Advances in Water Disaster Mitigation and Water Environment Regulation (WDWE2021)

水灾害防治与水环境调控

|研|究|进|展|

林鹏智　张建民　主编

项目策划：蒋　玙　唐　飞
责任编辑：蒋　玙
责任校对：唐　飞
封面设计：墨创文化
责任印制：王　炜

图书在版编目（CIP）数据

水灾害防治与水环境调控研究进展 = Advances in Water Disaster Mitigation and Water Environment Regulation (WDWE2021)：英文 / 林鹏智，张建民主编. — 成都：四川大学出版社，2021.7
ISBN 978-7-5690-4817-9

Ⅰ．①水… Ⅱ．①林… ②张… Ⅲ．①水灾－灾害防治－研究－英文②水灾－水环境－调控－研究－英文 Ⅳ．① P426.616

中国版本图书馆CIP数据核字（2021）第136260号

书　名	水灾害防治与水环境调控研究进展 Advances in Water Disaster Mitigation and Water Environment Regulation (WDWE2021)
主　编	林鹏智　张建民
出　版	四川大学出版社
地　址	成都市一环路南一段24号（610065）
发　行	四川大学出版社
书　号	ISBN 978-7-5690-4817-9
印前制作	四川胜翔数码印务设计有限公司
印　刷	成都金龙印务有限责任公司
成品尺寸	210mm×285mm
印　张	26
字　数	1040千字
版　次	2021年8月第1版
印　次	2021年8月第1次印刷
定　价	98.00元

◆ 版权所有 ◆ 侵权必究 ◆

◆ 读者邮购本书，请与本社发行科联系。
　电话：(028)85408408/(028)85401670/
　(028)86408023　邮政编码：610065
◆ 本社图书如有印装质量问题，请寄回出版社调换。
◆ 网址：http://press.scu.edu.cn

四川大学出版社
微信公众号

Advances in Water Disaster Mitigation and Water Environment Regulation（WDWE2021）

编写委员会

主 编

林鹏智　张建民

副主编

王　航　刘　超　安瑞冬

白瑞迪　陈　旻　崔宁博

韩　迅　胡兆永　司金华

宋春林　王　路　卫望汝

闫旭峰　姚慰炜　唐　恋

Contents

Full Paper

Hydraulics and Energy Dissipation on Stepped Spillways-prototype and Laboratory Experiences
.... Hubert Chanson(1)

Analysis of the Effect of Rainfall Station Spatial Distribution on Flash Flood Simulations
.... Zexing Xu, Xiekang Wang, Xufeng Yan(16)

Analysis of the Impact of Sediment on Flood Disaster in Shouxi River Basin
.... Tong Sun, Xiekang Wang, Xufeng Yan(24)

Channel Evolution Along the Yichang-Chenglingji Reach Downstream of the Three Gorges DAM
.... Hualin Wang, Shan Zheng, Guangming Tan, Lingyun Li(30)

Comparative Analysis of Comprehensive Evaluation Methods for Dam-break Consequence
.... Weiwei Sun, Qian Cai, Zhifei Long(38)

Comparisons Between SWAT and SWAT+ for Simulating Agriculture Management in a Large
  Watershed, a Case Study in Lijiang River Basin, China
.... Wuhua Li, Xiangju Cheng, Dantong Zhu, Jianhui Wen, Rui Xu(51)

Dynamic Flow Pattern Analysis Based on Energy Spectrum of Fluctuating Bottom Pressure
  in the Stilling Basin with a Negative Step
.... Wang Jia, Mingjun Diao, Lei Jiang, Guibing Huang(65)

Effects of Weir Height on Water Level and Bed Profile in a Degrading Channel
.... Weiming Wu, Lu Wang, Ruihua Nie, Xudong Ma, Xingnian Liu(75)

Experimental Study on Clear-water Scour Evolution Downstream of Rock Weirs
.... Wen Zhang, Lu Wang, Xingnian Liu, Ruihua Nie, Gang Xie(81)

Experimental Study on Flow Characteristics in Curve Open Channel with Suspended Vegetation
.... Qiaoling Zhang, Hefang Jing, Chunguang Li(88)

Experimental Study on Responses of Bed Morphology and Water Level to Unsteady Sediment
  Supply ········ Qihang Zhou, Lu Wang, Ruihua Nie, Xudong Ma, Xingnian Liu, Gang Xie(97)

Experiments on the Deterministic and Non-deterministic Effects of Water Inflow on Channelized
Configurations During the Lacustrine Delta Evolution
.... Xiaolong Song, Haijue Xu, Yuchuan Bai(105)

Free-Surface Characteristics of Undular Hydraulic Jumps in a Shallow Channel
.... Hong Hu, Ailing Cai, Ruidi Bai, Hang Wang(115)

Influence of Asymmetry on the Performance of a Breakwater Type Point Absorber Wave Energy
  Converter ········ Qi Zhang, Binzhen Zhou, Chaohe Chen, Peng Jin(124)

Influence of Sandspit Variability on Coastal Management Processes at the "Bouche du Roi" Inlet, Benin
.... Stephan Korblah Lawson, Hitoshi Tanaka, Keiko Udo, Nguyen Trong Hiep, Xuan Tinh(131)

Influence of Synchronization Time of Hybrid CPU/GPU 1D-2D Coupled Model for Urban Rainfall-runoff Process .... Donglai Li, Jingming Hou, Yangwei Zhang, Yongde Kang(139)

Longitudinal Distributions of Phytoplankton of the Pengxi River in Low Water Level
.... Jiaren Chen, Zhaowei Liu, Yongcan Chen, Xiao Chen, Jiabei Liu(150)

Long-term Water Security of Metropolitan Regions: Assessment of Water Supply Systems
.... Walter Manoel Mendes Filho, Wilson Cabral de Sousa Junior, Paulo Ivo Braga de Queiroz(159)

Modelling Sediment Transport Capacity of Loessial Slopes Based on Effective Stream Power
.... Chenye Gao, Jingwen Wang, Xile Liu, Shasha Han(168)

Non-invasive Measurements of Benthic Oxygen Exchanges of a Drinking Reservoir by Eddy Correlation Method
.... Yuanning Zhang, Bowen Sun, Chang Liu, Qingzhi Zong, Xiaobo Liu, Xueping Gao(174)

Real-time Water Depth Prediction of Multiple Manholes Based on Spatio-temporal Correlation
.... Changxun Zhan, Ting Zhang, Shuqi Li(184)

Reoccured Paleo-glacier Dammed Lakes and Outburst Floods in Langcang River
.... Xiwen Tang, Niannian Fan, Chengshan Wang, Xingnian Liu(192)

Research on Influence Factors of Sedimentation of Three Gorges Reservoir
.... Yue Zeng, Sichen Tong, Qi Shao(199)

Research Progress on the Formation Mechanism and Numerical Modeling of Cyanobacteria Blooms
.... Chenhui Xu, Sichen Tong, Guoxian Huang, Zhengze Lv, Qinghuan Zhang(207)

Risk Assessment of Landslide Dams with Different Internal Structure
.... Chuke Meng, Zhipan Niu, Yi Long(216)

Simulation of Colloids Transport and Retention by Pore-network Model: Influence of Particle Size and Flow Velocity
.... Dantong Lin, Scott Bradford, Liming Hu, Xinghao Zhang, Irene Lo(224)

Study on Relationship of Sediment Delivery Ratio with Inflow and Operational Water Level of the Three Gorges Reservoir
.... Ying Zhang, Sichen Tong, Xiaoping Long, Guoxian Huang(233)

Study on the Scheme of Diverting Clean Water Operation in the City Proper of Changzhou
.... Shousheng Mu, Jingxiu Wu, Ziwu Fan, Yang Liu, Guoqing Liu, Chen Xie(240)

The Analysis of Flood Limit Water Level in Longyangxia Reservoir During Different Flood Stages Based on Fuzzy Theory .... Song Yu, Kong Dezhi, Liu Qiang, Li Xinjie(246)

The Characteristics of Air Flow Driven by Free Surface of Open Channel
.... Jing Gong, Jun Deng, Wangru Wei, Weiwei Li(253)

The Impact of Climate and Land-use Changes on Freshwater Ecosystem Services Flows in Lianshui River Basin, China .... Yang Zou, Dehua Mao(261)

The Influence of Atmospheric Pressure Decrease on Cavity Characteristics of the Drop-step Aeration Facilities .... Yameng Wang, Jun Deng, Wangru Wei(275)

Water Quality Management(Sediment and Nutrients Run Off) at Brisbane Catchment
.... Fangrui Dong, Yongping Wei, Jinghan Li, Hui Li, Miao Zhang(282)

Abstract of Keynote Speech

A Particle-based Model for Simulation of Nonpoint Source Pollutant Dynamics Induced by Rainfall
.... Qiuhua Liang, Jinghua Jiang, Xue Tong(291)

An Experimental Study of Turbulent Structures in a Combined Fishway of Central Orifice and Vertical Slot ...... Zhiyong Dong, Zeyang Yan, Zhou Huang, Junpeng Yu, Jianli Tong(293)

Assessing the Hydraulics Safety of Existing Dams—A View to Dam Rehabilitation
.... Arturo Marcano(294)

Balancing Ecosystem Recovery and Water Demands in California and Chesapeake Bay, USA
.... Peter Goodwin(296)

Characteristics of Tsunami Damping with Different Friction Factors
.... Hitoshi Tanaka, Nguyen Xuan Tinh(297)

Eco-hydraulics for Sustainable Hydropower .... David Z. Zhu(300)

Effects of Data Temporal Resolution on Simulating Water Flux Extremes at the Field Scale
.... Lianhai Wu(301)

Field and ExperimentalModeling of Scour Induced by Turbulent Bores around Structures
.... Ioan Nistor, Razieh Mehrzad, Colin Rennie(302)

Flood Control Ability of the Three Gorges Reservoir and the Upstream Cascade Reservoirs in a Catastrophic Flood .... Shanghong Zhang, Zhu Jing, Wenjie An, Rongqi Zhang(303)

Flow and Sediment Resuspension in Vegetated Channels
.... Chao Liu, Yuqi Shan, Kejun Yang, Xingnian Liu(304)

Future Changes in Storm Surge due to Climate Change in the Western Pacific
.... Nobuhito Mori(305)

Grade Controlin Degrading Channels .... Ruihua Nie, Lu Wang, Xingnian Liu(307)

Granular Column Collapses: Implications for Studies of Landslides and Other Geo-Hazards
.... Ling Li(308)

How Does the Free Jump Dissipate Energy Efficiently? .... Sung-Uk Choi(309)

Hydraulics and Energy Dissipation on Stepped Spillways-prototype and Laboratory Experiences
.... Hubert Chanson(311)

Hydrodynamic Performance of a Bragg Reflection Breakwater
.... L. F. Chen, S. B. Zhang, D. Z. Ning(312)

Hyporheic Exchange of Reactive Substances
.... Guangqiu Jin, Zhongtian Zhang, Hongwu Tang, Qihao Jiang, Wenhui Shao(314)

Methane Emission from Inland Waters and Its Response to Climate Warming
.... Xinghui Xia, Liwei Zhang, Gongqin Wang(315)

Numerical Simulation of Turbulent Flow and Sediment Transport in Meandering Channel and River
.... Hefang Jing, Chunguang Li, Yakun Guo(316)

Responsive Strategies of Water Hazard Risk Under Global Change: Case Study in China
.... Jun Xia(319)
Sediment Transport in Vegetated Open Channel Flow .... Wenxin Huai(320)
The Key Technologies and Application of Intelligent Urban Drainage Management
.... Xiaohui Lei(321)
Theory and Practice of Water-soil Environmental Effects in the Process of Farmland
　Freezing-thawing Soil in Cold Regions .... Qiang Fu(322)
Understanding Hydro-eco-environment Changes in the Yangtze River
.... Dawen Yang, Ai wang, Ruijie Shi(323)
Unified Approach to Extreme Value Analysis of Coastal Storm Waves .... Zaijin You(324)
Using Nature-based Solutions for Integrated Water Management and Ecological Restoration:
　Lessons Learnt from the Netherlands .... W. Ellis Penning(325)
Velocity, Inundation and Run-up Measurements of Laboratory Generated Bores on a Planar Beach
.... Philip. L.-F. Liu, Lgnacio Barranco(326)

Extended Abstract

A GPU-based Two-phase SPH Model for Intense Sediment Transport in Geological-hazard Flows
.... Huabin Shi(327)
Accurate Short-term Prediction of the Water Levels Along the Yangtze Estuary by Using the
　NS_TIDE&AR Model .... Yongping Chen, Min Gan, Shunqi Pan(329)
Analysis of the Recirculation Length Downstream of Spur Dikes by Using a Two-dimensional
　Depth-averaged Mathematical Model .... Bingdong Li(331)
Assessing Impact Factors for Geological Hazards in Shaanxi Province Using Machine Learning
.... Shizhengxiong Liang, Dong Chen(333)
Biofilms as the "Architect": The Microbiological Mediation of Intertidal Sediment Behavior
.... Xindi Chen, Changkuan Zhang, Xiping Yu(335)
Changes in the Floodplain Channel Resistance in the Middle Yangtze River and Influencing
　Factors after the Impoundment of the Three Gorges Reservoir
.... Yong Hu, Yitian Li, Jinyun Deng(337)
Coupled 1D/2D Hydrodynamic Modelling and Hazard Risk Assessment
.... Boliang Dong, Junqiang Xia, Shanshan Deng, Meirong Zhou(339)
Data-driven Analysis and Forecasting for Sewer Flow of an Old Community in Ningbo, China
.... Biao Huang, Jiachun Liu, David Z Zhu(341)
Dynamic Simulation and Numerical Analysis of Coastal Flooding with Changes in Sea Level
　and Landscape .... Dongmei Xie, Qingping Zou, Yongping Chen (343)
Effect of a Single Air Bubble on the Migration Direction of a Cavitation Bubble
.... Jianbo Li, Yanwei Zhai(345)
Effect of Viscosity in Faraday Waves of Two-layer Liquids .... Dongming Liu, Yang Wu(347)
Effects of Ambient Temperature on Air-water Flow Properties of Free Surface Flow
.... Xiaohui Zheng, Wei Liu, Wei Sang, Shanjun Liu, Hang Wang, Ruidi Bai(349)

Effects of Hydrological Processes on Surface and Subsurface Nitrogen Losses from Purple Soil Slopes .... Meixiang Xie, Pingcang Zhang(351)

Error Analysis of PIV Based Pressure Measurement Method
.... Zhongxiang Wang, Qigang Chen, Yanchong Duan, Qiang Zhong(352)

Evolvement Mechanism of Bed Morphology Changes Around a Finite Patch of Vegetation
.... Fujian Li, Chao Liu(353)

Experimental Studies of Boundary-controlledrip Current Systems .... Sheng Yan, Zhili Zou(354)

Experimental Study on Cavitation Erosion Characteristics of Materials with Different Elasticities
.... Yanwei Zhai, Jianbo Li(356)

Flux Control Method for Uniform and Eco-friendly Fertilization Based on Differential Pressure Tank
.... Xinyu Hu, Xin Chen(358)

Hydrodynamics at the Channel Confluence with a Floodplain ...... Guanghui Yan, Saiyu Yuan(360)

Identification and Application of the Thresholds of Pre-release Indexes for Flood Control of a Reservoir
.... Sizhong He, Cao Huang, Xu Deng, Chuqi Zhou, Wan Jiang(362)

Imbalanced Stoichiometric Reservoir Sedimentation Regulates Methane Accumulation in China's Three Gorges Reservoir .... Zhe Li, Lunhui Lu(364)

In-situ Ecological Remediation of Urban Black-odor Water Bodies Based on Biological Carriers
.... Wenyao Jia, Xiaoyuan Qi, Kai He, Bingjun Liu(365)

ISPH Simulation of Wave Breaking with k-ε Turbulence Model
.... Dong Wang, Philip L.-F. Liu(367)

Mathematical Description of Formation and Emergency Response Process of Barrier Lake Based on Calculus Theory .... Jingwen Wang, Guangming Tan, Caiwen Shu, Shasha Han(369)

Microplastics in Freshwater River Sediments in Zhoushan, China
.... Yichen Sun, Cao Lu, Di Wu, Bin Zhou, Qiang Li(371)

Migration of Erosion Centers Along the Yichang-Chenglingji Reach Downstream of the Three Gorges Dam .... Hualin Wang, Shan Zheng, Guangming Tan(372)

Non-equilibrium Transport of Grouped Suspended Sediment and Its Effect on Channel Evolution in the Lower Yellow River ...... Yifei Cheng, Junqiang Xia, Meirong Zhou, Shanshan Deng (374)

Numerical Investigation of Flow in a Compound Flume with Rigid Vegetation by Lattice Boltzmann Method .... Hefang Jing, Weiwei Zhao, Weihong Wang(376)

Numerical Investigation of Two-phase Flow at a Chute Offset Aerator
.... Jijian Lian, Panhong Ren, Dongming Liu(378)

Numerical Investigation on Flow Around Two Side-by-side Patches of Emergent Vegetation
.... Jian Wang, Jingxin Zhang(379)

Numerical Simulation of Fluid-structure Interactions of Bridge Piers with Local Scours
.... Jing Chen, Yang Xiao, Wei Xu, Chentao Li, Zixuan Wang(381)

Numerical Simulation of Longdaohe River Water Quality Improvement .... Suiliang Huang(382)

Possible Impact of Climate Change on Tropical Cyclone Activities Based on a New Trajectory Model
.... Kaiyue Shan, Xiping Yu(385)

Research on the Influence of Rough Strip Energy Dissipator on Flow Characteristics of Channel Bend
　　·· Honghong Zhang, Zhenwei Mu, Fan Fan(387)
Research on the Producing Mechanism and Exchange Processes of Total Dissolved Gas
　　·· Bin Zhang, Xiaoli Fu(389)
Source Characteristics and Exacerbated Tsunami Hazard of the 2020 Mw 7.0 Samos Earthquake
　　in Eastern Aegean Sea
　　······ Gui Hu, Linlin Li, Wanpeng Feng, Yuchen Wang, Çagil KarakaŞ, Yunfeng Tian, Xiaohui He(390)
Spatial-temporal Characteristics of Hydrological Extremes in the Han River Basin
　　·· Yingying Feng, Maochuan Hu(392)
Study on Numerical Simulation and Emergency Control Strategy of Sudden Water Pollution
　　Scenarios in Dongzhang Reservoir ·························· Linlin Yan, Jijian Lian, Ye Yao(394)
Study on Water Wave Propagation Law of Submerged Jet with Rectangular Orifice
　　·· Shuguang Zhang, Jijian Lian(396)
The Climate Change of Summer Rainstorm and Water Vapor Budget in Sichuan Basin on the East
　　Side of Tibetan Plateau ······················· Dongmei Qi, Yueqing Li, Changyan Zhou(398)
The Influence of the BSISO Oscillation in the Summer Season on the Extreme Ocean Waves Off
　　Coast of China
　　····················· Xincong Chen, Xi Feng, Xiangbo Feng, Jingfu Peng, Yunshu Wu, Xinqi Zhang(400)
Water and Sediment Benefits Sharing Under Combined Operation of Cascade Reservoirs Along
　　the Yellow River Mainstream ······················ Fengzhen Tang, Yuanjian Wang, Xin Wang(402)

Hydraulics and Energy Dissipation on Stepped Spillways-prototype and Laboratory Experiences

Hubert Chanson

The University of Queensland, School of Civil Engineering, Brisbane QLD 4072, Australia

Abstract

Stepped spillways are used since more than 3,000 years. With the introduction of new construction materials(e.g. RCC, PVC coated gabions), the stepped spillway design has regained some interest for the last five decades. The steps increase significantly the rate of energy dissipation taking place along the chute, thus reducing the size of the required downstream energy dissipation basin. Stepped cascades are used also for in-stream re-aeration and in water treatment plants to enhance the air-water transfer of atmospheric gases and of volatile organic components. Yet the engineering design stepped spillways is not trivial because of the complicated hydrodynamics, with several possible, distinctly different flow regimes, complex two-phase air-water flow motions and massive rate of energy dissipation above the staircase invert. Altogether, the technical challenges in hydraulic engineering and design of stepped spillways are massive. This keynote lecture reviews the hydraulic characteristics of stepped channel flows and develops a reflection on nearly 30 years of active research including recent field measurements during major flood events. The writer aims to share his passion for hydraulic engineering, as well as some advice for engineering professionals and researchers.

Keywords: Stepped spillways; Hydraulic modelling; Field measurements; Energy dissipation and design; Multiphase air-water flows

1 Introduction

Extreme fluid flow events, and structures interacting with them, are critical for a wide range of areas, including reliability and design in engineering, as well as evaluation of cost-effectiveness of projects and optimisation of investments. Flooding is the most frequently occurring natural disaster in the world and has been increasing during the last two decades at an alarming rate(UNDRR 2019). It is expected that extreme flooding will occur even more frequently as a result of climate change(Rojas et al. 2013). In addition to the expected increase in the magnitude and frequency of future flooding, there will also be an increase in the world's population from the current level of 7.7 billion to 9.7 billion by 2050. Within this context, losses of life, and exorbitant aftermath damage costs will continue unless significant steps are taken to prevent catastrophic failures of critical infrastructure, in the face of pressures such as population growth, catchment development and climate change.

In spite of the continuous hydrological cycle of water, the rainfall pattern on our planet is not uniform, with more and more countries being subjected to hydrological extremes(floods, droughts) with high spatial and temporal variability. The catalogue of the world's maximum floods highlighted some astonishing record flow rates (IAHS 1984). In contrast, many regions have experienced recurrent long droughts for the last 300 years, e.g. Africa, Australia, Central Asia. Today, man-made reservoirs and dams are one of the most efficient means to deliver long-term reliable water reserves, as well as flood protection. During major flood events, the floodwaters must be passed safely above, beneath or beside the dam: this is achieved with a spillway system, a water system designed to discharge safely the extreme reservoir outflows(USBR 1965, Novak et al. 1996). In terms of fluid dynamics, the main functions of a spillway are the safe conveyance of floodwaters and energy dissipation. The dissipation of the spill's kinetic energy takes place down the steep chute and at the downstream end of the spillway system, and the amount of kinetic energy dissipation can be truly astonishing(Chanson 2015, pp. 5 − 8). Energy dissipation at dam spillways can take place in a hydraulic jump stilling structure, with a ski jump, a plunge pool, an impact dissipator, the installation of macro-roughness on the spillway chute, or a combination of the above(Vischer and Hager 1995, Chanson 2015). In modern times, the simplest type of macro-roughness is the installation of steps, that can be relatively easily accommodated during construction with several construction materials(Chanson 1995, 2001)(Fig. 1).

In this paper, it is argued that the hydraulic design stepped spillways is not trivial because of the complicated hydrodynamics and massive rate of energy dissipation above the staircase invert. The technical challenges are discussed, in the light of the state-of-the-art into stepped spillway hydraulic research and practice, including recent field observations during major flood events.

Figure 1. Stepped spillways

Note: (a) Gold Creek Dam(Australia, 1890) on 23 February 2015; (b) Les Olivettes Dam(France, 1987) in March 2003(Courtesy Mr & Mrs J. Chanson); (c) Hinze Dam Stage 3(Australia, 2011) on 26 September 2018; (d) Wadi Dayqah Dam(Oman, 2012) in January 2018(Courtesy Dr D. Wüthrich).

2 Stepped Spillway and Hydraulics

2.1 Presentation

Stepped spillways have been built and used for several millenia(Chanson 2000—2001). During the 19th century and early 20th century, a significant number of dams were equipped with a "staircase bye wash", i.e. a stepped spillway [Fig. 1(a)] (Wegmann 1911), until the mature design development of the hydraulic jump stilling basin at the toe of smooth chutes. In the last five decades, the advancement in construction technology, e.g. roller compacted concrete (RCC), and the geographic constraints of several dam sites, have driven a renewed interest in the stepped spillway design [Figs. 1(b)(c)(d) and 2]. The steps act as macro-roughness, increasing substantially the rate of kinetic energy dissipation along the spillway chute, in turn reducing the size and cost of the downstream stilling system(Sorensen 1985, Rajaratnam 1990).

A key feature of stepped spillway design is the broad diversity of applications, construction materials and design flow rates. The stepped invert design may be applied to re-aeration cascades, embankment overtopping protection systems, sabo check dams, storm waterways, low-head weirs as well as large concrete dam spillways. The construction of the steps maybe undertaken with timber cribs, gabions and Reno mattresses, precast concrete blocks, roller compacted concrete and conventional concrete blocks. There are further a very wide range of design unit discharges, from less than 0.1 m²/s to more than 100 m²/s! Importantly, each stepped chute operation is characterised by some complicated hydrodynamics which requires a high level of hydraulic expertise.

In the next sections, the author will share some thoughts in the challenges associated with the hydrodynamics and hydraulic design of stepped spillways, based upon his 30 years of experiences on the topic. These add an earlier very-relevant keynote paper(Matos and Meireles 2014) and a small number of relevant seminal papers(Bombardelli 2012, Chanson 2013a, Chanson et al. 2021). The writer initially developed an interest for stepped chute hydraulics for pedagogical interests, as part of the postgraduate teaching of hydraulic structures, before developing a strong research focus on physical modelling(Table 1), multiphase flow analyses and hydraulic design(Chanson 1995, 2001, Chanson et al. 2015).

Figure 2. Strong turbulence and intense free-surface aeration in a skimming flow on a stepped spillway
Note: Paradise Dam spillway operation on 5 March 2013 for $Q=2,320$ m³/s, $d_c/h=2.85$, $h=0.62$ m.

Table 1. Stepped spillway research at the University of Queensland. Characteristics of physical facilities(1995—2021)

$q(°)$	h(m)	B(m)	q(m²/s)	Step configuration	Others
2.6	0.143	0.25	0.07~0.14	Single drop	
3.4	0.0715~0.143	0.50	0.038~0.163	Flat horizontal	
3.4	0.0715~0.143	0.50	0~0.075	Flat horizontal	Dam break wave

(To be continued)

(Continue)

$q(°)$	$h(m)$	$B(m)$	$q(m^2/s)$	Step configuration	Others
15.9	0.05~0.10	1.0	0.020~0.26	Flat horizontal	
21.8	0.05~0.10	1.0	0.008~0.19	Flat horizontal	
21.8	0.10	1.0	0.10~0.219	Triangular vanes	
21.8	0.10	1.0	0.01~0.241	Rough steps	
26.6	0.10	0.52	0.005~0.241	Flat horizontal	
26.6	0.10	0.52	0.003~0.282	Pooled steps	
26.6	0.10	0.52	0.038~0.28	Gabion steps	
45	0.10	0.985	0.001~0.30	Flat horizontal	
45	0.10	0.985	0.083~0.216	Trapezoidal cavities	
45	0.10	0.985	0.0008~0.28	Inclined steps	
51.3	1.5	12.25	5.8~27.3	Flat horizontal	Hinze Dam Stage 3

Notes: q: longitudinal chute slope; h: vertical step height; B: chute width; q: unit discharge.

2.2 Stepped chute hydraulics and its challenges

The fluid mechanics of stepped spillway overflow is a very complicated topic (Horner 1969, Peyras et al. 1992, Chanson 1994). The challenges include (a) the strong turbulence and intense free-surface aeration, (b) the existence of markedly different flow regimes on a given stepped geometry, depending upon the unit discharge, and (c) the underlying limitations of both physical modeling and numerical modelling.

First, the stepped chute profile generates some intense turbulent dissipation during the spill associated with a significant kinetic energy dissipation, as well as strong self-aeration through the free-surface (Fig. 2) (Chanson and Toombes 2003, Boes and Hager 2003, Biethman et al. 2021). The steps are conducive to the development of strong cavity recirculation motion. At large discharges, the recirculation motion is maintained through the "pseudo-continuous" transfer of momentum from the main flow to the stepped cavity flow region, in the form of cavity ejection and replenishment process occurring irregularly (Chanson et al. 2002, Toro et al. 2017). The large-scale vortices are shed from the step cavities into the mainstream, before interacting with the free surface (Toro et al. 2017, Chanson 2021). The very strong free-surface turbulence induces a two-phase air-water free-surface mix, since neither gravity nor surface tension can maintain the free-surface cohesion (Ervine and Falvey 1987, Brocchini and Peregrine 2001, Chanson 2009). Visually, large amount of fluid ejections are seen, with complicated air-water structure ejections, and re-attachments further downstream, while the fluid ejections are accompanied with air entrapment. The violent process, i.e. 'eruptions' of air-water masses, is believed to be caused by the collision of tilted powerful large-scale streamwise vortices with the upper surface. In prototype, the upper free-surface may further be the locus of "surface waves", e.g. illustrated at Chinchilla weir (Australia) and Three-Gorges Project (China) (Toombes and Chanson 2007).

Second, the stepped chute flow may be one of several distinct flow regimes, for a fixed stepped geometry, depending upon the water discharge. For a rectangular prismatic channel equipped with flat horizontal impervious steps, the overflow is anappe flow at low unit discharges[Figs. 1(a)& 3(a)], a transition flow for a range of intermediate flow rates [Fig. 3(b)] or a skimming flow at larger discharges[Figs. 2 & 3(c)]. The nappe flow regime is typically designed for in older spillway structures built during the 19th century, and it corresponds to small unit discharges(Horner 1969, Chamani and Rajaratnam 1994). The transition flow regime is characterised by hydrodynamic instabilities and chaotic flow conditions, which should be avoided during wastewater spills unless at small flows(Chanson and Toombes 2004). The skimming flow is commonly observed in concrete gravity dam stepped spillways(Sorensen 1985, Matos 1999). Each flow regime has very different features and characteristics from one another(Fig. 3). More complex stepped geometries might have more than three flow regimes, e.g. four flow regimes were reported in gabion stepped chutes(Wüthrich and Chanson 2014), with stationary and instationary motions on pooled stepped spillways (Felder and Chanson 2014).

Third, the hydraulic modelling has been traditionally conducted based upon physical laboratory experiments, designed based upon a Froude similarity(Novak and Cabelka 1981, Sorensen 1985). The upscaling of the laboratory results may be affected by significant scaling issues, with the usage of small-size laboratory facilities. Recent works suggested that the laboratory testing must be conducted at large Reynolds numbers, i.e. $Re > 10^5$ to 5×10^5, to minimise potential scale effects, although scaling issues cannot be eliminated totally unless working at full scale(Chanson and Gonzalez 2005, Felder and Chanson 2009). In turn, the pre-requisite utilisation of large-size laboratory facilities operating with relatively large discharges may be economically expensive (Fig. 4). Free-surface aeration in such large laboratory flumes imply the needs for multiphase flow metrologies, with expensive fine instrumentation, operated by expert researchers. The requirement in human resources cannot be under-stated nor its cost under-estimated, considering that civil engineering undergraduate and postgraduate students are not taught about multiphase fluid dynamics. As an illustration, the author would educate and train a new research student for 2 to 6 months, before he/she can conduct reliable air-water flow measurements of acceptable quality.

Fourth, another form of modelling is the computational fluid dynamics CFD numerical modelling. CFD modelling is based upon the numerical integration of the time-dependent Navier-Stokes equations (Rodi et al. 2013, Rodi 2017). For the last three decades, some novel works developed some multiphase-dedicated CFD models (Prosperetti and Tryggvason 2009, Lubin 2021). The general equations for two-phase flows may be applied to hydraulic structures and stepped spillways (Bombardelli 2012). Yet, not all the results are equal, not all the outputs correct. Too often, the CFD modelling is not validated or incorrectly validated, leading to quantities of physically meaningless results. (As a journal editor and senior expert reviewer, the author has read too many erroneous CFD studies with wholly meaningless data outputs.). Although the AIAA and ASME developed clear guidelines for verification and validation(AIAA 1994, Roache 1998, Rizzi and Vos 1998), these are rarely applied rigorously in hydraulic engineering! There is further a lack of detailed laboratory data sets of high quality and of field observations, most relevant to CFD validation. This is not trivial. Indeed, the physical experiments do not provide all the detailed informations at all spatio-temporal scales, required for CFD model validation. There is no doubt that the potential applications of CFD to stepped spillway hydrodynamics offers a large interest, once it will be properly validated. Further, nearly all the CFD modelling is conducted with limited CPU resources, able only to model a laboratory-size chute. The correct CFD modelling of a full-scale stepped spillway would impose CPU and memory requirements that would be economically un-justifiable at present. Beyond the intrinsic numerical errors and bias, the current CFD results have to be upscaled to full-scale spillway with the implicit upscaling errors, already discussed with physical modelling

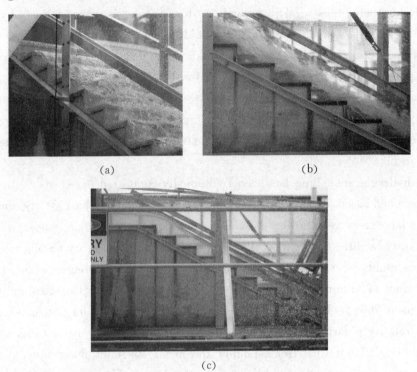

Figure 3. Flow regimes on a stepped spillway model: 1V : 2H(q=26.6°), h=0.10 m(University of Queensland)
Note: (a) Nappe flow(d_c/h=0.133); (b) Transition flow(d_c/h = 0.8); (c) Skimming flow(d_c/h=1.5).

Figure 4. Research student undertaking air-water flow measurements in a 1-m wide stepped spillway flume with a 1V : 1H longitudinal slope and 0.1 m high steps at the University of Queensland

3 Advancement in Stepped Spillway Hydraulics

3.1 Present advances

Laboratory measurements of air-water flows in a stepped chute are not trivial(Ruff and Frizell 1994, Matos 1999), but some advances in metrology combined with advanced post-processing and analyses may provide some in-depth characterisation of the turbulent aerated flows, in large-size facilities under controlled conditions. In self-aerated chute flows, the void fraction C ranges from very low values next to the invert to 100% in the atmosphere, with the mass and momentum fluxes typically conserved in the air-water column within $0<C<0.95$(Cain and Wood 1981, Wood 1985). In this gas-liquid region($C<0.95$), the high-velocity self-aerated flow motion is a quasi-homogenous mixture and the two phases, i.e. air and water, travel at identical speed with negligible slip(Rao and Kobus 1974, Chanson 1997). In the air-water flow, the fine characterisation of the turbulent gas-liquid flow necessitates a large number of parameters, including the void fraction, bubble count rate, bubble and drop size distributions, and flow fragmentation properties (Chanson 2013a). Some dedicated instruments were developed, especially the phase-detection needle probes, and more recently the application of the optical flow(OF) technique. The needle probes are designed to pierce the bubble and droplet interfaces, and they are well-suited to track air-water interfaces(Fig. 5). Figure 5 present high-shutter speed photographs of interactions between dual-tip phase-detection needle probes and air-water interface. Although developed in the early 1960s (Neal and Bankoff 1963, 1965), the applications of the needle probe system have drastically expanded over the last two decades, with the successful development of novel advanced signal processing and analysis techniques(Table 2). Further developments might include unbiased signal processing based upon the entire signal processing, as attempted in Zhang and Chanson(2019).

The optical flow(OF) is non-intrusive imaging technique, developed as a set of tools detecting the flow motion between consecutive frames. The OF technique was recently applied to self-aerated stepped spillway flows(Bung and Valero 2016). Sideview OF data were presented, showing great details on the cavity recirculation motion(Bung and Valero 2016, Zhang and Chanson 2018). But some detailed comparison with phase-detection probe data indicated that (a) the OF velocity data underestimated the centreline velocities by 10% to 30%, mostly as a result of sidewall effects, and(b) the OF outputs were meaningless in the upper air-water column, i.e. $C>0.3$ to 0.5(Zhang and Chanson 2018, Arosquipa Nina et al. 2021). Recently, some top view OF velocity data were also extracted(Arosquipa Nina et al. 2021). Newer developments are underway.

In-depth discussions on the various air-water flow signal post-processing are listed in Table 2. Table 2 summarises the key advancements in air-water flow signal analyses at the University of Queensland(2000—2021). The list focuses on a number of novel signal analyses for self-aerated air-water measurements in stepped chute flows, beyond the basic processing of void fraction and interfacial velocity in steady air-water flows(Jones and Delhaye 1976, Cartellier and Achard 1991).

All in all, some fairly major advances have been achieved during the last two decades with respect to detailed physical modelling of two-phase air-water flow in stepped channels. The combination of large-size facilities (Table 1) and complementary advanced instrumentation and signal processing (Table 2) can deliver a fine characterisation of the gas-liquid motion and internal structure down to the sub-millimetric scales.

Figure 5. Dual-tip phase-detection probes(DT-PDPs) piercing air-water interface in self-aerated flows-Probe systems designed and manufactured at the University of Queensland

Table 2. Advances in air-water flow signal processing and analyses with a focus on novel signal analyses in self-aerated stepped spillway flows developed at the University of Queensland(2000—2021)

Signal analysis	Outputs	Instrumentation	Reference(s)
Individual bubble detection technique	V, Tu	DT-PDP	Chanson(2005), Wang & Chanson(2015)
Interfacial turbulence intensity	Tu	DT-PDP	Chanson & Toombes(2002, 2003)
1D bubble clustering	N_c, Nb bubbles per cluster, ⋯	ST-PDP	Chanson & Toombes (2002), Chanson(2002)
Relationship between bubble count rate and void fraction	$F/F_{max} = f(C)$	ST-PDP	Toombes (2002), Toombes & Chanson(2008)
Spectral analyses		ST-PDP	Chanson & Gonzalez (2004), Zhang & Chanson(2019)
Integral turbulent time & length scales	T_{xx}, T_{xy}, L_{xy}, L_t, T_t	PDPA	Chanson & Carosi(2007a)
Inter-particle arrival time		ST-PDP	Chanson & Carosi(2007a)
2D bubble clustering	N_c, Nb bubbles per cluster, ⋯	PDPA	Sun & Chanson(2013)
Triple decomposition	C, F, T_{xx}, V, Tu	ST-PDP & DT-PDP	Felder & Chanson(2014)

(To be continued)

(Continue)

Signal analysis		Outputs	Instrumentation	Reference(s)
Total pressure & water turbulence intensity		P_t, Tu_p	M-TPP & ST-PDP	Zhang et al. (2016)
Optical flow(OF)	Turbulent velocity field		UHSC	Zhang & Chanson(2017, 2018)[1]
	Surface velocity field		UHSC & HD-VC	Arosquipa Nina et al. (2021), Chanson(2021)
Adaptive correlation technique		V	DT-PDP	Kramer et al. (2019)
Single-bubble event detection technique		V, Tu	DT-PDP	Shi et al. (2020)

Notes: DT-PDP: dual-tip phase-detection probe; M-TPP: miniature total pressure probe; PDPA: phase-detection probe array; ST-PDP: single-tip phase-detection probe; UHSC: ultra-high-speed camera; HD-VC: high-definition video camera.

[1] the seminal work by Bung and Valero(2016) must be acknowledged.

3.2 The near-future: hybrid modelling(Mark Ⅰ)

In hydraulic engineering, the term "hybridmodelling" is commonly used to design an integrated modelling technique combining physical modelling and computational fluid dynamics(CFD) modelling. The approach has been successfully applied to several hydraulic engineering projects, including box culvert hydrodynamics, unsteady breaking bores, and non-aerated skimming flows on stepped spillways(Meireles et al. 2014, Lopez et al. 2017).

More recently, some fascinating hybridmodelling results were achieved in the non-aerated skimming flow on a stepped spillway(Toro et al. 2016, 2017, Zabaleta et al. 2020). Fig. 6 illustrates an example. The three-dimensional CFD modelling used between 5.4×10^6 and 8.85×10^6 elements, with length about 0.024 to 0.02 times the step height. The results focused on the vorticity patches shed by the stepped cavities until they interacted with the free-surface and "break". The instantaneous vorticity field showed the formation of complex turbulent structures which detached from the stepped invert and were convected in the water column, ultimately interacting with the surface(Toro et al, 2017).

On another hand, detailed 3D CFD modelling are not yet as successful in the air-water skimming flow region downstream of the inception point of free-surface aeration. Current CFD models are not able to reproduce qualitatively and quantitatively the violent air-water ejections, re-attachment and air engulfment observed in prototype stepped spillways(Chanson 2013b, 2021). The modelling of the strong free-surface turbulence is extremely difficult and has to be resolved at the required level of details for further progresses, in the author's opinion.

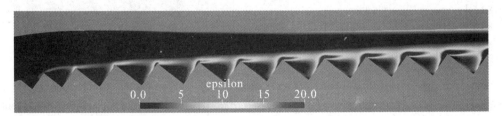

Figure 6. Computational Fluid Dynamics(CFD) modelling of the non-aerated flow region of skimming flow on a stepped chute($q = 51.3°$, $h = 0.05$ m, $d_c/h = 2.14$): field of dissipation rate of turbulent kinetic energy (TKE)(in m^2/s^3) by Toro et al. (2016) with flow direction from left to right(Courtesy of Prof. F. Bombardelli)

3.3 The near-future: hybrid modelling(Mark Ⅱ)

A different form of modelling may combine laboratory experiments and field observations (Fig. 7). Namely, the approach is based upon some complementary observations with stepped configurations that are geometrically scaled based upon a Froude similarity. Fig. 6 illustrates a recent application, combining optical observations in a prototype stepped spillway and imaging measurements in a relatively large-size laboratory facility. The left side of Fig. 7 presents the field observations. The right side shows some laboratory results. The middle(bottom) graph documents a comparison in terms of the streamwise surface velocity in the non-aerated flow region; the graph shows some good agreement between field observations(thick lines), laboratory observations(cross symbols) and ideal fluid flow theory(tick red dashed line). In practice, the most difficult component is the field observations. In that application(Fig. 7), the author documented several major flood events between 2013 and 2021 at the same facility(Chanson 2021).

The use of optical techniques in prototype spillways is still in the very early stages. Both LSPIV and OF techniques may provide some valuable information on the velocity field(Hauert A, 2021, *Pers. Comm.*, Chanson 2021). To date, the various experiments showed a number of non-trivial challenges. Any imaging technique relies upon physical and optical access, adequate atmospheric and light conditions, experienced operators, and high-resolution high-speed camera equipment. The adequacy of all these conditions is most challenging during a major flood event, particularly during natural disaster situations.

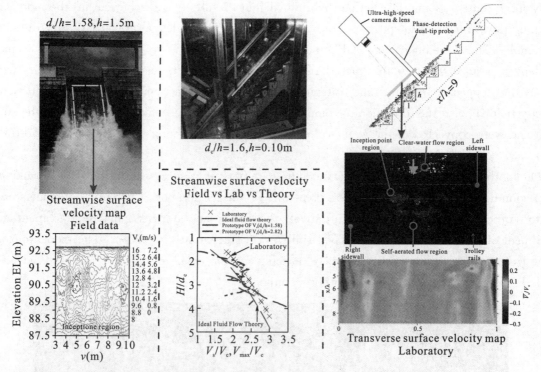

Figure 7. Hybrid modelling(Mark Ⅱ) of a stepped spillway: combining complementary field and laboratory observations-Left: time-averaged streamwise surface velocity map; Middle(bottom): dimensionless comparison of streamwise surface velocity on the channel centreline in the non-aerated flow region, between Prandtl-Pitot tube laboratory data, field OF observations and ideal fluid flow theory; Right: time-averaged transverse surface velocity map in laboratory(Data: Arosquipa Nina et al. 2021, Chanson 2021)

4 Conclusions

The stepped spillway design is well-suited to the stability of gravity dams, while its simplicity of shape and energy dissipation potential are conducive to cost-effective spillway systems, in particular in confined environments. The steps increase significantly the rate of energy dissipation taking place on the steep chute, thus reducing the size of the required downstream stilling structure(s) and the risk of scour. However, hydraulic engineers must analyse carefully the hydraulic operation of stepped spillways and the design is not trivial. The estimation of energy dissipation performances relies heavily upon an accurate prediction of the hydrodynamic properties, including strong turbulence and intense free-surface aeration taking place above the steps. Both strong turbulence and self-aeration are two physical processes always observed in prototype stepped spillways. The hydraulic challenges are amplified to a number of various possible flow regimes, with markedly different flow patterns and hydrodynamic properties.

The hydraulic modelling of stepped spillways may be undertaken physically and/or numerically. Both approaches have their own limitations, discussed above(section 2.2). Laboratory experiments must be conducted in large-size facilities, with multiphase-flow instrumentation. CFD numerical modelling requires some detailed validation, although current practice does not allow a full-scale modelling. All in all, the limitations and costs of hydraulic modelling should not be under-estimated, nor the demands in human resources, i.e. engineers and researchers with a high level of expertise in hydraulic engineering. Recent developments encompass some large-size physical modelling with advanced complementary metrologies, as well some hybrid modelling. Two types of hybrid modelling, Mark I and Mark II, have the potential to deliver a rational, science-based framework of accurate energy dissipation predictions on stepped spillways, and in turn robust and economically-viable spillway designs, to provide our society with reliable flood protection. Any future research on the topic should transcend traditional discipline boundaries by exploiting knowledge derived from state-of-the-art hydraulic modelling and instrumentation techniques for assessing the performances of full-scale prototype structures.

Acknowledgements

The author wishes to acknowledge the exchanges with and contributions of many individuals, including numerous undergraduate research students, his Masters research students(G. Carosi, S. Felder, P. Guenther, D. Wüthrich, Y. Arosquipa Nina), his Ph.D. students(L. Toombes, C. Gonzalez, S. Felder, G. Zhang, R. Shi), and many research collaborators and friends (F. Bombardelli, D. Bung, J. Matos, I. Ohtsu, M. Takahashi, D. Valero, D. Wüthrich, Y. Yasuda). The financial support of the School of Civil Engineering at the University of Queensland(Australia) is acknowledged.

References

[1] AIAA. Editorialpolicy statement on numerical accuracy and experimental uncertainty[J]. AIAA Journal, 1994, 32 (1): 3-3.

[2] Arosquipa Nina Y, Shi R, Wuthrich D. Intrusive and non-Intrusive air-water measurements on stepped spillways with inclined steps: a physical study on air entrainment and energy dissipation. Hydraulic Model Report No. CH121/21

[R]. Brisbane: The University of Queensland, 2021.

[3] Biethman B, Ettema R, Thornton C. Air entrained in flow along a steep-stepped spillway: data and insights from a hydraulic model[J]. Journal of Hydraulic Engineering, 2021, 147(5): 05021001.

[4] Boes R M, Hager W H. Two-phase flow characteristics of stepped spillways[J]. Journal of Hydraulic Engineering, 2003, 129(9): 661−670.

[5] Bombardelli F A. Computational multi-phase fluid dynamics to address flows past hydraulic structures [C]. Proceedings of the 4th IAHR International Symposium on Hydraulic Structures, Porto, Portugal. 2012: 9−11.

[6] Brocchini M, Peregrine D H. The dynamics of strong turbulence at free surfaces. Part 1. Description[J]. Journal of Fluid Mechanics, 2001, 449: 225−254.

[7] Bung D B, Valero D. Optical flow estimation in aerated flows[J]. Journal of Hydraulic Research, 2016, 54(5): 575−580.

[8] Cain P, Wood I R. Measurements of self-aerated flow on aspillway[J]. Journal of the Hydraulics Division, 1981, 107(11): 1425−1444.

[9] Cartellier A, Achard J L. Local phase detection probes in fluid/fluid two-phase flows [J]. Review of Scientific Instruments, 1991, 62(2): 279−303.

[10] Chamani M R, Rajaratnam N. Jet flow on stepped spillways[J]. Journal of Hydraulic Engineering, 1994, 120(2): 254−259.

[11] Chanson H. Hydraulics of skimming flows over stepped channels and spillways[J]. Journal of Hydraulic Research, 1994, 32(3): 445−460.

[12] Chanson H. Hydraulic design of stepped cascades, channels, weirs and spillways[M]. Oxford: Pergamon, 1995.

[13] Chanson H. Air bubble entrainment in free-surface turbulent shear flows[M]. London: Academic Press, 1997.

[14] Chanson H. Historical development of stepped cascades for the dissipation of hydraulic energy[J]. Transactions of the Newcomen Society, 2000, 72(2): 295−318.

[15] Chanson H. The hydraulics of stepped chutes and spillways[M]. Lisse: Balkema, 2001.

[16] Chanson H. Air-water flow measurements with intrusive, phase-detection probes: Can we improve their interpretation?[J]. Journal of Hydraulic Engineering, 2002, 128(3): 252−255.

[17] Chanson H. Air-water and momentum exchanges in unsteady surging waters: an experimental study [J]. Experimental Thermal and Fluid Science, 2005, 30(1): 37−47.

[18] Chanson H. Turbulent air-water flows in hydraulic structures: dynamic similarity and scale effects [J]. Environmental fluid mechanics, 2009, 9(2): 125−142.

[19] Chanson H. Hydraulics of aerated flows: qui pro quo?[J]. Journal of Hydraulic Research, 2013, 51(3): 223−243.

[20] Chanson H. Interactions between a developing boundary layer and the free-surface on a stepped spillway: Hinze Dam spillway operation in January 2013[C]. ICMF, 2013.

[21] Chanson H. Energy dissipation in hydraulic structures[M]. Boca Raton: CRC Press, 2015.

[22] Chanson H. Stepped spillway prototype operation, spillway flow and air entrainment: the hinze dam, australia. hydraulic model report No. CH123/21[R]. Brisbane: The University of Queensland, 2021

[23] Chanson H, Bung D, Matos J. Stepped spillways and cascades [M]. Boca Raton: CRC Press, 2015: 45−64.

[24] Chanson H, Carosi G. Turbulent time and length scale measurements in high-velocity open channel flows [J]. Experiments in Fluids, 2007, 42(3): 385−401.

[25] Chanson H, Carosi G. Advanced post-processing and correlation analyses in high-velocity air-water flows [J]. Environmental Fluid Mechanics, 2007, 7(6): 495−508.

[26] Chanson H, Gonzalez C A. Interactions between free-surface, free-stream turbulence and cavity recirculation in open channel flows: measurements and turbulence manipulation [C]. Proceeding of 5th International Conference on Multiphase Flow, ICMF, Yokohama, 2004: 1−14.

[27] Chanson H, Gonzalez C A. Physical modelling and scale effects of air-water flows on stepped spillways[J]. Journal of Zhejiang University-Science A, 2005, 6(3): 243−250.

[28] Chanson H, Leng X, Wang H. Challenging hydraulic structures of the twenty-first century-from bubbles, transient turbulence to fish passage[J]. Journal of Hydraulic Research, 2021, 59(1): 21−35.

[29] Chanson H, Toombes L. Air-water flows down stepped chutes: turbulence and flow structure observations[J]. International Journal of Multiphase Flow, 2002, 28(11): 1737−1761.

[30] Chanson H, Toombes L. Strong interactions between free-surface aeration and turbulence in an open channel flow [J]. Experimental Thermal and Fluid Science, 2003, 27(5): 525−535.

[31] Chanson H, Toombes L. Hydraulics of stepped chutes: the transition flow[J]. Journal of Hydraulic Research, 2004, 42(1): 43−54.

[32] Chanson H, Yasuda Y, Ohtsu I. Flow resistance in skimming flows in stepped spillways and its modelling[J]. Canadian Journal of Civil Engineering, 2002, 29(6): 809−819.

[33] Ervine D A, Falvey H T. Behaviour of turbulent water jets in the atmosphere and in plunge pools[J]. Proceedings of the Institution of Civil engineers, 1987, 83(1): 295−314.

[34] Felder S, Chanson H. Turbulence, dynamic similarity and scale effects in high-velocity free-surface flows above a stepped chute[J]. Experiments in Fluids, 2009, 47(1): 1−18.

[35] Felder S, Chanson H. Triple decomposition technique in air-water flows: application to instationary flows on a stepped spillway[J]. International Journal of Multiphase Flow, 2014, 58: 139−153.

[36] Horner M W. An analysis of flow on cascades of steps[D]. Birmingham: University of Birmingham, 1969.

[37] Rodier J A, Roche M. World catalogue of maximum observed floods[J]. IAHS-AISH publication, 1984(143).

[38] JonesJr O C, Delhaye J M. Transient and statistical measurement techniques for two-phase flows: a critical review [J]. International Journal of Multiphase Flow, 1976, 3(2): 89−116.

[39] Kramer M, Valero D, Chanson H. Towards reliable turbulence estimations with phase-detection probes: an adaptive window cross-correlationtechnique[J]. Experiments in Fluids, 2019, 60(1): 1−6.

[40] Lopes P, Leandro J, Carvalho R F. Alternating skimming flow over a stepped spillway[J]. Environmental Fluid Mechanics, 2017, 17(2): 303−322.

[41] Lubin P. Wave breaking and air entrainment[J]. Advanced Numerical Modelling of Wave Structure Interactions, 2021: 69−85.

[42] Matos J. Emulsionamento de ar e dissipação de energia do escoamento em descarregadores em degraus. (Air entrainment and energy dissipation in flow over stepped spillways.)[D]. Lisbon: Instituto Superior Técnico, 1999.

[43] Matos J, Meireles I. Hydraulics of stepped weirs and dam spillways: engineering challenges, labyrinths of research [C]//11th National Conference on Hydraulics in Civil Engineering & 5th International Symposium on Hydraulic Structures: Hydraulic Structures and Society-Engineering Challenges and Extremes. Engineers Australia, 2014: 330.

[44] Meireles I C, Bombardelli F A, Matos J. Air entrainment onset in skimming flows on steep stepped spillways: an analysis[J]. Journal of Hydraulic Research, 2014, 52(3): 375−385.

[45] Neal L G, Bankoff S G. A high resolution resistivity probe for determination of local void properties in gas-liquid flow[J]. AIChE Journal, 1963, 9(4): 490−494.

[46] Neal L G, Bankoff S G. Local parameters in cocurrent mercury-nitrogen flow: Parts I and II[J]. AIChE Journal, 1965, 11(4): 624−635.

[47] Novák P, Čabelka J. Models in hydraulic engineering: physical principles and design applications[M]. Pitman Publishing, 1981.

[48] Novak P, Moffat, A I B, Nalluri C. Hydraulic structures[M]. London: E & FN Spon, 1996.

[49] Peyras L, Royet P, Degoutte G. Flow and energy dissipation over stepped gabion weirs[J]. Journal of hydraulic Engineering, 1992, 118(5): 707−717.

[50] Prosperetti A, Tryggvason G. Computational methods for multiphase flows[M]. Cambridge: Cambridge University Press, 2009.

[51] Rajaratnam N. Skimming flow in stepped spillways[J]. Journal of Hydraulic Engineering, 1990, 116(4): 587−591.

[52] Rao N S L, Kobus H, Barczewski B. Characteristics of self-aerated free-surface flows (Water and waste water current research and practice)[M]. Berlin, Bielefeld, Munichen: E. Schmidt, 1975.

[53] Roache P J. Verification of codes and calculations[J]. AIAA journal, 1998, 36(5): 696−702.

[54] Rodi W. Turbulence modeling and simulation in hydraulics: A historical review[J]. Journal of Hydraulic Engineering, 2017, 143(5): 03117001.

[55] Rodi W, Constantinescu G, Stoesser T. Large-eddy simulation in hydraulics[M]. Leiden: Crc Press, 2013.

[56] Rojas R, Feyen L, Watkiss P. Climate change and river floods in the European Union: Socio-economic consequences and the costs and benefits of adaptation[J]. Global Environmental Change, 2013, 23(6): 1737−1751.

[57] Ruff J F, Frizell K H. Air concentration measurements in highly-turbulent flow on a steeply-sloping chute[C]// Hydraulic Engineering. ASCE, 1994: 999−1003.

[58] Shi R, Wüthrich D, Chanson H. Introducing a single bubble event detection technique for air-water interfacial velocity measurements in unsteady turbulent bore[C]. Proceedings of 22nd Australasian Fluid Mechanics Conference AFMC20202020, Brisbane: The University of Queensland, 2020: 20−24.

[59] Sorensen R M. Stepped spillway hydraulic model investigation[J]. Journal of hydraulic Engineering, 1985, 111(12): 1461−1472.

[60] Sun S, Chanson H. Characteristics of clustered particles in skimming flows on a stepped spillway[J]. Environmental fluid mechanics, 2013, 13(1): 73−87.

[61] Tombes L. Experimental study of air-water flow properties on low-gradient stepped cascades[D]. Brisbane: The University of Queensland, 2002.

[62] Toombes L, Chanson H. Surface waves and roughness in self-aerated supercritical flow[J]. Environmental Fluid Mechanics, 2007, 7(3): 259−270.

[63] Toombes L, Chanson H. Interfacial aeration and bubble count rate distributions in a supercritical flow past a backward-facing step[J]. International Journal of Multiphase Flow, 2008, 34(5): 427−436.

[64] Toro J P, Bombardelli F A, Paik J. Characterization of turbulence statistics on the non-aerated skimming flow over stepped spillways: a numerical study[J]. Environmental Fluid Mechanics, 2016, 16(6): 1195−1221.

[65] Toro J P, Bombardelli F A, Paik J. Detached eddy simulation of the nonaerated skimming flow over a stepped spillway[J]. Journal of Hydraulic Engineering, 2017, 143(9): 04017032.

[66] UNDRR. Global assessment report on disaster risk reduction 2019[R]. Geneva: United Nations Office for Disaster Risk Reduction, 2019.

[67] USBR. US Department of the interior, bureau of reclamation, Denver CO, USA, 1st edition, 3rd printing[J]. Design of Small Dams, 1965.

[68] Vischer D, Hager W H. IAHR Hydraulic Structures Design Manual No. 9: Hydraulic Design Considerations[J]. Energy Dissipators, 1995.

[69] Wang H, Chanson H. Experimental study of turbulent fluctuations in hydraulic jumps[J]. Journal of Hydraulic Engineering, 2015, 141(7): 04015010.

[70] Wegmann E. The design and construction of dams[M]. New York: John Wiley & Sons, 1911.

[71] Wood I R. Air water flows[C]. Proceding of 21st IAHR Congress, Melbourne, Australia, Keynote address, 1985: 18−29.

[72] Wüthrich D, Chanson H. Hydraulics, air entrainment, and energy dissipation on a Gabion stepped weir[J]. Journal of Hydraulic Engineering, 2014, 140(9): 04014046.

[73] Zabaleta F, Bombardelli F A, Toro J P. Towards an understanding of the mechanisms leading to air entrainment in the skimming flow over stepped spillways[J]. Environmental Fluid Mechanics, 2020, 20(2): 375−392.

[74] Zhang G, Chanson H. Application of local optical flow methods to high-velocity air-water flows: validation and application to skimming flows on stepped chutes. hydraulic model report No. CH105/17[R]. Brisbane: The University of Queensland, 2017.

[75] Zhang G, Chanson H. Application of local optical flow methods to high-velocity free-surface flows: Validation and

application to stepped chutes[J]. Experimental Thermal and Fluid Science, 2018, 90: 186−199.

[76] Zhang G, Chanson H. On void fraction and flow fragmentation in two-phase gas-liquid free-surface flows[J]. Mechanics Research Communications, 2019, 96: 24−28.

[77] Zhang G, Chanson H, Wang H. Total pressure fluctuations and two-phase flow turbulence in self-aerated stepped chute flows[J]. Flow Measurement and Instrumentation, 2016, 51: 8−20.

Analysis of the Effect of Rainfall Station Spatial Distribution on Flash Flood Simulations

Zexing Xu, Xiekang Wang, Xufeng Yan

State Key Laboratory of Hydraulics and Mountain River Engineering,
Sichuan University, Chengdu 610065, China

Abstract

Flash floods are directly and solely related to the precipitation in a river basin, so reliable rainfall data is the key to accurately simulate flash flood events. Uncertainty may be introduced when point rainfall of a certain number is used to represent the rainfall variation of the whole basin because the spatial-temporal distribution of rainfall is uneven. The objective of this paper is to evaluate the effect of the amount and location of rainfall stations on hydrological simulations. A total of 10 rainfall stations with records were taken into account in the simulation of flood events in Zhongdu River Basin, Sichuan Province. Four conditions (close to the outlet, away from the outlet concentrate in central region, and scatter in the basin) of the spatial distribution of stations are considered by reducing the number of rain gauges, which were designed to study how the rain gauge density and distribution affect the results of flash flood simulation. The results show that the uncertainty of rainfall estimates arising from the amount and location of rainfall stations has a considerable impact on model simulation. The distance of rain gauge to the outlet of basin has a direct influence on peak discharge and runoff volume. Reasonable spatial distribution of rain gauge over the forecast basins was one of the main determinants of simulate accuracy during a flash flood event.

Keywords: Flash flood; Rainfall stations; Hydrologic simulation; Uncertainty analysis

1 Introduction

Precipitation is the key driving force of flash food in mountain watershed and one of the essential meteorological inputs that affect the performance of the hydrological model(Fraga et al., 2019; Yuan et al., 2019). Rainfall data is generally measured using rain gauges, radars, or satellites(Sun et al., 2018). Ground-based rain gauges accurate rainfall reaching the ground at a certain location while radars and satellites provide high spatio-temporal resolution areal rainfall but less accurate(Marra et al., 2017). Currently, the rainfall data obtained through reasonably distributed networks of rain gauge is still the most common input of hydrological modeling(Zhang et al., 2018). However, rainfall records are often incomplete because the recommendation about minimum rain gauge densities has not been fully followed by instrument malfunction, funding shortage, and poor geographic location(Tan and Yang, 2020). Obviously, the spatial distribution of precipitation plays a key role in flood simulation. Therefore, it is essential to study the influence of rainfall station distribution on the hydrological model. Several studies have been focused on the effects of rain station density and position on hydrologic modeling. Zeng et al. (2018) found that the model performance of a lumped model no longer improves when rain gauge density reaches a certain threshold. Xu et al. (2013) also raise a similar conclusion that rain gauge density has an optimal threshold for hydrological model performance. Tan and Yang(2020) reported that distance of rainfall stations to the streamflow station has a considerable impact on SWAT modelling and at least four stations are required to accurately simulate monthly streamflow. Hohmann et al. (2020) applied a physically based hydrological model to identify the impact of station density and interpretation schemes on runoff simulations and point out that the influence of station density is more sensitive to the small watershed and short-term rainfall.

Previous studies were concentrated on the influence of rainfall distribution on long-term runoff, whereas the short-duration rainstorms have been rarely discussed, particularly in mountain watersheds. The objective of this study is to evaluate the effect of the spatial distribution of the rainfall station(station number and station position) on flash flood simulations. The semi-distribution model based on excess storage-excess infiltration and geomorphologic instantaneous unit hydrograph theory was applied for the flash flood simulation to evaluate runoff and described the spatial distribution of rainfall stations. And the study is carried out for Zhongdu river basin, China.

2 Study Area and Data

The Zhongdu river basin, located in southern Sichuan province in China, was selected as the study area. The catchment area above the Longshan stream gauge is approximately 376 km² with the main river length of 36 km, and the absolute elevations range from 400 m to 1900 m(Fig. 1). The climate for the basin is characterized as a humid subtropical monsoon zone. The average annual rainfall is about 800 mm and short-duration, high-intensity storm events that occur in the summer season (from July to August) account for 60% of annual rainfall. The western part of the basin is mostly covered by forest, while in the east near the catchment outlet, the land use type is mainly cropland and grassland along the streams. The majority of land cover in the basin is yellow and purplish soil. Consequently, the saturation-excess is the main form of runoff generation in Zhongdu river basin.

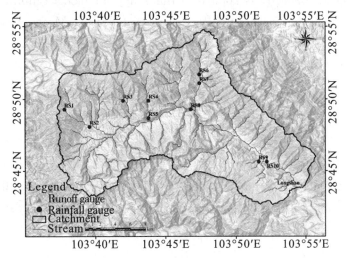

Figure 1. Study area and the distribution of rain gauges

With the effect of climate change, the increase in rainfall intensity may lead to more severe flash floods in Zhongdu river basin. The last significant flood event in the basin occurred in August 2018, caused flooding downstream where 600 acres of farmland were occupied, at least 4 people missed, and several bridges and hundreds of houses were destroyed. We select this rainstorm event to analyze the effect of the amount and location of rainfall stations on simulation of the flash flood process. Hourly rainfall data for August 2018 was obtained from the excerpt data of 11 rainfall stations in Zhongdu river basin, (Fig. 1). The streamflow data at the Longshan stream gauge for the same period were also collected from the same department for hydrologic model calibration and validation.

A 30 m Digital Elevation Model (DEM) of Advanced Space-borne Thermal Emission and Reflection Radiometer (ASTER) was used for river formation, basin and sub-basins delineation in hydrologic modeling. And the geomorphic parameters of the basin, including river lengths and slopes, sub-basin area and slopes, are obtained from DEM using ArcGIS.

3 Methodology

3.1 Model setup and calibration

The semi-distribution model based on excess storage-excess infiltration and geomorphologic instantaneous unit hydrograph theory is used in this study. The model consists of the following three major processes: evapotranspiration calculation, runoff production and separation, and flow routing. In this model, three layers of soil were used to represent the evaporation capacity and order of soil, and then potential evapotranspiration is calculated by pan evaporation or other meteorological data. When the precipitation at the present moment is greater than the sum of potential evapotranspiration and water storage capacity or exceeds the infiltration capacity, runoff will be generated and divided into three types including surface flow, interflow and groundwater runoff based on mechanism of excess storage-excess infiltration. Surface flow and interflow are further routed to the catchment outlet through kinematic-wave-based geomorphologic instantaneous unit hydrograph with manning formula and Darcy formula, ground flow is routed to a catchment outlet through a linear reservoir.

Multi-objective automatic optimization algorithm, the non-dominated sorting genetic algorithm II (NSGA-II; see Deb et al., 2002), is applied in this study for model calibration. Three commonly used objective functions, the Nash-Sutcliffe efficiency (NSE), Mean Absolute Relative Error (MARE), and Flood Peak Error (FPE) are used to evaluate the performance of model. The functions of the three indices are expressed as follows:

$$NSE = 1 - \frac{\sum (Q_{obs}^t - Q_{sim}^t)^2}{\sum (Q_{obs}^t - \overline{Q}_{obs})^2} \quad (1)$$

$$MARE = \frac{\sum |Q_{obs}^t - Q_{sim}^t|}{\sum Q_{obs}^t} \quad (2)$$

where Q_{obs}^t and Q_{sim}^t is the observed and simulated runoffs; $\overline{Q}_{obs}$ is mean of the observed runoff series. NSE represents the fitting performance between the observed and simulated data. MARE reflects the degree of dispersion of results.

3.2 Spatial rainfall scenarios

In this study, different scenarios considering rainfall station scarcity were designed according to the actual situation of the watershed. To quantify the impacts of the spatial distribution of stations, the scenarios were divided into different station numbers and different station positions. and one baseline(S0) and 26 scenarios(S1~S26) with different station numbers and different station positions are shown in Table 1. The baseline scenario includes all the 10 rain gauges and others is then designed by gradually reducing the number of rainfall stations and their spatial locations. The 9 scenarios named S1 to S9 considered the reduced number of rain gauges starting with the station located farthest from the outlet, while another 9 scenarios named S10 to S18 firstly removed the station closest to the outlet. The S19 to S22 scenarios were designed by removing the station nearest to the boundary of the drainage basin, and The S23 to S26 scenarios represent the scattered distribution of rainfall stations using the maximum dissimilarity algorithm (MDA) to assign 2, 6, and 8 of rainfall stations for calculating areal rainfall. The details about all the 27 scenario settings can be found in Table 1.

Table 1. The scenario design of rainfall station distribution

Scenario	Category	Number	Detailed description
S0	Baseline	10	Retained a complete 10 stations
S1	Reduce the number of rainfall stations by removing the station farthest from the outlet	9	Remove RS1
S2		8	Remove RS1, RS2
S3		7	Remove RS1, RS2, RS3
S4		6	Remove RS1, RS2, RS3, RS4
S5		5	Remove RS1, RS2, RS3, RS4, RS5
S6		4	Remove RS1, RS2, RS3, RS4, RS5, RS6
S7		3	Remove RS1, RS2, RS3, RS4, RS5, RS6, RS7
S8		2	Remove RS1, RS2, RS3, RS4, RS5, RS6, RS7, RS8
S9		1	Remove RS1, RS2, RS3, RS4, RS5, RS6, RS7, RS8, RS9
S10	Reduce the number of rainfall stations by removing the station nearest from the outlet	9	Remove RS10
S11		8	Remove RS9, RS10
S12		7	Remove RS8, RS9, RS10
S13		6	Remove RS7, RS8, RS9, RS10
S14		5	Remove RS6, RS7, RS8, RS9, RS10
S15		4	Remove RS5, RS6, RS7, RS8, RS9, RS10
S16		3	Remove RS4, RS5, RS6, RS7, RS8, RS9, RS10
S17		2	Remove RS3, RS4, RS5, RS6, RS7, RS8, RS9, RS10
S18		1	Remove RS2, RS3, RS4, RS5, RS6, RS7, RS8, RS9, RS10
S19	Reduce the number of rainfall stations by removing the station closest to the boundary	8	Remove RS1, RS10
S20		6	Remove RS1, RS2, RS9, RS10
S21		4	Remove RS1, RS2, RS3, RS6, RS9, RS10
S22		2	Remove RS1, RS2, RS3, RS4, RS6, RS7, RS9, RS10
S23	Reduce the number of rainfall stations by maximum dissimilarity algorithm	8	Remove RS7, RS9
S24		6	Remove RS4, RS7, RS8, RS9
S25		4	Remove RS2, RS3, RS4, RS7, RS8, RS9
S26		2	Remove RS2, RS3, RS4, RS5, RS6, RS7, RS8, RS9

4 Result and Discussion

4.1 Effect of rain gauge number

The complete rainfall station data(S0 scenario) is used as the meteorological input of model for the calibration and verification period. The two flood events that occurred in the Zhongdu river basin on August 1, 2018 and August 10, 2018 were selected as calibration periods, and the flood event on August 16, 2018 was selected as verification period and then used for streamflow simulations under the 26 rainfall scenarios. The baseline scenario simulation had a good performance in both the calibration(NSE=0.9 and MARE=0.27) and validation(NSE=0.8 and MARE=0.31) period.

The evaluation results of different spatial scarcity conditions of rainfall stations are shown in Fig. 2. It is worth noting that the number of rainfall stations has a significant impact on the performance of the model. An obvious increasing relationship between the NSE of simulated runoff and the number of rainfall stations can be found in Fig. 2(a) and an obvious decreasing relationship between the MARE of simulated runoff and the number of rainfall stations can be found in Fig. 2(b). In comparison, the mean MARE of the 8-station scenario is 27% lower than that of the 2-station scenario. This indicates that the model performance gradually improved when the number of rainfall stations increased, and previous studies have reached similar conclusions(Xu et al., 2013; Zeng et al., 2018). For a small watershed, at least 4 stations can achieve the passing performance(NSE>0.6) and at least 8 stations can achieve the desired performance(NSE>0.8). Although the NSE of S12 of only three stations was greater than 0.6, the performance decreased as the stations increased in the situation far away from the outlet. The poor performance of the scenarios with less than 3 rainfall stations can be observed, such as S9, S18, S22 and S26[Fig. 3(a)(b)]. The negative values for NSE indicate that single or two rainfall stations will cause great uncertainty of rainfall and are insufficient for flash flood simulations. For example, the result of S22 greatly underestimated the rainfall in the basin because there is almost no rainfall near the station, while S22 greatly overestimates the rainfall because the rainfall center is concentrated in the middle of the basin[Fig. 3(b)].

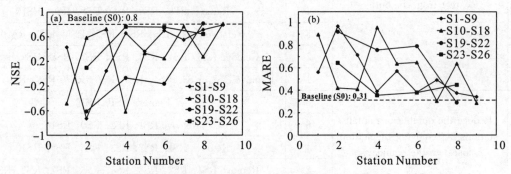

Figure 2. Evaluation of observed streamflow and simulated streamflow from 27 scenarios

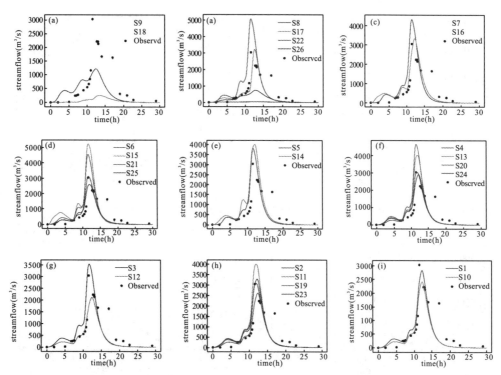

Figure 3. streamflow simulations for different number of rain gauges

Note: (a) 1-gauge scenarios; (b) 2-gauge scenarios; (c) 3-gauge scenarios; (d) 4-gauge scenarios; (e) 5-gauge scenarios; (f) 6-gauge scenarios; (g) 7-gauge scenarios; (h) 8-gauge scenarios; (i) 9-gauge scenarios.

4.2 Effect of rain gauge distribution

The distribution of different rainfall stations also caused huge differences in the performance of the model. Thus, the impacts of scarcity data location in the basin were further analyzed based on a concept of gauge influence sphere by comparing different rainfall station distribution schemes. Through the comparison of rain gauge distributions that lead to different model performance, it seems that the distribution scheme with the rainfall stations near the outlet and far away from the outlet has greater fluctuations than the other two schemes(Fig. 4). The performance of the model for scheme of rain gauge close to the outlet becomes more stable as the number of stations increases, while the scheme of keeping the rain gauge away from the outlet has obvious uncertainty(Fig. 1). This show that at least one station should be close to the outlet, which can greatly improve the performance of the model. When rainfall stations are concentrated in the center of the basin, the simulated flow will be larger than the actual flow[Fig. 4(c)], which means that the center of the storm is concentrated in the middle of the basin, and this strategy will cause the areal rainfall of the basin to be overestimated. It can be found that the scheme with the smallest fluctuation range is scattered distribution [Fig. 4(d)], and a limited number of rainfall stations can also reflect the flood process in the basin. Therefore, the scattered distribution of rain gauges is more representative to the rainfall of the whole basin than other schemes, which can reflect the non-uniformity of rainfall distribution in small mountain watersheds and accurately simulate the flood process.

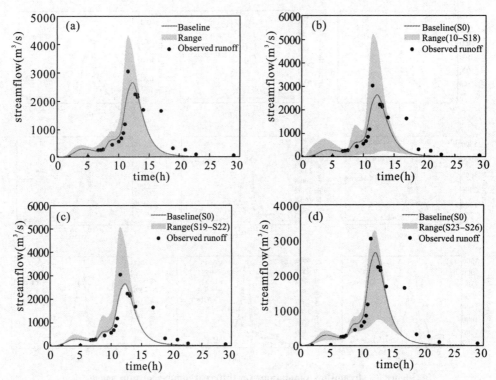

Figure 4. Streamflow ranges for 27 different combinations of rainfall stations

Note: (a) Rainfall stations are close to the outlet: S1-S9; (b) Rainfall stations are far away from outlet: S10-S18; (c) Rainfall stations are concentrated in the center of the basin: S19-S22; (d) rainfall stations are scattered in the basin: S23-S26.

5 Conclusions

In this study, an evaluation was made to find out the impacts of rain gauge density and distribution on flash flood simulation in Zhongdu river basin. 27 scenarios with different locations and density of rainfall stations were carefully designed to reflect the uncertainty of rainfall spatial distribution. The study focuses on the short-duration flood process in small mountain watershed and adopts a hydrological model based on excess storage-excess infiltration and geomorphologic instantaneous unit hydrograph theory for hydrological simulation.

The results reveal that the number of rainfall stations is an important factor in the simulation effect of the model as its performance improved with the increase of the number of stations. At least 4 rainfall stations are required to achieve the desired performance of a model, and single or two stations will cause unestimable uncertainty. The study shows that location effect or spatial distribution of rainfall stations has a considerable influence on the assessment of rainfall in the basin, and then affects the flood simulation. It is recommended to set up rain gauges near the outlet in the basin, because the scheme of rain gauges close to the outlet has a more stable improvement on the effect of the model. In addition, the scheme of distributed rain gauge stations is recommended. The scheme of distributed rainfall stations is more reasonable and can reflect the unevenness of rainfall in small mountain basins.

However, some of the results of this study should be applied cautiously to other watersheds because the rainfall characteristics of each watershed may be different. Future research should consider more catchments with different climatic conditions and different geomorphological characteristics to make the results more representative.

Acknowledgements

This research was supported by the National Key R&D Program of China(2019YFC1510702), the National Natural Science Foundation of China(51639007), and the Open Foundation project (SKHL1913).

References

[1] Deb K, Pratap A, Agarwal S, et al. A fast and elitist multiobjective genetic algorithm: NSGA-II[J]. IEEE transactions on evolutionary computation, 2002, 6(2): 182−197.

[2] Fraga I, Cea L, Puertas J. Effect of rainfall uncertainty on the performance of physically based rainfall-runoff models [J]. Hydrological Processes, 2019, 33(1): 160−173.

[3] Hohmann C, Kirchengast G, Rieger W, et al. Runoff sensitivity to spatial rainfall variability: A hydrological modeling study with dense rain gauge observations[J]. Hydrology and Earth System Sciences Discussions, 2020: 1−28.

[4] Marra F, Morin E, Peleg N, et al. Intensity-duration-frequency curves from remote sensing rainfall estimates: comparing satellite and weather radar over the eastern Mediterranean[J]. Hydrology and Earth System Sciences, 2017, 21(5): 2389−2404.

[5] Sun Q, Miao C, Duan Q, et al. A review of global precipitation data sets: data sources, estimation, and intercomparisons[J]. Reviews of Geophysics, 2018, 56(1): 79−107.

[6] Tan M L, Yang X. Effect of rainfall station density, distribution and missing values on SWAT outputs in tropical region[J]. Journal of Hydrology, 2020, 584: 124660.

[7] Xu H, Xu C Y, Chen H, et al. Assessing the influence of rain gauge density and distribution on hydrological model performance in a humid region of China[J]. Journal of Hydrology, 2013, 505: 1−12.

[8] Yuan W, Liu M, Wan F. Study on the impact of rainfall pattern in small watersheds on rainfall warning index of flash flood event[J]. Natural Hazards, 2019, 97(2): 665−682.

[9] Zeng Q, Chen H, Xu C Y, et al. The effect of rain gauge density and distribution on runoff simulation using a lumped hydrological modelling approach[J]. Journal of hydrology, 2018, 563: 106−122.

[10] Zhang A, Shi H, Li T, et al. Analysis of the influence of rainfall spatial uncertainty on hydrological simulations using the bootstrap method[J]. Atmosphere, 2018, 9(2): 71.

Analysis of the Impact of Sediment on Flood Disaster in Shouxi River Basin

Tong Sun, Xiekang Wang, Xufeng Yan

State Key Laboratory of Hydraulics and Mountain River Engineering,
Sichuan University, Chengdu 610065, China

Abstract

Affected by the Wenchuan earthquake, a large number of loose deposits left on the hillsides in Shouxi River Basin, which provided abundant material sources for geological disasters and flash floods. On August 20, 2019, a serious flood and sediment disaster occurred in this area caused by a rainstorm. In this paper, we calculated the distribution and volume of landslides in the basin by TRIGRS model. The result indicated that landslide disasters occurred frequently in this area under the influence of rainstorms. Moreover, many potential landslides are mainly distributed near the river. When a large amount of sediment flows into the river, it will have a huge impact on the early warning of flash flood disaster. To reduce the hazards of flash floods in Shouxi River Basin, the sediment caused by geological disasters should be fully considered.

Keywords: Flash Flood; TRIGRS; Landslide Volume; Mountain River

1 Introduction

In China, flash floods are a very common natural disaster. The Southwest mountain area is the most serious area of flash flood disaster in China. With the increase of people's utilization of mountain resources, the mountain torrent disaster is becoming more and more serious(Duan et al., 2016; Tu et al., 2020). According to the distribution of landslides, many scholars have done relevant research. Tang et al.(2012) interpreted the remote sensing images to obtain the corresponding landslide area in Wenchuan area after the Wenchuan earthquake. Chang et al.(2015) interpreted the remote sensing images of different periods in Longxihe area, and obtained the corresponding landslide area change with time. Chiu et al.(2019) estimated the landslide risk area in Taiwan based on the AHP(analytic hierarchy process) and the empirical power law formula of area and volume. There are many landslides in nature, which are connected by many landslides. In the process of landslide volume estimation, the existence of landslide clusters will lead to the overestimation of landslide volume. In order to reduce the influence of landslide group on the calculation results, Fan et al.(2019) cut the landslide group based on the flow direction. Calculating landslide volume through TRIGRS is also affected by landslide clusters.

Affected by the Wenchuan earthquake, a large number of loose deposits left on the hillsides in Shouxi River Basin, which provided abundant material sources for geological disasters and flash floods. In order to better early warning of mountain flood disaster in Shouxi River Basin, it is necessary to simulate the size and distribution of landslides in the basin.

In this study, we simulated the risk coefficient of landslide by TRIGRS model. At the same time, slope-units are used as calculation units to deal with the risk coefficient. Then we use slope-units to cut the landslide clusters. When FS <1, the slope-unit is considered as a dangerous unit. We use slope-units with FS values less than 1 as hazard units. Finally, we use the empirical power law formula to calculate the volume of each hazard unit.

2 Methods

2.1 TRIGRS model

TRIGRS(the transient rainfall infiltration and grid-based regional slope-stability model) is a FORTRAN program used to calculate the spatio-temporal variation of safety factor distribution caused by a rainfall(Baum et al., 2008). The model is based on the calculation and analysis of the transient pore water pressure changes caused by the grid unit. The model is suitable for the simulation of slope stability in saturated and unsaturated soil. The model mainly includes three modules: infiltration model, hydrological model and slope stability model. The infiltration model is an instantaneous rainfall infiltration model based on time and space. The model uses a series of Heaviside step functions to solve the proposed sum of the original solutions for constant intensity rainfall proposed by Iverson to represent a general time-varying sequence of surface fluxes of different intensity and duration (Iverson, 2000). The generalized solution used in TRIGRS is as follows:

$$\psi(Z,t) = (Z-d)\beta + k_1 + k_2 \tag{1}$$

$$k_1 = 2\sum_{n=1}^{N} \frac{I_{nZ}}{K_s} H(t-t_n) D_1(t-t_n)^{\frac{1}{2}}$$

$$\sum_{m=1}^{\infty} \left\{ ierfc\left[\frac{(2m-1)d_{LZ} - (d_{LZ}-Z)}{2 D_1(t-t_n)^{\frac{1}{2}}}\right] + ierfc\left[\frac{(2m-1)d_{LZ} + (d_{LZ}-Z)}{2D_1(t-t_n)^{\frac{1}{2}}}\right] \right\} \tag{2}$$

where ψ is the groundwater pressure head, t is the time, Z is the soil thickness, d is the depth of the water table, K is the saturated hydraulic vertical conductivity, I_{nZ} is the surface flux at time n, N is the total number of time intervals, d_{LZ} is the boundary of the impervious base, and m is a coefficient showing an infinite series of odd terms.

$ierfc$ is the complementary error function(Saulnier et al., 1997), which is defined as the following equation:

$$ierfc(\eta) = \frac{1}{\sqrt{\pi}}\exp(-\eta^2) - \eta erfc(\eta) \tag{3}$$

2.2 Calculation method

We first collected relevant parameters of the watershed through field investigation and literature review, including DEM, soil depths, soil parameters, and rainfall data. These parameters are input into the TRIGRS model to calculate the risk coefficient map of the basin. the obtained coefficient graph is segmented by slope-units. Slope-units of the watershed are made according to the curvature. The average of the safety factors on all grid cells within each slope cell is the value of all grid cells within that slope cell. When the safety factor FS<1, the slope-unit is considered a dangerous unit.

The area of each dangerous slope unit is sorted out. The landslide volume of each dangerous slope-unit is obtained through the formula. The relationship was exponential, established by the scale index γ, and intercept α. The relationship is as follows:

$$V = \alpha A^{\gamma} \tag{4}$$

Parker et al. (2011) revised the α and γ for the Wenchuan earthquake region based on field measurements. We set the parameters according to the research results. The total landslide volume in the basin is obtained by adding the landslide volumes of all dangerous slope-units.

3 Study Area and Data

3.1 Study area and rainfall data

The basin studied in this paper is Shouxi River Basin, a tributary of Minjiang River in China, which is located in the south of Wenchuan County, Aba Tibetan and Qiang Autonomous Prefecture, Sichuan Province. The Shouxi River, The Xihe River, which originates in Datang Mountain, is the main stream of this basin. It joins the two tributaries of the Zhonghe River and the Heishi River at Sanjiang, and then flows through Baishi and Shuimo towns to Xuankou, where it flows into the Minjiang River. The catchment area of the basin is about 554 km², with a geographical location between 103°02′04″~ 103°26′56″E, 30°47′42″~ 31°02′19″N, and an altitude of 895~4952 m. The basin is a typical mountain basin with complex terrain, criss-crossing gullies and steep slopes. Basin is located in western Sichuan rainy central area, frequent rainstorms, a total of two hydrological stations in the basin, respectively, is the Sanjiang hydrological station and Guojia dam hydrological station, according to the hydrological station measured data that the average precipitation in the basin for many years is about 1125 mm. Since the Wenchuan earthquake in 2008, a large number of landslides have been produced in the basin. Landslides and loose deposits in the channel provide abundant sediment sources for the flood disaster. In 2019, a large-scale mountain flood disaster broke out in the Shouxi river. The flash flood has caused a lot of damage to the villages on both sides of the riverbank.

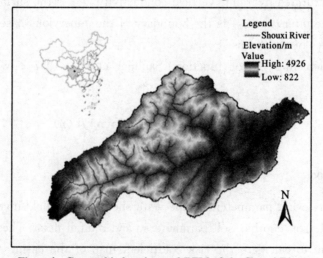

Figure 1. Geographic location and DEM of the Shouxi River

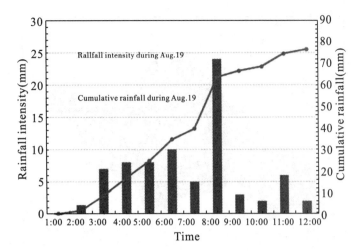

Figure 2. Rainfall data on 13 August 2010

3.2 Soil parameters

Rawls et al. used multiple linear regression to estimate BC parameters from a large database of approximately 2540 soil layers(Rawls 1982). Carsel and Parrish(1988) statistically processed the results to obtain the probability density function of VG parameters. The water Fflow parameter module of Hydrus 1D summarizes the results and obtains soil mechanical parameters of 12 soil types. Combined with the local soil type, the specific parameters are in the table below:

Table 1. The specific parameters

Residual soil water content θ_s	0.067
Saturated soil water content θ_r	0.45
Parameter α in the soil water retention function	0.02
Saturated hydrological conductivity K_S (m/s)	1.25×10^{-6}
Background infiltration rate I_{ZLT} (m/s)	1.25×10^{-8}

3.3 Soil thickness

Soil thickness is the deepest possible failure surface of a landslide. Due to the lack of local data, the empirical function associated with the slope defined by Saulnier et al. was used to estimate the simulated soil thickness(Saulnier et al., 1997).

$$D_{IZ} = Z_{max}\left[1 - \frac{\tan\delta - \tan\delta_{min}}{\tan\delta_{max} - \tan\delta_{min}}\left(1 - \frac{Z_{min}}{Z_{max}}\right)\right] \tag{5}$$

where Z_{max} and Z_{min} are the maximum and minimum values of soil thickness. δ_{min} and δ_{max} are the maximum and minimum values of the slope. According to the field investigation, Z_{max} is set at 4 m and Z_{min} at 0.4 m. The water table is considered to be the same as the soil thickness.

Figure 3. The soil depth in Shouxi River

4 Results and Discussion

Through the method, we get the safety factor distribution map of the basin. We divided the risk of the area into four levels based on the Fs value. The resulting hazard distribution is shown in Fig. 4. The results show that there are a lot of potential landslides in the Shouxi River Basin. The calculation shows that there are 2002 dangerous units in the basin. The potential landslide area in the basin is 4.96×10^7 m², which is accounting for 8.3% of the total basin. The potential landslide volume in the basin is 3.1×10^8 m³.

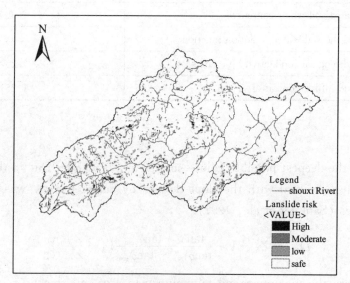

Figure 4. Landslide distribution

The landslides in the basin are mainly concentrated in the upstream of the basin, with a large slope. In order to explore the influence of slope on landslide, the paper analyzes the slope distribution in the basin and the slope distribution of landslide. When rainstorm happens, landslide disaster will occur. Landslide body enters river channel, which provides a lot of sediment supply for flood disaster.

5 Conclusion

In this paper, we calculated the distribution and volume of landslides in the basin by TRIGRS model. We use slope-units to cut the landslide clusters, which makes the result more reliable. The result indicated that landslide disasters occurred frequently in this area under the influence of rainstorms. Moreover, many potential landslides are mainly distributed near the river. When a large amount of sediment flows into the river, it will have a huge impact on the early warning of flash flood disaster. To reduce the hazards of flash floods in Shouxi River Basin. the sediment caused by geological disasters should be fully considered.

Acknowledgements

This research was supported by the National Key R&D Program of China(2019YFC1510702), the National Natural Science Foundation of China(51639007), and the Open Foundation project (SKHL1913)

References

[1] Baum R L, Savage W Z, Godt J W. TRIGRS—A Fortran Program for Transient Rainfall Infiltration and Grid-Based Regional Slope-Stability Analysis, Version 2.0[R]. US Geological Survey Open-File Report, US Department of the Interior, 2008.

[2] Carsel R F, Parrish R S. Developing joint probability-distributions of soil-water pretention characteristics[J]. Water Resources Research, 1998, 24(5): 755−769.

[3] Chang M, Tang C, Ni H. Evolution process of sediment supply for debris flow occurrence in the Longchi area of Dujiangyan City after the Wenchuan earthquake[J]. Landslides, 2015, 12(3): 611−623.

[4] Chiu Y J, Lee H Y, Wang T L. Modeling sediment yields and stream stability due to sediment-related disaster in shihmen reservoir watershed in Taiwan[J]. Water, 2019, 11(2): 332.

[5] Duan W L, He B, Nover D. Floods and associated socioeconomic damages in China over the last century[J]. Journal of the International Society for the Prevention and Mitigation of Natural Hazards, 2016, 82(1): 401−413.

[6] Fan R L, Zhang L M, Shen P. Evaluating volume of coseismic landslide clusters by flow direction-based partitioning [J]. Engineering Geology, 2019, 260(3):105238.

[7] Iverson R M. Landslide triggering by rain infiltration[J]. Water Resources Research, 2000, 36(7): 1897−1910.

[8] Parker R N, Densmore A L, Rosser N J. Mass wasting triggered by the 2008 Wenchuan earthquake is greater than orogenic growth[J]. Nature Geoscience, 2011, 4(7): 449−452.

[9] Rawls W J, Brakensiek D L, Saxtonn K E. Estimating soil water properties[J]. Transactions of the ASAE, 1982, 25, 1316−1320.

[10] Saulnier G M, Beven K, Obled C. Including spatially variable effective soil depths in TOPMODEL[J]. Journal of Hydrology, 1997, 202(1−4): 158−172.

[11] Tang C, Zhu J, Chang M. An empirical-statistical model for predicting debris-flow runout zones in the Wenchuan earthquake area[J]. Quaternary International, 2012, 250: 63−73.

[12] Tu H W, Wang X K, Zhang W S. Flash flood early warning coupled with hydrological simulation and the rising rate of the flood stage in a mountainous small watershed in Sichuan Province, China[J]. Water, 2020, 12(1): 255.

Channel Evolution Along the Yichang-Chenglingji Reach Downstream of the Three Gorges DAM

Hualin Wang[1], Shan Zheng[2], Guangming Tan[3], Lingyun Li[4]

[1,2,3] State Key Laboratory of Water Resources and Hydropower Engineering Science,
Wuhan University, Wuhan 430072, China

[4] River Research Department, Changjiang River Scientific Research Institute, Wuhan 430012, China

Abstract

The operation of the Three Gorges Dam has caused profound morphology adjustment downstream of channel reaches since 2002. In this study, based on the data of water discharge, sediment load and cross-sectional surveys during 2002—2018, we studied the morphological adjustments of the Yichang-Chenglingji Reach downstream of the TGD. The channel reaches were divided into 32 sub-reaches according to the planform. The vertical, lateral and bankfull channel changes were calculated. The result indicated that the channel downstream of the TGD experienced fairly strong erosion during the initial impoundment of the Three Gorges Reservoir(from 2003—2007)—weakened erosion after the reservoir was operated with normal pool level of 175 m(2008—2012)—enhanced erosion after the cascade reservoirs upstream of the TGD started to operate(2012—2018).

Keywords: Three Gorges Dam; Yichang-Chenglingji Reach; Morphological evolution; Erosion centers; Spatial clustering

1 Introduction

Large dam emplacement can profoundly modify the natural flow and sediment regimes, which usually leads to important adjustments in downstream channels(Williams and Wolman, 1984; Ma et al., 2012). These adjustments usually present in different dimensions, including the river bed incision and armoring, the variation of longitudinal concavity of channel profile, lateral channel shift, and planform variation(Smith et al., 2016).

Similar phenomenon of the downstream adjustments have also been reported for the channel reaches downstream of the Three Gorges Dam(TGD), the largest dam in the world. For instance, the bed material had armored obviously at Zhicheng Station ~100 km downstream of TGD(Li et al., 2020). Lyu et al. (2019) revealed that the residual depth increased nonlinearly in the reaches downstream of the TGD. Bank erosion in the lower Jingjiang reach had also been widely reported (Deng et al, 2019). Besides, the channel adjustment is unevenly distributed(Williams and Wolman, 1984). Dong et al., (2019) studied the erosion and deposition processes from the channel reach between Yichang to Hukou(955 km downstream the TGD), and argued that the channel reach downstream of Chenglingji(~450 km downstream the TGD) started to be eroded obviously after 2012. Xu et al.(2021) proposed the area of coarse sediment supplement had moved downstream from Yichang-Zhicheng reach(38 km downstream the TGD) to Shashi-Jianli reach(~160 km downstream the TGD) based on studies of sediment exchange.

In this study, the morphological adjustment of the 408-km-long channel reach between Yichang and Chenglingji just downstream of the TGD during 2002—2018 is studied based on the data of water discharge, sediment load, bed material size and cross-sectional resurveys provided by the Changjiang Water Resources Commission(CWRC). The objectives of this paper are to (1) analyze the vertical, lateral and erosion/aggradation adjustments of the channel reaches, (2) compare erosion/aggradation characteristics at channel reaches with different distance to the dam and in different time periods.

2 Study Area and Data Source

2.1 Study area

The study channel reach is between Yichang and Chenglingji(denoted as YCR) downstream of the TGD, with a river length of ~408 km(Fig. 1). It is generally divided into the Yichang-Zhicheng reach (~60 km) and the Jingjiang reach(~348 km). The later can be further divided into the upper and lower reaches, with Ouchikou being the dividing point. The river bed surface layer is mainly composed of gravel, pebbles, and sand from Yichang to Yangjianao, which is located in Jing25 (Fig. 1). The riverbed of the upper Jingjiang reach downstream of Yangjianao mainly consists of moderately fine sand. The banks and bed of the lower Jingjiang reach are composed of loose fine sand (Lyu et al. 2019). The water is diverted into the Dongting Lake through three outlets (named Songzikou, Taipingkou and Ouchikou) on the right bank of the channel, and flows into the Yangtze River from Donging Lake at Chenglingji(Fig. 1).

According to the channel planform, we divided the YCR into 32 sub-reaches, each of them contains about 5 cross-sections and the length of them range between about 8~14 km(Fig. 1). Considering the single- or multi-thread patterns and the sinuosity, these sub-reaches were divided into six river patterns(i.e. meandering, slightly curved, straight, meandering branching, slightly curved branching and straight branching patterns, Fig. 1). according to the follows rule(Xie, 1997):

$$\Omega = \frac{L_0}{L} \begin{cases} \Omega > 1.2: \text{meandering} \\ 1.05 < \Omega \leqslant 1.2: \text{slightly curved} \\ \Omega \leqslant 1.05: \text{straight} \end{cases} \quad (1)$$

where Ω= the sinuosity of the sub-reach; L_0, L= the river length of the sub-reach, and the distance between the starting and ending point of the sub-reach, respectively.

The TGD started to impound water from 2003 with the initial pool level 135 m. Since 2008, the reservoir has been operated with 175 m as the normal pool level(SPT, 2013). The cascade reservoirs upstream of TGD, such as Xiluodu and Xiangjiaba reservoirs, have been put into operation after 2012. Considering the different operation modes of the TGD and the construction of the upstream cascade reservoirs, we divide the study time period into three sub-periods, i.e. Ⅰ. 2003—2007: the initial impoundment of TGD, Ⅱ. 2008—2012: normal operation of TGD, and Ⅲ. 2013—2018: post-construction of cascade dams upstream of the TGD.

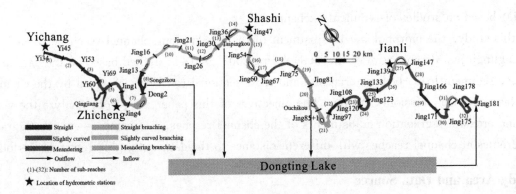

Figure 1. Yichang-Chenglingji Reach and the 32 sub-reaches

2.2 Data source

Our data was provided by Changjiang Water Resources Commission (CWRC), including 163 cross-sectional resurveys and the hydrological data at hydrometric stations of Yichang, Zhicheng, Shashi and Jianli (Table 1). The annual-averaged water discharge generally increased, while the annual-averaged sediment load decreased in the three time periods (Table 2). The water discharge increased by 2% and 4% in the 2nd time period and the 3rd time period at Yichang station, respectively. The annual sediment load reduced by 55% and 63% in 2nd time period and 3rd time period at Yichang station, respectively, and the reducing extent decreased in the downstream direction. The ratio of water discharge and sediment load transported during flood seasons to those in years both declined with time at the four hydrometric stations.

Table 1. Sources of water and sediment regimes

Data type	Location	Period of record	Sources
Daily discharge	Yichang, Zhicheng, Shashi, Jianli	2002—2017	CWRC
Daily sediment concentration	Yichang, Zhicheng, Shashi, Jianli	2002—2017	CWRC
Cross-sectional resurveys	163 cross-sections within the YCR	2002—2018	CWRC

Table 2. Water and sediment conditions at the hydrometric stations

Period	Station	2003—2007	2008—2012	2013—2018
Water discharge (billion m³)	Yichang	394	402	419
	Zhicheng	402	417	425
	Shashi	372	380	388
	Jianli	356	370	377
Sediment discharge (million t)	Yichang	67	30	11
	Zhicheng	82	35	13
	Shashi	93	46	24
	Jianli	102	60	41

3 Methods

Based on the 163 cross-sectional surveys, we calculated the thalweg elevation, lateral migration rates of thalweg, and bankfull area at the cross-sections. For the cross-sections with multiple branches, the thalweg elevation was determined as the lower elevation between the left and right thalweg. The bankfull area refers to the cross-sectional area below the bankfull elevation, which is taken as the elevation of the lower lip between the left and right bankfull lips. An empirical relationship between bankfull elevation and distance to the dam was obtained based on cross-sections with obvious bankfull lips. This relationship was used to estimate the bankfull elevation at those without obvious bankfull lips. After the bankfull elevation was determined, we calculated the bankfull width W and bankfull depth H according to the cross-sectional profile.

The lateral migration distance of thalweg ΔX (m) at a cross-section of a single-thread pattern is calculated as follows:

$$\Delta X = |x_{t+1} - x_t| \qquad (2)$$

where x_t, x_{t+1} = the distance to the monument of thalweg in t^{th} and $(t+1)^{th}$ year, respectively.

The lateral migration distance of thalweg at a cross-section with two branches is taken as the sum of the lateral migration distance of thalweg in each branch.

$$\Delta X = |x_{t+1}^1 - x_t^1| + |x_{t+1}^2 - x_t^2| \qquad (3)$$

where x_t^1, x_{t+1}^1 = the distance to the monument of the thalweg of the left branch at a cross-section in t^{th} and $(t+1)^{th}$ year, respectively. x_t^2, x_{t+1}^2 = the distance to the monument of the thalweg of the right branch at a cross-section in t^{th} and $(t+1)^{th}$ year, respectively.

Based on change in the thalweg elevation ΔZ, lateral migration distance of thalweg ΔX, bankfull area ΔA and bankfull depth ΔH at each cross-section, the adjustment of these morphological variables in reach-scale may be calculated by:

$$\bar{G} = \frac{\sum_{1}^{M-1}(G_j + G_{j+1})/2 \times L_{j,j+1}}{L_{1,M}} \qquad (4)$$

where $\bar{G}$ = the reach-scale changes in the morphological variables. G_j and G_{j+1} = the changes in the morphological variables at cross-sections j and $(j+1)$, respectively. $L_{j,j+1}$ and $L_{1,M}$ = the river length between the cross-sections j and $(j+1)$, and the river length of the calculated channel reach. M = the number of cross-sections in the calculated channel reach.

Based on bankfull area of 163 cross-sections, we calculated the erosion/deposition volume at the 32 sub-reaches (Zheng et al. 2014),

$$V = \sum_{j=1}^{M-1} \frac{\Delta A_j + \Delta A_{j+1}}{2} \times L_{j,j+1} \qquad (5)$$

where ΔA_j and ΔA_{j+1} = changes of bankfull area of cross-sections j and $(j+1)$, respectively. $L_{j,j+1}$ = river length between the cross-sections j and $(j+1)$. M = number of cross-sections in the calculated channel reach.

4 Results

4.1 Longitudinal adjustments

The variation of the thalweg elevation and bankfull depth showed that the YCR was generally eroded from 2003 to 2018(Fig. 2). The erosion rate was the greatest in the 1st time period(2003—2007), it decreased in the 2nd time period(2008—2012), and then increased again in the 3rd time period (2013—2018). For instance, the sum of the length of the eroded sub-reaches accounted for 93%, 69% and 88% of the total length of YCR in the three time periods, respectively. The average change rate of thalweg elevation was -0.41, -0.17, -0.2 m/yr. in the three time periods, respectively. The change rate of bankfull depth was smaller than that of thalweg elevation, and was 0.2, 0.12 and 0.14 m/yr. in the three time periods, respectively. Although the erosion rate increased after 2012, it was still smaller than that in the first time period.

4.2 Lateral migration

The lateral migration rates of thalweg decreased in both single- and multi-thread channel, and the lateral activity of single-thread channel decreased slightly, whereas the lateral migration of multi-thread channel obviously declined(Fig. 3). The lateral migration rates of single-thread sub-reaches were 70, 63 and 51 m/yr. in the three time periods, respectively, whereas the migration rates of multi-thread sub-reaches were 176, 125 and 57 m/yr. in the three time periods, respectively. The reasons are various for thalweg migration in single- and multi-thread channel. For single-thread channel, the bank erosion in the convex bank may induce the lateral migraion of thalweg[Fig. 4(a)]. Thalweg migration of the multi-thread channel may be caused by the greater migration of thalweg in the main branch with relatively large river width, such as the right branch of Jing 143[Fig. 4(b)].

4.3 Erosion and deposition volume

The erosion rates in the Yi-Zhi, upper and lower Jingjiang reaches were 2.78, 3.78 and 2.69 million m³/km from 2003 to 2018, respectively, indicating the greatest erosion rate in the upper Jingjiang reach[Fig. 5(a)]. The erosion rate of the Yi-Zhi reach reduced nonlinearly induced by the riverbed armoring probably. However, the erosion rate of the upper Jingjiang reach tended to increase after about 2010, which may be ascribed to the reduction of sediment supply in Yi-Zhi reach [Fig. 5(a)]. The erosion rate of Yi-Zhi reach was smaller than that of the upper Jingjiang reach, which may because that the river bed is mainly composed of gravel, pebbles, and sand in Yi-Zhi reach with higher anti-erodibility, whereas the river surface layer of the upper Jingjiang reach mainly consists of moderately fine and fine sand with lower anti-erodibility. Therefore, the erosion rates decreased gradually in the YCR from 2003 to 2009, and increased again when the upper Jingjiang reach became the main sediment supply reach after 2010.

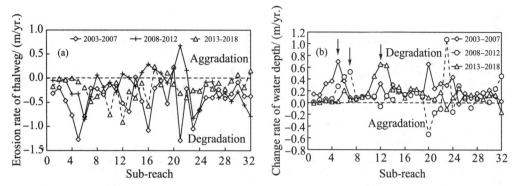

Figure 2. Change rate of (a) thalweg elevation, and (b) water depth

The erosion/deposition rates at the sub-reaches showed three "erosion peaks" at sub-reaches No. 5, No. 7 and No. 13 during the three time periods, respectively [Fig. 5(b)]. The erosion rates were 700, 650 and 1000 thousand m^3/km/yr of the three erosion peaks., respectively. The erosion peaks tended to migrate downstream as time elapsed. Interestingly, bankfull depth also increased obviously at the three erosion peaks in the sub-reaches [Fig. 2(b)], implying that bed incision was significant in the channel erosion processes downstream of dams (Lyu et al., 2019).

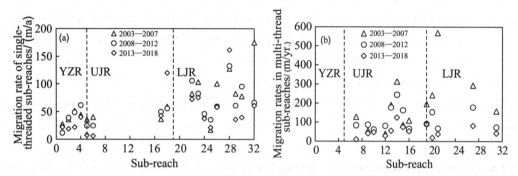

Figure 3. Lateral migration rates of (a) single- and (b) multi-thread sub-reaches in three time periods

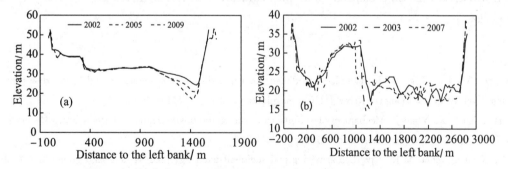

Figure 4. Typical cross-sections in single- and multi-thread sub-reaches

Note: (a) Yi 63 in sub-reach No. 4; (b) Jing 143 in sub-reach No. 27.

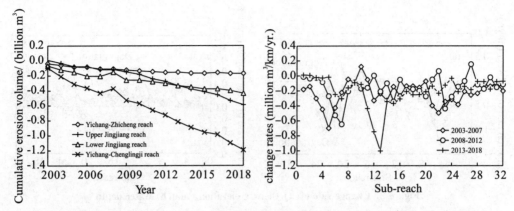

Figure 5. (a) Cumulative erosion volume of channel reaches, (b) Erosion/deposition rates

5 Conclusion

Based on the data of water discharge, sediment load, bed material size and cross-sectional surveys, we analyzed the vertical, lateral and bankfull channel geometry changes in the Yichang-Chenglingji Reach downstream of the TGD. The main conclusions are as follows:

(1) The Yichang-Chenglingji reach downstream of the TGD experienced fairly strong erosion during the initial impoundment of the Three Gorges Reservoir(from 2003—2007)—weakened erosion after the reservoir was operated with normal pool level of 175 m(2008—2012)—enhanced erosion after the cascade reservoirs upstream of the TGD started to operate(2012—2018).

(2) The lateral migration of single- and multi-thread channel weakened drastically and changed unobviouly after 2013, respectively; The migration rates in lower Jingjiang were smaller than that in Yi-zhi Reach.

(3) The erosion/deposition rates at sub-reaches presented three "erosion peaks", which tended to migrate downstream as time elapsed, and the bankfull depth also increased obviously at the three erosion peaks.

References

[1] Deng S S, Xia J Q, Zhou M R. Coupled two-dimensional modeling of bed evolution and bank erosion in the upper jingjiang reach of middle Yangtze River[J]. Geomorphology, 2019, 344(1): 10−24.

[2] Dong B J, Xu Q X, Yuan J. Mechanism of serious scour along the downstream of Three Gorges Reservoir in recent years[J]. Journal of Sediment Research, 2019,44(5): 42−47.

[3] Li L L, Xia J Q, Zhou M R. Riverbed armoring and sediment exchange process in a sand-gravel bed reach after the Three Gorges Project operation[J]. Acta Geophysica, 2020, 68(1).

[4] Lyu Y W, Zheng S, Tan G M. Effects of three gorges dam operation on spatial distribution and evolution of channel thalweg in the yichang-chenglingji reach of the middle Yangtze River, China[J]. Journal of Hydrology, 2018, 565: 429−442.

[5] Ma Y X, Huang H Q, Nanson G C. Channel adjustments in response to the operation of large dams: the upper reach of the lower Yellow River[J]. Geomorphology, 2012, 147−148(1−4): 35−48.

[6] Smith N D, Morozova G S, Pérez-Arlucea M. Dam-induced and natural channel changes in the Saskatchewan River below the E. B. Campbell Dam, Canada[J]. Geomorphology, 2016, 269: 186−202.

[7] Xie J H. Riverbed evolution and regulation[M]. Beijing: China water resources and electric power press, 1997.

[8] Xu Q X, Li S X, Yuan J. Analysis of equilibrium sediment transport in the middle and lower reaches of the Yangtze

River after the impoundment of Three Gorges reservoir[J]. Journal of Lake Sciences, 2021(3): 806−818.

[9] Williams G P, Gordon W M. Downstream effects of dams on alluvial rivers[M]. Washington: U. S. Government Printing Office, 1984.

Comparative Analysis of Comprehensive Evaluation Methods for Dam-break Consequence

Weiwei Sun[1,2], Qian Cai[1,2], Zhifei Long[1,2]

[1] Nanjing Hydraulic Research Institute, Nanjing 210029, China
[2] Dam Safety Management Center of the Ministry of Water Resources, Nanjing 210029, China

Abstract

Dam failure consequence assessment is one of the core elements in the calculation of dam risk. The accuracy of dam failure consequence assessment is directly related to the degree of dam risk. Dam-break consequence assessment is a complex and systematic work, which needs to carry out four aspects of assessment, including life loss, economic loss, environmental impact and social impact, and the four aspects of assessment results are synthesized to get the comprehensive assessment results. In this paper, five methods, linear weighted comprehensive evaluation, fuzzy mathematics comprehensive evaluation, matter-element comprehensive evaluation, grey correlation degree comprehensive evaluation and principal component analysis comprehensive evaluation, are applied to evaluate the dam-break consequences of five reservoirs which are Changlong, Xialan, Shibikeng, Longshan and Lingtan respectively, and the severity ranking of the dam-break consequences of five reservoirs is obtained. The calculation results and the characteristics of the five evaluation methods are compared and analyzed. It is concluded that when evaluating the severity ranking of dam-break consequences for multiple reservoirs, the linear weighted comprehensive evaluation method is preferred, followed by the matter-element comprehensive evaluation method.

Keywords: Dam-break consequence; Evaluation method; Comparative analysis; Severity ranking

1 Introduction

Dam risk is a combination of the probability of dam failure and the consequences of dam failure (Lou J K, 2000). The consequence of dam failure is the core content of dam risk analysis and evaluation. At present, many foreign countries, such as the United States, Canada, Australia, Western Europe, South Africa, etc. have carried out early research on the technology of dam failure assessment. The assessment of dam-break consequence is one of the core elements for calculating the risk of a dam. Whether the assessment of the consequences of dam-breaking losses is directly related to the magnitude of the dam risk. The assessment of the consequence of dam-break is a complicated, tedious and systematic work. It needs to carry out four aspects of assessment such as life loss, economic loss, environmental impact and social impact.

There are several methods for estimating life loss: Brown-Graham method of the US Bureau of Reclamation(Dekay M L, and Mcclelland G H, 1993), Dekay-McCleland method(Brown C A, and Graham W J, 1988), Graham method, Finland's RECDAM method(Reiter P, 2001), Canada's Assaf method Utah State University Act(Aboelata M., Bowles D.S., and McClelland D.M., 2002), etc.

In China, Li L et al. (Li L, and Zhou K F, 2006) summarized several key parameters affecting life loss estimation, namely the population at risk, the severity of flood, the alarm time, and the public's understanding of the severity of dam-break events. Zhou K F(Zhou K F, Li L, and etc., 2007) analyzed the research results of the loss of life of eight ruptured dams, analyzed the basic laws and characteristics of the loss of life of ruptured dams in China, summarized the range of mortality of the risk population, and made recommendations. Reference values, in terms of dam-break life loss estimate, put forward some indirect influencing factors, including the proportion of young and middle-aged adults in the risk population composition, weather, dam-break occurrence time, distance from the dam, emergency plan implementation, dam height, the storage capacity, the descent of the dam downstream, and the buildings' impact resistance, etc., and based on this, a Li-Zhou model that is more suitable for the actual situation of dam failure in China is proposed.

For the estimation of dam-break economic loss, some research work has been carried out at home and abroad, but China's research results are more operable and can be directly used for estimation. Dam-break economic losses include both the damage caused by reservoir engineering damage and damage caused by dam-break flood.

In terms of dam-break social impact assessment, Li L et al. (Li L, Wang R Z, Sheng J B and etc., 2006) believed that the social impacts of reservoir dam-breaking mainly include political influences(that is, adverse effects on national and social stability), and physical and mental health caused by injuries or mental stress. Injuries, degradation of daily living standards and quality of life, and other irreparable losses of cultural relics, art treasures and rare animals and plants, so the impact factors are divided into risk population, important cities, important Facilities, as well as cultural relics, art treasures and rare animals and plants, are used to comprehensively estimate the social impact of reservoir dam-break.

In this paper, five methods including linear weighted comprehensive evaluation, fuzzy mathematical comprehensive evaluation, matter-element comprehensive evaluation, gray correlation comprehensive evaluation, and principal component analysis comprehensive evaluation are applied to the five reservoirs of Changlong, Xialan, Shibikeng, Longshan, and Lingtan respectively. A case evaluation of the dam-break consequences was carried out, and the severity ranking of the dam-break consequences of the five reservoirs was obtained. According to the calculation results, the characteristics and results of the five evaluation calculation methods are compared and analyzed. It is concluded that when ranking the severity assessment of dam-break consequences of multiple reservoirs, the linear weighted comprehensive evaluation method is recommended first, followed by the matter-element comprehensive evaluation law.

2 Comorehensive Evaluation Methods

2.1 Linear weighted comprehensive evaluation method

Linearity refers to the proportional and linear relationship between quantity and quantity. The working attribute between the output quantity and the input quantity conforms to the principle of superposition. It can be mathematically understood as a function whose first derivative is constant(Shi Z K, 2008). Comprehensive evaluation of the dam-break consequences(outputs), the evaluation indicators(inputs) which are loss of life, economic loss, environmental impact and social impact.

$$L = \sum_{j=1}^{4} w_j F_j \qquad (1)$$

In the formula, L is the comprehensive severity coefficient, w_j is the weight of each evaluation index, and F_j is the severity coefficient.

The linearly weighted comprehensive evaluation expression of the dam break consequence is as follows:

$$L = \sum_{j=1}^{4} w_j F_j = 0.636F_1 + 0.091F_2 + 0.136F_3 + 0.136F_4 \qquad (2)$$

In the formula, w_j weight coefficients for each evaluation index, namely, life loss, economic loss, environmental impact, and social impact, are 0.636, 0.091, 0.136, and 0.136, respectively; F_1, F_2, F_3, F_4 are life loss, economic loss, and environment, respectively. Impact and social impact severity coefficients are calculated based on comprehensive evaluation function L of Li Lei et al. for the consequences of dam failure(Li L, Wang R Z and Sheng J B, 2006).

2.2 Comprehensive evaluation method of fuzzy mathematics

Fuzzy theory mainly uses the basic concepts of fuzzy sets and membership function theory(Kerre E E, Huang C F, and Ruan D, 2004), which can be roughly divided into five branches: fuzzy mathematics, fuzzy systems, fuzzy decision making, uncertainty and information, fuzzy logic and artificial intelligent. Fuzzy mathematics is a new branch of mathematics that uses fuzzy sets as the basic theoretical basis. It uses mathematical methods to study and deal with fuzzy phenomena. It provides a new method for dealing with uncertainties and inaccuracy. A powerful tool for information. Calculation steps of comprehensive evaluation model of dam-break consequences based on fuzzy mathematics are as follow.

2.2.1 *Determination of the set of factors for the thing being evaluated*

The set of factors that determine the thing to be evaluated is $U = \{u_1, u_2, u_3, u_4\}$, u_1, u_2, u_3, u_4 which respectively mean loss of life, economic loss, environmental impact, and social impact.

2.2.2 *Determination of the evaluation level set*

Determine the evaluation level set $V = \{v_1, v_2, v_3, v_4\}$, v_1, v_2, v_3, v_4 which mean general accident, major accident, major accident, especially major accident.

2.2.3 *Determination of the weight distribution set among various factors*

Determine the weight distribution set among various factors $W = \{w_1, w_2, w_3, w_4\}$, w_1, w_2, w_3, w_4 which respectively represent the weight of the loss of life, economic loss, environmental impact, and social impact in the comprehensive evaluation of the consequences of dam failure, which are 0.636, 0.091, 0.136, and 0.136.

2.2.4 *Determine the fuzzy evaluation matrix*

For each thing being evaluated, the relationship between the evaluation elements and the evaluation level, that is, the fuzzy relationship from to, can be described by a fuzzy evaluation matrix, which is expressed by $R = \begin{bmatrix} r_{11} & r_{12} & r_{13} & r_{14} \\ r_{21} & r_{22} & r_{23} & r_{24} \\ r_{31} & r_{32} & r_{33} & r_{34} \\ r_{41} & r_{41} & r_{41} & r_{41} \end{bmatrix}$ The element in R ; r_{ij} ($i = 1, 2, \cdots, 4, j = 1, 2, \cdots, 4$)

indicates the degree of membership of v_j in which the thing being evaluated can be rated from the perspective of factors u_i.

According to the theoretical analysis, the trapezoid membership function is selected for the comprehensive evaluation of the consequences of dam-break.

The membership function for x rank is

$$r_{ij}(x) = \begin{cases} 1 & c_1 < x \leqslant c_2 \\ \dfrac{c_3-x}{c_3-c_2} & c_2 < x \leqslant c_3 \\ 0 & c_3 < x \end{cases} \tag{3}$$

The membership function of x for B、C ranks is

$$r_{ij}(x) = \begin{cases} 0 & x \leqslant c_{i-1} \\ \dfrac{x-c_{i-1}}{c_i-c_{i-1}} & c_{i-1} < x \leqslant c_i \\ \dfrac{c_{i+2}-x}{c_{i+2}-c_{i+1}} & c_{i+1} < x \leqslant c_{i+2} \\ 0 & x > c_{i+2} \end{cases} \tag{4}$$

The membership function of x for D ranks is

$$r_{ij}(x) = \begin{cases} 0 & x \leqslant c_3 \\ \dfrac{x-c_3}{c_4-c_3} & c_3 < x \leqslant c_4 \\ 1 & c_4 < x \leqslant c_5 \end{cases} \tag{5}$$

In the formula, x is the value of the evaluation index of dam-breaking consequences; c_i is the standard value corresponding to the level i of the accident; $r_{ij}(x)$ is the degree of membership of the index i corresponding to the level j.

Four levels of A, B, C, D corresponding to general accidents are major accidents, major accidents, particularly major accidents. Using the membership function, the degree of membership of the four evaluation indicators to the four types of accident levels is obtained, and the membership matrix R.

2.2.5 *Solution of comprehensive evaluation results*

Using ordinary matrix multiplication, the comprehensive evaluation result B of the thing being evaluated can be obtained, that is

$$B = W \cdot R = (w_1, w_2, w_3, w_4) \cdot \begin{bmatrix} r_{11} & r_{12} & r_{13} & r_{14} \\ r_{21} & r_{22} & r_{23} & r_{24} \\ r_{31} & r_{32} & r_{33} & r_{34} \\ r_{41} & r_{42} & r_{43} & r_{44} \end{bmatrix} = (b_1, b_2, b_3, b_4) \tag{6}$$

Further normalize the result of $B = (b_1, b_2, b_3, b_4)$, and get the final result as $B' = (b'_1, b'_2, b'_3, b'_4)$.

2.3 Matter-element comprehensive evaluation method

Matter-element theory is a new discipline that is developed on the basis of classical mathematics and fuzzy mathematics. Matter-element analysis theory consists of matter-element theory and extension set theory. The basic unit of intricate things is called matter-element, and the possibility of things change is called the extension of things. It describes the internal structure of things and the various external relationships. Matter-element analysis theory takes matter-element as the basic unit, and applies matter-element transformation to the contradiction problem as a compatibility problem. The basis of matter-element transformation is extension(Yang C Y, and Cai W, 2007).

2.3.1 Establishment of matter-element matrix

The consequences of dam-break N need to be described by four characteristics c_1, c_2, c_3, c_4, namely loss of life, economic loss, environmental impact, social impact, and the corresponding magnitude x_1, x_2, x_3, x_4. It is called a 4-dimensional matter element and is expressed by a matrix as

$$R = (N, c_i, x_i) = \begin{bmatrix} N & c_1 & x_1 \\ & c_2 & x_2 \\ & c_3 & x_3 \\ & c_4 & x_4 \end{bmatrix} \tag{7}$$

2.3.2 Matter-element matrix of classical domain and node domain

The matter-element matrix of the classical domain can be expressed as

$$R_{oj} = (N_{oj}, c_i, x_{oij}) = \begin{bmatrix} N_{oj} & c_1 & x_{o1j} \\ & c_2 & x_{o2j} \\ & c_3 & x_{o3j} \\ & c_4 & x_{o4j} \end{bmatrix} = \begin{bmatrix} N_{oj} & c_1 & \langle a_{o1j} & b_{o1j} \rangle \\ & c_2 & \langle a_{o2j} & b_{o2j} \rangle \\ & c_3 & \langle a_{o3j} & b_{o3j} \rangle \\ & c_4 & \langle a_{o4j} & b_{o4j} \rangle \end{bmatrix} \tag{8}$$

In the formula: N_{oj} is the level j ($j=1,2,3,4$) of things, which are four levels of general accident, major accident, major accident, and particularly serious accident; c_i is the feature i of the level j of things; x_{oij} is the magnitude range of N_{oj} on c_i, that is, the classic domain $< a_{oij}, b_{oij} >$ of each level on corresponding features.

2.3.3 Determination of the matter element to be evaluated

The matter-element matrix of the classical domain can be expressed as

$$P_k = (P_k, c_i, x_i) = \begin{bmatrix} P_k & c_1 & x_1 \\ & c_2 & x_2 \\ & c_3 & x_3 \\ & c_4 & x_4 \end{bmatrix} \tag{9}$$

In the formula, P_k is the thing to be evaluated ($k = 1, 2\cdots, l$); x_i is the magnitude of P_k about c_i, that is the actual data of each feature.

2.3.4 Calculation of correlation function and correlation degree

According to the theoretical analysis, $k_1(x)$-type correlation function formula is used for the comprehensive evaluation of the dam-break consequences. The correlation function indicates the extent of compliance with the requirements when the value of the matter element is a point on the real axis. The relevance of the object P_k to the level j is

$$k_j(P_k) = \sum_{i=1}^{4} a_i k_j(x_i) \tag{10}$$

In the formula: a_i is the weight coefficient.

2.3.5 Determination of the weight distribution set

Determine the weight distribution set $W = \{w_1, w_2, w_3, w_4\}$ among various factors, w_1, w_2, w_3, w_4 respectively indicate the weight of the loss of life, economic loss, environmental impact, and social impact in the comprehensive evaluation of the consequences of dam failure is 0.636, 0.091, 0.136, and 0.136.

2.3.6 Evaluation levels and standards

According to the principle of maximum membership, the maximum correlation function value is sought in $k_j(P_k)$.

$$k'_j(P_k) = \max[k_1(P_k), k_2(P_k), k_3(P_k), k_4(P_k)] \tag{11}$$

Then the thing to be evaluated P_k should belong to the level j.

2.4 Comprehensive evaluation method of grey correlation

Grey system theory is in the 1980s, and it is an applied mathematics subject whose research information is partially clear, partially unclear, and with uncertain phenomena (Deng J. L., 1992). The comprehensive evaluation of dam-break consequences based on the grey correlation degree method is mainly to calculate the degree of correlation between the indicators of dam-breaking consequences of each reservoir to be evaluated and its standard series. The severity of the consequences of the dam failure of each reservoir to be evaluated is comprehensively compared by the correlation degree. The analysis and calculation steps are as follows.

2.4.1 Determination of the evaluation standard sequence

There are m evaluation indicators $x_1, x_2, \cdots, x_m$. The consequences of dam failure are four evaluation indicators: loss of life, economic loss, environmental impact, and social impact; there are n reservoirs to be evaluated; and there are P dam collapse consequence levels, which are general accidents, major accidents, major accidents, and particularly serious accidents.

The evaluation standard matrix is

$$x = \begin{bmatrix} x_{11} & x_{12} & \cdots & x_{1m} \\ x_{21} & x_{12} & \cdots & x_{2m} \\ \vdots & \vdots & \vdots & \vdots \\ x_{p1} & x_{p2} & \cdots & x_{pm} \end{bmatrix} = \begin{bmatrix} x_{11} & x_{12} & \cdots & x_{14} \\ x_{21} & x_{22} & \cdots & x_{24} \\ \vdots & \vdots & \vdots & \vdots \\ x_{41} & x_{42} & \cdots & x_{44} \end{bmatrix} \tag{12}$$

2.4.2 Dimensionless Processing of data

The dimensionless matrix of evaluation criteria is

$$x = \begin{bmatrix} X_{11} & X_{12} & \cdots & X_{1m} \\ X_{21} & X_{12} & \cdots & X_{2m} \\ \vdots & \vdots & \vdots & \vdots \\ X_{p1} & X_{12} & \cdots & X_{pm} \end{bmatrix} = \begin{bmatrix} X_{11} & X_{12} & \cdots & X_{14} \\ X_{21} & X_{22} & \cdots & X_{24} \\ \vdots & \vdots & \vdots & \vdots \\ X_{41} & X_{42} & \cdots & X_{44} \end{bmatrix} \tag{13}$$

2.4.3 Determination of the correlation coefficient

There are n reservoirs to be evaluated. When each reservoir is evaluated separately, i is a fixed value and the range is 1-n. Get the correlation coefficient matrix:

$$E = \begin{bmatrix} \varepsilon_{11} & \varepsilon_{12} & \cdots & \varepsilon_{1m} \\ \varepsilon_{21} & \varepsilon_{12} & \cdots & \varepsilon_{2m} \\ \vdots & \vdots & \vdots & \vdots \\ \varepsilon_{p1} & \varepsilon_{12} & \cdots & \varepsilon_{pm} \end{bmatrix} = \begin{bmatrix} \varepsilon_{11} & \varepsilon_{12} & \cdots & \varepsilon_{14} \\ \varepsilon_{21} & \varepsilon_{22} & \cdots & \varepsilon_{24} \\ \vdots & \vdots & \vdots & \vdots \\ \varepsilon_{41} & \varepsilon_{42} & \cdots & \varepsilon_{44} \end{bmatrix} \quad (14)$$

2.4.4 *Determination of the weight distribution set among various factors*

Determine the weight distribution set $W = \{w_1, w_2, w_3, w_4\}$ for each factor, w_1, w_2, w_3, w_4 respectively represents: the weight of the loss of life, economic loss, environmental impact, and social impact in the comprehensive evaluation of the consequences of dam failure, namely 0.636, 0.091, 0.136, 0.136.

2.4.5 *Calculation of relevance and rank*

As the correlation coefficient only reflects the degree of correlation between the reservoir to be evaluated and the evaluation standard samples at all levels, the $r_k = \sum_{j=1}^{m} w_j \varepsilon_{kj} = 0.636 \times \varepsilon_{k1} + 0.091 \times \varepsilon_{k2} + 0.136 \times \varepsilon_{k3} + 0.136 \times \varepsilon_{k4}$ is used to calculate the comprehensive correlation between the dam break consequences to be evaluated and the evaluation standard samples at all levels.

The maximum $r_k^* = \max\limits_{1 \leqslant k \leqslant p} \{r_1, r_2, \cdots, r_p\} = \max\limits_{1 \leqslant k \leqslant 4} \{r_1, r_2, \cdots, r_4\}$ calculated from the above formula.

The corresponding dam break consequence level is the dam break consequence level of the reservoir to be evaluated.

2.5 Principal component analysis comprehensive evaluation method

The principal component analysis comprehensive evaluation method is a statistical method for comprehensive evaluation of multiple indicators. Through appropriate mathematical transformations, the new variables become linear combinations of the original variables, and the principal component is used to comprehensively evaluate the consequences of dam-break(Zhang P, 2004). Calculation steps of comprehensive evaluation model of dam-break consequences based on principal component analysis.

$$X = \begin{bmatrix} x_{11} & x_{12} & \cdots & x_{1p} \\ x_{21} & x_{22} & \cdots & x_{2p} \\ \vdots & \vdots & \vdots & \vdots \\ x_{n1} & x_{n2} & \cdots & x_{np} \end{bmatrix} = \begin{bmatrix} x_{11} & x_{12} & \cdots & x_{14} \\ x_{21} & x_{22} & \cdots & x_{24} \\ \vdots & \vdots & \vdots & \vdots \\ x_{n1} & x_{n2} & \cdots & x_{n4} \end{bmatrix} \quad (15)$$

In the formula, n is the number of samples, that is, the number of reservoirs to be evaluated; p is the number of variables, that is, four variables: loss of life, economic loss, environmental impact, and social impact.

2.5.1 *Establishing a correlation matrix of variables*

$$R = (r_{ij})_{p \times p} = (r_{ij})_{4 \times 4} \quad (16)$$

In the formula, r_{ij} is the correlation coefficient between indicator X_i and indicator X_j, note:

$$r_{ij} = \frac{1}{n-1} \sum_{k=1}^{n} x_{ki}^* x_{kj}^* \quad (17)$$

2.5.2 *Calculating characteristic equation*

Solve the characteristic root of R, $\lambda_1 \geqslant \lambda_2 \geqslant \cdots \geqslant \lambda_P > 0$, in this case that: $\lambda_1 \geqslant \lambda_2 \geqslant \cdots \geqslant \lambda_4 > 0$, and its corresponding unit feature vector. The greater the variance, the greater the contribution to the total variance.

2.5.3 *Determination of principal components*

Calculate the variance contribution rate of each component.

$$e_j = \lambda_j / \sum_{j=1}^{p} \lambda_j \times 100\% \tag{18}$$

The number of principal components selected depends on the cumulative variance contribution rate of principal components.

$$\sum_{j=1}^{m} e_j m = 1,2,\cdots,p, \text{ in this case that } m = 1,2,3,4 \tag{19}$$

Write the main components:

$$Y_j = a_{1j}X_1 + a_{2j}X_2 + \cdots + a_{pj}X_p \quad j = 1,2,\cdots,p, \text{ in this case that } j = 1,2,3,4 \tag{20}$$

2.5.4 *Calculation of comprehensive score*

$$z = \sum_{j=1}^{m} e_j Y_j \tag{21}$$

3 Basic Information of Five Reservoirs

3.1 Project overview

3.1.1 *Changlong Reservoir*

Changlong Reservoir is located in Changjing Village, Xingguo Town, Xingguo County, Jiangxi Province. It is located on the tea garden of the Pingjiang tributary of the Ganjiang River system. It has a rainfall area of 116 km² and a total storage capacity of 16.85 million m³. It is a medium-sized reservoir with irrigation-based, combined power generation, aquaculture and other comprehensive benefits. The dam is an earth-rock mixed dam with a dam top elevation of 216.30 m, a wave barrier top elevation of 216.94 m, a dam top width of 5 m and a length of 170 m, and a maximum dam height of 42.96 m.

The reservoir's geographical location is important. The downstream flood protection protects 200,000 people, and Xingguo County (17 km downstream), the Beijing-Kowloon Railway (300 m downstream), 319 National Highway (500 m downstream), "General Park", a patriotic education base, Zeng The safety of important towns and infrastructures, such as the Three Classes, will cost more than 3.5 billion yuan in the event of a crash.

3.1.2 *Xialan Reservoir*

Xialan Reservoir is located in Jicun Village, Xianxia Township, Yudu County. It is located on the Xianxia River, a tributary of the Meijiang River in the Ganjiang River. The reservoir has a rainfall area of 32 km² and a total storage capacity of 11.68 million m³. It is a medium-sized reservoir with mainly irrigation, flood prevention, power generation and comprehensive benefits such as breeding. The dam is a clay sloped rockfill dam. The height of the dam top is 209.20 m, the maximum dam height is 33.40 m, the width of the dam top is 5 m, and the length of the dam top is 141 m.

The geographical position of Xialan Reservoir is important. The provincial highway Yuning Highway is 4km downstream, and there are also important military facilities and the Long March First

Crossing. In the lower reaches of Xianxia and Chexi, there are 240,000 towns with a population of 240,000. In the event of a crash, the direct economic loss exceeds 2.5 billion yuan.

3.1.3 *Shibikeng Reservoir*

Shibikeng Reservoir is located in Xiaoba Village, Wenwu Town, Huichang County, Jiangxi Province. It is located on the Bankeng River, a tributary of the Gongjiang River in the Ganjiang River. It has a rain-collecting area of 164 km^2 and a total storage capacity of 58.6 million m^3. It is a medium-sized reservoir with mainly irrigation, flood control, power generation and comprehensive benefits such as breeding and urban water supply. The main dam is a core-wall earth dam, with a dam top elevation of 203.64 m, a maximum dam height of 36 m, a dam top length of 140 m, and a width of 5.5 m.

Shibikeng Reservoir has an important geographical location. 5km downstream of the reservoir are important towns such as Huichang County, Wenwuba Town, Zhuangkou Town, Zhuangbao Township, and Baie Township, as well as 206 National Road and Provincial Road and other infrastructure facilities. The downstream flood protection protects 250,000 people, once the project crashed, the direct economic loss exceeded 4.1 billion yuan.

3.1.4 *Longshan Reservoir*

Longshan Reservoir is located in Chetou Village, Rentian Town, Ruijin City, Jiangxi Province. It is located on the Sanguan River, a tributary of the Mianjiang River in the Ganjiang and Gongjiang River systems. It has a rainfall area of 80 km^2 and a total storage capacity of 28.1 million m^3. It is a medium-sized reservoir with comprehensive benefits such as power generation and breeding. The main dam is a clay core wall dam, with a dam top elevation of 245.80 m, a maximum dam height of 33.50 m, a dam top width of 5.00 m and a length of 287.36 m.

The geographical location of the reservoir is very important. Downstream there are the urban area of Ruijin City, the former Soviet Central Government Yeping site, Rentian Town, Yeping Township, Huangbai Township, Xianghu Town, Shazhouba Town, Zeqin Township, Wuyang Town, Xiefang Town, Ganlong Railway, National Highway 206, 319, 323 and other important facilities. The downstream flood protection protected 240,000 people. Once the project crashed, the direct economic loss exceeded 2.5 billion yuan.

3.1.5 *Lingtan Reservoir*

Lingtan Reservoir is located in Changling Village, Ziyang Township, Shangyou County, Jiangxi Province. It is located on Ziyang Water, a tributary of the Zhangjiang River, a third branch of the Ganjiang River. It has a rain-collecting area of 26.5 km^2 and a total storage capacity of 15 million m^3. It is a medium-sized reservoir with comprehensive benefits such as power generation and breeding. The main dam is a masonry block gravity dam with a dam top elevation of 397.72 m, a wave wall top elevation of 398.22 m, a dam top width of 4.3 m, a length of 132.5 m, and a maximum dam height of 38.4 m.

Lingtan Reservoir has an important geographical location. There are 5 towns and towns in Shangyou County and Nankang City, important military facilities, and Hunan-Jiangxi inter-provincial highway in the downstream. The downstream flood protection protected 180,000 people. Once the project crashed, the direct economic loss exceeded 2 billion yuan.

3.2 Loss data

Statistics of life loss, economic loss, environmental impact and social impact of five reservoirs are shown in Table 1.

Table 1. Loss data of dam-break of five reservoirs

Name of reservoir	Changlong	Xialan	Shibikeng	Longshan	Lingtan
Loss of life(person)	454	735	975	887	1709
Economic loss(100 million yuan)	35	25	41	25	20
Environmental impact(coefficient)	1.43	1.43	1.43	1.43	1.43
Social impact(coefficient)	19.28	13.82	9.68	34.85	7.71

4 Evaluation Results and Analysis

4.1 Evaluation result

See Table 2 for the evaluation results of dam break consequences of five reservoirs in Jiangxi Province by using five comprehensive evaluation methods.

Table 2. Summary of evaluation results of five methods

Name of reservoir	Changlong	Xialan	Shibikeng	Longshan	Lingtan
Linear weighted comprehensive evaluation method	Extremely serious accident⑤ 0.7813	Especially significant Accident③ 0.7840	Especially significant Accident④ 0.7837	Especially significant Accident① 0.8147	Especially significant Accident② 0.7862
Comprehensive evaluation method of fuzzy mathematics	Extremely serious accident⑤ 0.4119	Especially significant Accident④ 0.4202	Especially significant Accident② 0.4301	Especially significant Accident③ 0.4267	Especially significant Accident① 0.4512
Matter element comprehensive evaluation method	Extremely serious accident⑤ 0.0279	Especially significant Accident③ 0.0533	Especially significant Accident④ 0.0513	Especially significant Accident① 0.1414	Especially significant Accident② 0.1026
Comprehensive evaluation method of grey correlation degree	Extremely serious accident① 0.6445	Extremely serious accident④ 0.5170	Extremely serious accident② 0.5488	Extremely serious accident⑤ 0.5147	Extremely serious accident③ 0.5374
Comprehensive evaluation method of principal component analysis	Extremely serious accident① 0.2641	Extremely serious accident③ 0.0789	Extremely serious Accident ④ −0.0492	Extremely serious accident ② 0.1492	Extremely serious Accident⑤ −0.4430

Note: the numbers in the table indicate the ranking of severity of dam break consequences corresponding to each evaluation method, and the ranking of severity of accidents is from strong to weak.

4.2 Analysis of evaluation results

(1) The data of the dam-break consequences of the five reservoirs in Jiangxi based on the five

methods are the same. The dam-break consequences of each of the reservoirs are particularly significant accidents. The evaluation conclusions are consistent and are in line with objective reality. Therefore, it is feasible to use these five methods to evaluate the dam-break consequences of a single reservoir.

(2) Comparing the ranking of the severity of the dam-break effects of the five reservoirs in Jiangxi by the five methods, it can be seen that the ranking results are not completely consistent, and the ranking results obtained by the linear weighted comprehensive evaluation method and the matter-element comprehensive evaluation method are completely Consistent, in order of severity from strong to weak: Longshan, Lingtan, Xialan, Shibikeng, Changlong. Through comparison and synthesis of one-way indicators of life loss, economic loss, environmental impact, and social impact of the five reservoirs considering the weight of the four indicators, it is consistent with the comprehensive ranking results determined from the objective analysis of the data. Therefore, the ranking results obtained by these two methods are considered to be more reasonable.

The linear weighted comprehensive evaluation method starts directly from the consequence severity coefficient, and its evaluation index data is clear and intuitive. The significance of the correlation function of the matter-element comprehensive evaluation method mainly involves the meaning of distance and distance. It is relatively easy to understand and the difficulty is average. The two methods are more suitable for the severity of the consequences of dam breaks in multiple reservoirs.

(3) The fuzzy-math comprehensive evaluation method uses the minimum-maximum algorithm. Through analysis, the ranking results obtained by the minimum-maximum algorithm are completely determined by the loss of life. Therefore, in terms of the comprehensive evaluation method of fuzzy mathematics, the weighting matrix algorithm is considered to be more reasonable than the minimum and maximum algorithm when it is used to rank the severity of dam break consequences of multiple reservoirs.

(4) The grey correlation analysis method reached the conclusion that the consequences of dam failure of Changlong Reservoir are the most serious, and there are at least 454 people in the five reservoirs whose lives were lost, and there is a large gap with the number of dam failures in the other four reservoirs Contrary to this, the ranking of severity is quite unreasonable. This is because, although the calculation method of the gray correlation degree reflects the integrity of the data of all the dam failure consequences of the reservoirs to be evaluated, the meaning of the gray correlation degree with its corresponding level is more ambiguous.

(5) Although the principal component analysis comprehensive evaluation method uses three principal components, the results are quite different compared with the results of other methods. The severity of the dam-break results obtained is ranked from strong to weak in order: Changlong, Longshan, Xialan Reservoir. The loss of life in Shibikeng, Lingtan, and Changlong Reservoir is at least 454 among the five reservoirs, and the loss of life in Lingtan Reservoir is at most 1709 among the five reservoirs. Environmental impact and social impact, but the weight of life loss in the comprehensive evaluation is 0.636, indicating that it still occupies a very important position, and its ranking results are obviously not in line with objective reality. The weight determination method in this method is different from other evaluation methods. It is determined according to the variance contribution rate, and its value changes with data changes. Such changes may have a greater impact

on the evaluation conclusions. Considering that the method is difficult, the calculation is complicated, the meaning of the principal components is not clear, and the weight determination method is not reasonable, so its applicability should be further explored.

5　Conclusions

(1) Five comprehensive evaluation methods, including linear weighted comprehensive evaluation, fuzzy mathematical comprehensive evaluation, matter-element comprehensive evaluation, gray correlation comprehensive evaluation, and principal component analysis comprehensive evaluation, are selected in this paper to carry out dam-breaking losses of five typical reservoirs. The calculation analysis and ranking are done, and the ranking results are basically reasonable.

(2) If the severity of the dam-break consequences of a single reservoir is evaluated, five methods are available. But if the severity of the dam-break consequences of multiple reservoirs is ranked and comprehensive evaluation is performed, the linear weighting comprehensive evaluation method is recommended first in this paper, and followed by matter-element comprehensive evaluation method.

(3) Analysis of the consequences of dam-break losses involves many factors such as life, economy, environment, society, etc. Although some researches have been made in this area, the research results obtained are still based on statistics with semi-qualitative and semi-quantitative stage. So, there is still much more work to do in the research of risk loss assessment.

Acknowledgements

Research Funds of Nanjing Hydraulic Research Institute (Y720009, Y719010); National Key R&D Program of China(2018YFC0407104).

References

[1] Aboelata M, Bowles D S, McClelland D M. A Model for Estimating Dam Failure Life Loss[C]. Processdings of Australian Committee on Large Dams Risk Workshop, Launceston, Tasmania, Australia, 2005.

[2] Brown C A, Graham W J. Assessing the threat to life from Dam failure[J]. Water Resource Bulletin, 1988, 24(6): 1303−1309.

[3] Dekay M L, Mcclelland G H. Risk Analysis Predicting loss of life in cases of dam failure and flash flood[J]. Risk Analysis,1993, 13(2): 193−205.

[4] Deng J L. Grey prediction and decision[M]. Wuhan: Huazhong University of Science and Technology Press, 1992.

[5] Kerre E E, Huang C F, Ruan D. Fuzzy set theory and approximate reasoning, MSc Thesis[M]. Wuhan: Wuhan University Press, 2004.

[6] Li L, Zhou K F. Research status of estimation methods of life loss caused by dam collapse[J]. Advances in Science and Technology of Water Resources and Hydropower, 2006, 26(2): 76−80.

[7] Li L, Wang R Z, Sheng J B, et al. Dam risk evaluation and risk management[M]. Beijing: China Water Resources and Hydropower Press, 2006.

[8] Li L, Wang R Z, Sheng J B. Study on evaluation model of severity of dam break consequence[J]. Journal of Safety and Environment, 2006, 6(1): 1−4.

[9] Lou J K. Dam safety risk management of Canadian BC Hydro[J]. DAM & SAFETY,2000, 9(4): 7−11.

[10] Reiter P. Loss of life caused by dam failure. The RESCDAM LOL method and its application to Kyrkosjarvi Dam in Seinajoki. Helsinki: PR Water Consulting Ltd.

[11] Shi Z K. Linear system theory[M]. Beijing: Science Press, 2008.

[12] Yang C Y, Cai W. Extension Engineering[M]. Beijing: Science Press, 2007.
[13] Zhang P. Research on Comprehensive Evaluation Based on Principal Component Analysis, MSc Thesis[D]. Nanjing: Nanjing University of Science and Technology, 2004.
[14] Zhou K F, Li L, et al. Preliminary study on evaluation model of dam loss life loss in China. Journal of Safety and Environment, 2007, 7(3): 145−149.

Comparisons Between SWAT and SWAT+ for Simulating Agriculture Management in a Large Watershed, a Case Study in Lijiang River Basin, China

Wuhua Li[1], Xiangju Cheng[2], Dantong Zhu[3], Jianhui Wen[4], Rui Xu[5]

[1,2,3] State Key Laboratory of Subtropical Building Science,
South China University of Technology, Guangzhou 510640, China
[4] Guilin Environmental Monitoring Center, Guilin 541002, China
[5] School of Computer Science and Information Security,
Guilin University of Electronic Technology, Guilin 541004, China

Abstract

Non-point source(NPS) pollution is becoming the main sources of water pollution. Due to its characteristics of decentralization and path uncertainty, distributed hydrological models are widely used in the study of NPS pollution, and Soil and Water Assessment Tool(SWAT) is one of the typical tools. However, it still cannot probably reflect the crop growth process in the catchment, making the model less representative in the cultivated catchment dominated by agricultural NPS pollution. SWAT+, a completely reconstructed version of the SWAT model, is possible to simulate mixed cropping by defining plant communities. In this paper, taking Lijiang River Basin as a case, SWAT and SWAT+ models were constructed under the default setting and detailed management setting, and the simulated Leaf Area Index(LAI) values, an indicator representing plant growth, were used for comparison with remote sensing LAI values and calculation of their correlation coefficient, so as to compare the simulation effects of the two models on crop growth under the two settings. The results showed that:(1) SWAT+ was more capable of covering the growing period than SWAT, and LAI values of SWAT+ would not decline to zero after the growing period, which was more consistent with actual LAI changes;(2) after implementing management operation, both SWAT and SWAT+ can reflect two growing seasons and the simulation of SWAT+ was closer to the actual situation;(3) the correlation coefficients of simulated LAI values and remote sensing LAI values of SWAT+ were higher than those of SWAT in both cases, and the correlation coefficients increased after adding detailed management operations, which quantitatively indicated the advantages of SWAT+ in simulating crop growth.

Keywords: SWAT+; NPS pollution; Agricultural management; Crop growth; LAI

1 Introduction

Water pollution can be classified into point source(PS) pollution and non-point source(NPS) pollution. In some developed areas, PS pollution can be well controlled due to the state or city's attention and treatment. However, water pollution is still very serious in many areas. NPS pollution has become the main source of water pollution and agriculture is becoming a major contributor, accounting for more than 85% (Ongley et al., 2010; FAO, 2013). In the process of agricultural production, due to excessive fertilization or unreasonable fertilization methods, the nitrogen and phosphorus in the fertilizer cannot be absorbed by plants, and they will enter the surface or groundwater with runoff when it rains (Green & van Griensven, 2008). The increase of nutrient concentration in water can lead to eutrophication (Ongley et al., 2010), resulting in algae blooms, water hypoxia, fish death, and water quality deterioration, which might reduce available water resources and cause potential public health problems. Especially in China's subtropical areas, where there is abundant rainfall and runoff, improper agricultural production has a more significant impact on water quality. Unlike PS pollution, NPS pollution sources are dispersed and have no clear migration path. Moreover, it can be greatly affected by environmental and human factors such as weather, runoff, topographic conditions, and agricultural management, which makes it more difficult to control and treat NPS pollution.

With the development of computer and 3S technology, distributed hydrological models have been effectively used in the study of NPS pollution. SWAT (Soil and Water Assessment Tool), a semi-distributed hydrological and water quality model based on physical processes, is one of the representative tools (Arnold et al., 1998). SWAT was originally developed to predict the effects of different soil, land use and management practices on water quality and sediment in a catchment (Kiniry et al., 2008). It can simulate not only hydrology, but also processes such as crop growth, nutrient migration and management practices in the catchment. In addition, SWAT has a variety of built-in plant and management databases, allowing users to easily invoke or modify relevant parameter settings. SWAT is recommended as the most suitable tool for long-term simulation in agriculture-dominated watersheds (Dechmi, 2012; Ha et al., 2018) and it has been broadly applied to various regions, including in the USA (White et al., 2017), Canada (Amon-Armah et al., 2015), South Africa (Mengistu et al., 2019), China (Liu et al., 2016), and etc.

For both practitioners and researchers, modeling needs to truly reflect the real world (Vieux, 2004). Over the last 20 years, numbers of attempts have been made to meet the adaptability requirements of different regions (Bieger et al., 2017; Hoang, 2017). For example, some scholars developed modified SWAT models suitable for specific environmental conditions, including karst areas (Reza Eini et al., 2020) and tropic areas (Strauch & Volk, 2013); some coupled with other models to expand more functions of SWAT, including MODFLOW (Park et al., 2019; Liu et al., 2020; Sabzzadeh & Shourian, 2020), HSPF (Sarkar et al., 2019) and WASP (Chueh et al., 2021).

SWAT was originally developed for temperate regions (Strauch & Volk, 2013) and monoculture crops (Krysanova & Arnold, 2008). The seasonality of the vegetation is simulated by scheduling operations based on dates or heat units (Baffaut et al., 2015). Although SWAT can often achieve good performance in hydrology and water quality, it does not mean that plant growth can be well simulated within the model, because most studies do not reflect the suitability to simulate vegetation dynamics (Strauch & Volk, 2013), especially in tropical and subtropical areas where mixed cropping and multiple growing seasons are common. As mentioned above, agriculture is an important source of NPS pollution. Therefore, for watersheds with large agricultural areas, effective simulation of crop growth and its management measures are the keys to simulate the distribution of NPS pollution in these areas and better modelling of crop growth and development helps to improve simulation of hydrological, soil erosion, and nutrient transport process(Sinnathamby et al., 2017).

The simulations of multicultural plant communities and multiple growing seasons are two weaknesses of SWAT (Krysanova & Arnold, 2008), and SWAT+ model has been improved accordingly. SWAT+ is a completely revised version of SWAT, offering greater flexibility and extensibility. It absorbed many years research foundation and user needs, and differs from SWAT in many different concepts and methods. Mixed cropping now is possible in SWAT+ using the plant community file. Timing of management operations and land use dynamic can be scheduled based on user-defined criteria using decision tables, which allows to take several conditions(e.g., nutrient stress, soil moisture and precipitation) into account(Bieger et al., 2017).

Nksawa et al. (2020) analyzed seasonal land use maps pixel by pixel to derive land use trajectories and then implement in SWAT and SWAT+ model. And the result demonstrated the advantage of SWAT+ over SWAT in simulating seasonal land use dynamics in a cultivated catchment. The study area was 240 km^2 and focus on the HRU for specific crops. Many regions are larger and do not have such detailed data on phenology, land use map, etc. There are only agricultural statistics at the county level and the oral experience of local farmers available. To compare the performance of SWAT and SWAT+ models in simulating the general agricultural management in large scales, we constructed SWAT and SWAT+ models of Lijiang River Basin (6358 km^2), set up general agricultural managements and compared the simulated LAI values and remote sensing LAI values.

2 Study Area and Data

2.1 Lijiang River Basin(LRB), Guangxi

The Lijiang River Basin(LRB) is a 6358 km^2 watershed located in Guangxi Province, China, ranging from latitude 24°40′ N to 25°50′ N and longitude 110°05′ E to 110°45′ E. The elevation of the study region varies from 9 m on the southern edge to 2110 m at the highest point on the northern edge. The study area is located in mid-subtropical monsoon climate, with hot and rainy summer. The annual average temperature is 16.5℃ ~ 20.0℃, and the annual precipitation ranges from 1367.5 to 1932.9 mm/year. The wet season is from March to August every year while the dry season is from September to February of the following year.

As shown in Fig.1, land use is predominantly characterized by forest (68% of area) and cultivation(26% of area), with minor land covers of residential areas(2%) in the mid-stream of the watershed. In SWAT, the various attributes of different land uses are saved in the *plant.dat* and *urban.dat* files. The document uses four English letters to indicate a certain land use type. For example, the SWAT code AGRL represents agricultural land-generic, which is different with the GLC30 code, and we should use a lookup table to link them. The land use classification description of GLC30 and its connection with SWAT database are shown in Table 1.

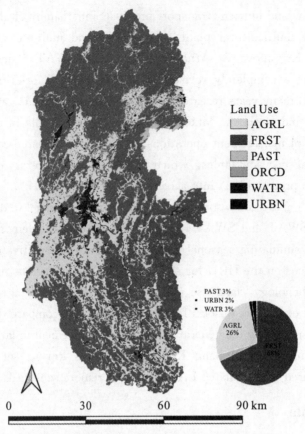

Figure 1. Lijiang River Basin and land use distribution

Table 1. Global Land Cover 30 description

GLC30 Code	SWAT Code	Land Use	Description
10	AGRL	Cultivated Land	Lands used for agriculture, horticulture and gardens
20	FRST	Forest	Lands covered with trees, with vegetation cover over 30%
30	PAST	Grassland	Lands covered by natural grass with cover over 30%
50	ORCD	Wetland	Lands covered with wetland plants and water bodies
60	WATR	Water	Water bodies in the land area
80	URBN	Artificial Surface	Lands modified by human activities

Various kinds of crops are grown in the Lijiang River basin, therefore involving many agricultural management measures. According to agricultural statistics, double cropping rice is the main crop in paddy fields, while single cropping rice and sweet potato are planted in some fields in the basin. In addition to paddy fields, dryland crops are mainly vegetables, and corn, soybean and peanut intercropping is common.

2.2 Preparation of data

Spatial data and meteorological data need to be prepared to build the basic model, as shown in Table 2. The spatial data used in constructing models for the study including a digital elevation model (DEM) with resolution of 90 meters which was provided by Geospatial Data Cloud site, Computer Network Information Center, Chinese Academy of Sciences. These data were used to delineate the watershed. Soil distribution and properties were obtained from the Harmonized World Soil Database developed by FAO(Food and Agriculture Organization) with a resolution of 1 km. The land use map data with a resolution of 30 meter was obtained from Global Land Cover 30. The spatial data were projected using $WGS\_1984\_UTM\_Zone\_49N$ projected coordinate system. Because the original land use and soil data are different from the model code, lookup tables of land use and soil were imported from csv file, and *usersoil* database was also been built and imported before constructing the models.

The climate data were required for simulating hydrological processes in the watershed to drive the model. Daily weather data were acquired from the China metrological Administration including precipitation, temperatures (maximum and minimum), relative humidity, wind speed and solar radiation(sunshine duration).

In order to evaluate the effect of agricultural management simulation, LAI data is used to compare with the simulated LAI value. The Normalized Differential Vegetation Index(NDVI) data at 1 km resolution obtained from Resource and Environment Science and Data Center and was further used to calculated LAI according a universal equation, as shown in Eq.(1)(Su, 1996).

Table 2. **Data sources and resolution for model datasets**

Data	Resolution	Year	Source
Topography	90 m	2000	Geospatial Data Cloud site
Land use	30 m	2010	Global Land Cover 30
Soil	1000 m	1995	HWSD v1.2
Climate	5 stations	1986—2015	China Meterological Agency
NDVI	1 km	2010—2015	Resource and Environment Science and Data Center

$$LAI = \sqrt{NDVI \frac{(1+NDVI)}{(1-NDVI)}} \qquad (1)$$

where NDVI refers to a raster value in NDVI dataset, and the corresponding LAI value of the grid can be estimated by the above equation.

3 Model Setup and Evaluation

3.1 Model description

The Soil and Water Assessment Tool(SWAT) has been developing for over 20 years and has been used effectively in hydrology and water management around the world. The model initially runs on ArcGIS platform called ArcSWAT. To meet more users' need, another version of SWAT runs on QGIS platform was introduced, which is called QSWAT. Compared with ArcSWAT, QSWAT adds functions like land use split and result visualization. QSWAT+ is a completely reconstructed version with greater flexibility and expandability, which was developed using Python and used various QGIS functionalities(Bieger et al., 2017).

The differences on platforms and software requirements of each SWAT version are shown in Table 3. All the versions are available on the SWAT official website at https://swat.tamu.edu/

software/. The SWAT+ has a modeling interface using QGIS plug-in and uses the SQLite database for data management. While the SWAT has an interface using ArcGIS extension and use Microsoft Access for database management. Both SWAT and SWAT+ have the rather similar interfaces and procedures, which mainly consist of four main steps. The main forms of the two programs and their editors are shown in Fig. 2.

Table 3. Comparison of different versions of SWAT and corresponding platforms

	ArcSWAT	QSWAT3	QSWAT+
GIS	ArcGIS 10.5	QGIS 3(32 bit)	QGIS 3(64 bit)
SWAT	SWAT 2012	SWAT 2012	SWAT+
Editor	A build-in Editor	SWAT Editor	SWAT+ Editor
Database	MS Access	MS Access	SQLite

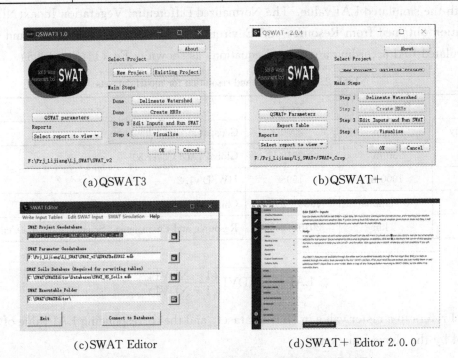

(a) QSWAT3　　(b) QSWAT+

(c) SWAT Editor　　(d) SWAT+ Editor 2.0.0

Figure 2. Interface comparisons between SWAT and SWAT+

For simulation and calculation, the model divides the watershed into several sub-basins based on topography, and further divides into a number of Hydrologic Response Units (HRUs) according to land use, soil and slope. HRU is the smallest calculation unit of the model, with unique land use, soil, slope and management combinations. The model calculates hydrological processes separately at the HRU level and aggregates them at the sub-basin level(Neitsch et al., 2005).

SWAT+ uses similar equations as SWAT for calculations. Weather input provides rainfall and energy to the model, which drives hydrological processes and crop growth. The hydrological processes base on the water balance equation, as shown in Eq. (2).

$$SW_t = SW_0 + \sum (R_{day} - Q_{surf} - E_a - w_{seep} - Q_{gw}) \times \Delta t \qquad (2)$$

where SW_t is the final soil water storage(mm H_2O), SW_0 is the initial soil water storage(mm H_2O), Δt is the time step(day) and R_{day}, Q_{surf}, E_a, w_{seep} and Q_{gw} represent the total amount of rainfall, surface runoff, evaporation, percolation volume and groundwater volume in the time step, respectively.

The plant growth component of SWAT is derived from EPIC model. It simulates plant development based on heat unit theory assuming that the heat demand of plants can be quantified and related to maturity time. Eq. (3) is used to calculate the total number of heat units required for plants to reach maturity.

$$PHU = \sum_{d=1}^{m} HU \qquad (3)$$

where PHU is the total heat unit required for plant maturity, HU is the number of heat units accumulated on day d and $d = 1$ means the day of planting.

Although it is convenient to simulate crop growth in terms of thermal units, it is sometimes inconsistent with the reality. For example, it is not possible to simulate multiple growing seasons, which does not correspond to the multiple growing seasons in tropical and subtropical regions. In this case, users need to set up operations by-date, and the model will calculate the cumulative heat units according to different operations.

Leaf Area Index(LAI) is defined as the area of green leaf per unit area(Watson, 1947). The leaf development is controlled by the optimal leaf area development curve and is related to the fraction of potential heat units accumulated for the plant on day d in the growing season. Therefore, the simulated LAI changes can reflect the plant growth process.

SWAT+ provides more flexibility in terms of management measures, spatial connectivity and decentralization, thus enabling more options to be set up to better represent the watershed situation. The SWAT+ model allows two or more plants or crops to grow at the same time by defining a plant community. In a plant community, different crops can be harvested at different time. This function can easily simulate the mixed planting scenario in reality, which cannot be realized in SWAT.

3.2 Model configuration

In this study, we constructed models using QSWAT3 and QSWAT+ and compare their simulation results. All the simulations were run on a laptop Intel © Core™ i7−9750H CPU @ 2.60 GHz, 16.0 GB RAM. The same spatial data (including DEM, land use, soil data), meteorological data, and tabular data were used to construct SWAT and SWAT+ models. Penman-Monteith method was used to calculate potential evaporation(PET), SCS curve number method was used to calculate surface runoff and variable storage method was used to calculate flow routing.

The threshold of catchment area was set at 6000 hectares to create streams, and the basin was divided into 67 sub-basins. No threshold was set for land use, soil and slope, so the maximum number of HRUs could be generated. The model constructed by SWAT and SWAT+ generated 3769 and 3891 HRUs, respectively. The difference in the number of HRUs could be attributed to the creation of HRUs through SWAT model sub-basins, while SWAT+ uses landscape units to create HRUs (Nkwasa et al., 2020).

Splitting land uses allows modeler to define more precise land uses than provided land use map. For example, if we know that in the study area the AGRL land use consists of 50% CORN and 50% RICE, we can use the function to further split the original land use and make sure the percentages of all sub-land uses summed to 100. And different agricultural management practices can then be applied to different sub-land uses.

According to local agricultural statistics, we divided the original AGRL land use into four categories, namely double cropping rice, single cropping rice, mixed planting and vegetables, and their proportions were 25%, 5%, 20% and 50%, respectively. The detailed management measures are shown in Table 4.

Table 4. Management and schedule of crops in the catchment

Management set	Percentage	Plant	Harvest and kill	Crop
Double cropping rice	25%	3~21	6~30	RICE
		7~21	11~1	RICE
Single cropping rice	5%	4~21	8~20	RICE
		8~22	11~30	SPOT
Mixed planting	20%	3~11	7~11	CORN
		3~21	7~1	SOYB
		3~21	7~21	PNUT
Vegetable	50%	—	—	CANP

By default, both SWAT and SWAT+ models simulate plant growth with a single growing season driven by accumulated heat units. This may not work for tropical and subtropical regions that may have multiple cropping seasons. In this study, SWAT and SWAT+ default models were set, that is, no crop management changes were made. The results were compared with the results of the model with more detailed management settings, so as to highlight the improvement effect of appropriate management settings on the model.

Crop rotation can be simulated by date in both SWAT and SWAT+, i.e., planting another kind of crop after harvest or starting the next round of crop planting. As shown in Table 4, in the management of double cropping rice, two rounds of rice are planted in March and July respectively; In single cropping rice management, sweet potatoes are planted after the rice harvest in August. Since SWAT can only simulate one crop at a time, CORN was used to simulate mixed planting. SWAT+ simulated mixed planting by defining corn, soybean and peanut as a plant community.

3.3 Model evaluation

The models simulated plant growth by leaf area development, light interception and converting the intercepted light into biomass. Therefore, the simulated LAI trend can reflect the simulation effect of plant growth. Before the model calibration, remote sensing LAI data can be used as the reference to compare the correlation between simulated LAI and observed LAI values, so as to evaluate the representativeness of management settings to the watershed.

The monthly remote sensing LAI values with a resolution of 1 km were calculated from NDVI data, and the raster average value of the whole watershed was calculated on the GIS platform. The models output simulation LAI values at HRU level. The monthly simulated LAI values at watershed level were calculated according to area weighting.

In the field of natural science, Pearson correlation coefficient is widely used to measure the degree of correlation between two variables, and its value is between -1 and 1. Pearson correlation coefficient between two variables is defined as the quotient of covariance and standard deviation between two variables, as the following equation:

$$R_{XY} = \frac{\text{Cov}(X, Y)}{S_X S_Y} \qquad (4)$$

where R_{XY} is the Pearson correlation coefficient between variables X and Y; S_X and S_Y is the standard deviations of variable X and variable Y, respectively. $\text{Cov}(X, Y)$ is their covariance.

In our study, the Pearson correlation coefficient analysis was applied to calculate the correlation between the models' LAI outputs and the LAI derived from NDVI, evaluating the simulation output for plant growth. Correlation analysis was performed on R platform.

4 Results and Discussion

4.1 Water balance check

Although both SWAT and SWAT+ models choose the same hydrological calculation method, there are differences in water balance. The annual mean water balance components between SWAT and SWAT+ were compared in Table 5. The water components simulated by SWAT and SWAT+ are quite different, reflecting the uncertainty of SWAT and SWAT+ in water volume simulation.

The average annual rainfalls are slightly different between the two models, which can be attributed to the way precipitation stations distributed in different models(Nkwasa et al., 2020). The average annual evapotranspiration simulated by SWAT+ was 183.78 mm(31%) more than that by SWAT, while the surface runoff simulated by SWAT+ was 507.36 mm(71%) less than that by SWAT. The percolation in SWAT+ is much bigger than SWAT, with a difference of 391.23 mm (87%). Although the lateral flow component is the smallest in both SWAT and SWAT+, the lateral flow simulated by SWAT+ is only 0.786 mm, the SWAT simulated value differs by two orders of magnitude.

Table 5. Comparison of annual water balance components between SWAT and SWAT+ by default(unit: mm)

Water component	Default		
	SWAT	SWAT+	Difference
Precipitation	1829.2	1824.85	−4.35
Evapotranspiration	585.6	769.38	183.78
Surface runoff	715.93	208.57	−507.36
Percolation	451.93	843.16	391.23
Lateral flow	60.88	0.786	−60.10
Mass balance	−14.86	−2.95	11.9

The changes of water balance components after adding detailed management measures can be seen from Table 6, reflecting the influence of management operations. The water balance components changed little. However, the variation trend of water components in the two models is not consistent. The evaporation of SWAT model increased by 4 mm, while SWAT+ model decreased by 2 mm. The surface runoff, percolation and lateral flow components of SWAT model decreased while the corresponding components of SWAT+ increased.

Table 6. Comparison of water balance components in each model after adding detailed managements(unit: mm)

Water balance component	SWAT			SWAT+		
	Default	Detailed	Difference	Default	Detailed	Difference
Precipitation	1829.2	1829.2	—	1,824.85	1,824.85	0
Evapotranspiration	585.6	589.3	4	769.38	767.38	−2
Surface runoff	715.93	714.41	−0.77	208.57	208.96	0.39
Percolation	451.93	449.69	−3.3	843.16	844.73	1.57
Lateral flow	60.88	60.87	−0.02	0.785	0.785	0
Mass balance	−14.86	−14.95	−0.09	−2.95	−2.99	−0.04

4.2 LAI comparison

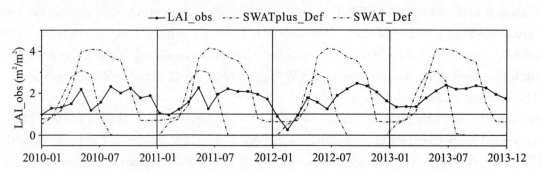

Figure 3. The monthly changes of observed LAI and simulated LAI(2010—2013)

Comparison of remote sensing LAI values with simulated LAI values of SWAT and SWAT+ is shown in Fig. 3, reflecting the simulation effect of different models on plant growth under default settings. The remote sensing LAI value is within 0.25~2.6. It can be seen that there are two main growing seasons(May and August to November), which can roughly indicate the planting schedule of local crops. Besides the growing season, the LAI values are still relatively high in other times. The simulated LAI values are the monthly mean values of all land use types in the whole basin. The watershed covers a wide area of forest which is mainly broad-leaved forest, so there is a certain degree of vegetation coverage after crops harvested. The LAI values in February 2012 dropped to close to 0, which may be due to convective weather affecting the observation effect(Strauch & Volk, 2013).

Both the LAI values simulated by SWAT and SWAT+ are higher than the peak values of remote sensing LAI and vary in a wider range. For example, the LAI range simulated by SWAT was 0~3.19, while the LAI range simulated by SWAT+ was about 0.63~4.16.

Although both default SWAT and SWAT+ models simulated crop growth based on heat units, SWAT shows an earlier and shorter growing season. The growth season of SWAT model is mainly from May to July, while the growth season of SWAT+ is approximately from June to October.

The LAI values of SWAT model drop to 0 value after the growing season, which is inconsistent with the actual situation. There is still some vegetation cover within the basin and there is a large area of forest besides agricultural land.

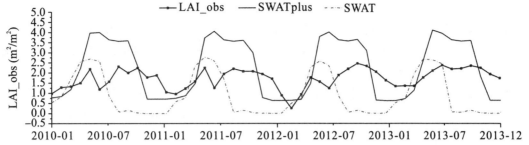

Figure 4. The monthly changes of observed LAI and simulated LAI after adding detailed managements(2010—2013)

Fig. 4 shows the monthly changes of observed LAI and simulated LAI after adding detailed managements during 2010—2013. It can be seen that in the two detailed models, both SWAT and SWAT+ could reflect two growing seasons. However, the second growing season in SWAT model is far from the first growing season, while the second growth season simulated by SWAT+ is slightly lower than the first one. It is clear that the growing seasons simulated by SWAT+ are more in line with the actual situation.

The peak LAI values simulated by SWAT and SWAT+ model decreased after setting detailed management operation, but the value of SWAT model went down more, making it closer to the remote sensing value.

The analysis above reflects the uncertainty of plant growth simulated by SWAT and SWAT+ models, so the simulation performance is further quantitatively evaluated through correlation analysis.

4.2 Correlation analysis

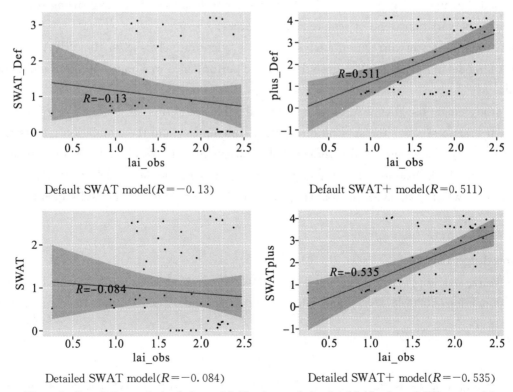

Figure 5. Correlation analysis of observed LAI values and simulated LAI values of different models

As shown in Fig. 5, through correlation analysis between simulated LAI values and remote sensing LAI values, we can quantitatively evaluate the degree of correlation between them and we found that:

(1) Both by default and adding detailed operations, the correlation coefficients of the SWAT+ model are larger than those of the SWAT model, indicating that the simulated LAI values by SWAT+ can better reflect the LAI changes.

(2) Although both SWAT and SWAT+ can capture two growing seasons in the figure, the correlation between simulated LAI values by SWAT and remote sensing LAI values decreases after adding management operation (from −0.13 to −0.084), while for SWAT+, the correlation increased from 0.511 to 0.535.

5 Conclusions

In this paper, we constructed SWAT and SWAT+ models respectively in the Lijiang River Basin (6538 km^2), set default operation and detailed management operation, compared remote sensing LAI values with simulated LAI values, and calculated their Pearson correlation coefficients. The results were as follows:

(1) The numbers of HRUs generated by SWAT and SWAT+ models and the balance components simulated by them are quite different, which may be caused by the different discretization and aggregation way in SWAT+.

(2) By comparing the graphs of simulated LAI and remote sensing LAI, SWAT+ can better cover the actual vegetation growth period than SWAT by default; After adding the detailed management operation, both models could reflect multiple growing seasons in a year, but the growing seasons of SWAT+ had the similar degree, and LAI values would not drop to 0 after the growing season, which was closer to the variation trend of remote sensing values.

(3) Through Pearson correlation analysis, it was quantitatively shown that SWAT+ is better than SWAT model in reflecting plant growth, and the performance was improved after adding detailed operation.

Our study suggests that SWAT+ is more flexible in terms of simulation of agricultural management practices (e.g., the application of plant community) and the simulation results are more accord with reality. However, it also reflected the uncertainty of SWAT and SWAT+ model. It is recommended to check crop growth simulation effect after building the models to verify its realistic representation, which can avoid introducing too many parameters in the subsequent calibration process.

Acknowledgements

This work was supported by Guangxi Key R&D Program (AB18221108), China Postdoctoral Science Foundation (2020M672635) and Guangdong Basic and Applied Basic Research Foundation (2020A1515111152).

References

[1] Amon-Armah F, Yiridoe E K, Jamieson R, et al. Comparison of Crop Yield and Pollution Production Response to Nitrogen Fertilization Models, Accounting for Crop Rotation Effect[J]. Agroecology and Sustainable Food Systems, 2015, 39: 245−275.

[2] Arnold J G, Srinivasan R, Muttiah R S, et al. Large Area Hydrologic Modeling and Assessment Part I: Model

Development1[J]. Journal of the American Water Resources Association, 1998, 34: 73—89.

[3] Baffaut C, Dabney S, Smolen M, et al. Hydrologic and water quality modeling: Spatial and temporal considerations[J]. American Society of Agricultural and Biological Engineers, 2015, 58.

[4] Bieger K, Arnold J G, Rathjens H, et al. Introduction to SWAT+, A Completely Restructured Version of the Soil and Water Assessment Tool. Journal of the American Water Resources Association, 2017, 53: 115—130.

[5] Chueh Y Y, Fan C, Huang Y Z. Copper concentration simulation in a river by SWAT-WASP integration and its application to assessing the impacts of climate change and various remediation strategies[J]. Journal of Environmental Management, 2021, 279:111613.

[6] Dechmi F. SWAT application in intensive irrigation systems: Model modification, calibration and validation[J]. Journal of Hydrology, 2012, 470—471: 227—238.

[7] FAO. Guidelines to Control Water Pollution from Agriculture in China[M]. FAO, 2013.

[8] Green C H, van Griensven, A. Autocalibration in hydrologic modeling: Using SWAT 2005 in small-scale watersheds[J]. Environmental Modelling & Software, 2008, 23: 422—434.

[9] Ha L T, Bastiaanssen W G M, Van Griensven A, et al. Calibration of Spatially Distributed Hydrological Processes and Model Parameters in SWAT Using Remote Sensing Data and an Auto-Calibration Procedure: A Case Study in a Vietnamese River Basin[J]. Water, 2018, 10: 212.

[10] Hoang L. Enhancing the SWAT model for simulating denitrification in riparian zones at the river basin scale[J]. Environmental Modelling, 2017, 17.

[11] Kiniry J R, Macdonald J D, Kemanian A R, et al. Plant growth simulation for landscape-scale hydrological modelling[J]. Hydrological Sciences Journal, 2008, 53: 1030—1042.

[12] Krysanova V, Arnold J G. Advances in ecohydrological modelling with SWAT—a review[J]. Hydrological Sciences Journal, 2008, 53: 939—947.

[13] Liu R, Xu F, Zhang P, et al. Identifying non-point source critical source areas based on multi-factors at a basin scale with SWAT[J]. Journal of Hydrology, 2016, 533: 379—388.

[14] Liu W, Bailey R T, Andersen H E, et al. Quantifying the effects of climate change on hydrological regime and stream biota in a groundwater-dominated catchment: A modelling approach combining SWAT-MODFLOW with flow-biota empirical models[J]. Science of The Total Environment, 2020, 745:140933.

[15] Mengistu A G, van Rensburg L D, Woyessa Y E. Techniques for calibration and validation of SWAT model in data scarce arid and semi-arid catchments in South Africa[J]. Journal of Hydrology: Regional Studies, 2019, 25:100621.

[16] Neitsch S, Arnold J, Kiniry J, et al. SWAT theoretical documentation[J]. Grassland, 2005, 494: 234—235.

[17] Nkwasa A, Chawanda C J, Msigwa A, et al. How Can We Represent Seasonal Land Use Dynamics in SWAT and SWAT+ Models for African Cultivated Catchments? [J]. Water, 2020, 12:1541.

[18] Ongley E D, Xiaolan Z, Tao Y. Current status of agricultural and rural non-point source Pollution assessment in China[J]. Environmental Pollution, 2010, 158: 1159—1168.

[19] Park S, Nielsen A, Bailey R T, et al. A QGIS-based graphical user interface for application and evaluation of SWAT-MODFLOW models[J]. Environmental Modelling & Software, 2019, 111: 493—497.

[20] Reza Eini M, Javadi S, Delavar M, et al. Development of alternative SWAT-based models for simulating water budget components and streamflow for a karstic-influenced watershed[J]. CATENA, 2020, 195:104801.

[21] Sabzzadeh I, Shourian M. Maximizing crops yield net benefit in a groundwater-irrigated plain constrained to aquifer stable depletion using a coupled PSO-SWAT-MODFLOW hydro-agronomic model[J]. Journal of Cleaner Production, 2020, 262:121349.

[22] Sarkar S, Yonce H N, Keeley A, et al. Integration of SWAT and HSPF for Simulation of Sediment Sources in Legacy Sediment-Impacted Agricultural Watersheds[J]. Journal of the American Water Resources Association, 2019, 55:497—510.

[23] Sinnathamby S, Douglas-Mankin K R, Craige, C. Field-scale calibration of crop-yield parameters in the Soil and Water Assessment Tool(SWAT)[J]. Agricultural Water Management, 2017, 180: 61—69.

[24] Strauch M, Volk M. SWAT plant growth modification for improved modeling of perennial vegetation in the tropics[J]. Ecological Modelling, 2013, 269: 98−112.

[25] Su Z. Remote sensing applied to hydrology: the Sauer river basin study [D]. Ruhr University Bochum, Bochum, Germany, 1996.

[26] Vieux B E. Distributed Hydrologic Modeling Using GIS[M]. 2nd ed. Boston: Kluwer, Dordrecht, 2004.

[27] Watson D J. Comparative Physiological Studies on the Growth of Field Crops: I. Variation in Net Assimilation Rate and Leaf Area between Species and Varieties, and within and between Years[J]. Annals of Botany, 1947, 11: 41−76.

[28] White M, Beiger K, Gambone M, et al. Development of a Hydrologic Connectivity Dataset for SWAT Assessments in the US[J]. Water, 2017, 9: 892.

Dynamic Flow Pattern Analysis Based on Energy Spectrum of Fluctuating Bottom Pressure in the Stilling Basin with a Negative Step

Wang Jia, Mingjun Diao, Lei Jiang, Guibing Huang

State Key Laboratory of Hydraulics and Mountain River Engineering,
Sichuan University, Chengdu 610065, China

Abstract

The hydrodynamic pressure pulsation may cause the self-vibration of the bottom of the stilling pool, which seriously threatens the safety of the building. In this paper, the pulsating pressure on the bottom plate of the conventional stilling basin and stilling basin with a negative step was tested under different flow rates. First, the pulsating pressure signal is decomposed by EDM to obtain several eigenmode functions, and the stability of the pulsating pressure of the water flow is tested. On the basis that the pulsating pressure signal is a stationary random process, the energy eigenvalues of each order of IMFs are extracted, and the relationship between the energy distribution, conversion and flow regime of IMFs is discussed. The research results show that the pulsating pressure acting on the bottom plate of the stilling basin with a negative step can be regarded as the state undergoing a stable and random process; the pulsating pressure intensity of the low-order IMFs component is much smaller than that of the high-order IMFs, and the EDM decomposition is from the high frequency The order of low frequency is gradually separated; the pulsating pressure energy on the bottom of the conventional stilling basin and stilling basin with a negative step is mainly concentrated in the front of the stilling basin; under different flow rates, the pulsating pressure on the central axis of the bottom of the stilling basin is high The proportion of frequency bands gradually decreases with the increase of x/L, and the proportions of middle and low frequency bands gradually increase with the increase of x/L.

Keywords: Stilling basin with a negative step; Pulsating pressure; EMD; IMFs; Energy

1 Introduction

With the continuous improvement of high dam engineering technology, energy dissipation and erosion prevention methods are no longer limited to traditional stilling basins. New types of energy dissipators continue to emerge(Zhivoderov and Tupikov, 1994). Stilling basin with a negative step is a widely used new type of energy dissipater. In China, many high dam projects have adopted step-down stilling pools as energy dissipation and erosion prevention methods, such as hydropower projects such as Xiangjiaba, Guandi and Liyuan. The stilling basin with a negative step is to set a reasonable height drop in the first section of the conventional stilling basin. Compared with the traditional stilling basin, the stilling basin with a negative step has a higher efficiency rate and effectively reduces the hydraulic index of the bottom of the stilling basin. Causes the advantages of small atomization(Kazemi et al.,

2016). Fig. 1 shows a schematic diagram of a conventional stilling basin and a stilling basin with a negative step.

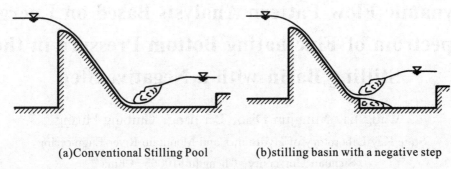

(a)Conventional Stilling Pool   (b)stilling basin with a negative step

Figure 1.

In recent years, the bottom plates of the stilling pools of a number of water conservancy projects have been damaged to varying degrees, and some damages even caused engineering accidents. The pressure pulsation of the bottom of the stilling pool is an important reason for the damage of the bottom of the stilling pool(Tang et al.). The pressure pulsation may cause the self-vibration of the bottom of the stilling pool, which seriously threatens the safe operation of the stilling pool. Predecessors have made a lot of research results on pulsating pressure(Yan et al., 2006). At present, the analysis methods of pulsating pressure mainly include mathematical statistics and spectrum analysis. When using these two methods to analyze the pulsating pressure signal, it must be assumed that the pulsating pressure belongs to a stable and random process experienced by various states.(Lu et al., 2021) In the experiment, we think If the sampling time is long enough, the signal can be considered as a stationary signal. This article will use the empirical mode decomposition(EMD) method to test the stationarity of the pulsating pressure signal. On this basis, the energy distribution of IMFs will be analyzed for the purpose of conventional stilling pools. And the relationship between energy distribution and flow regime in stilling basin with a negative step(Huang et al., 1998).

In this paper, through model tests, under different flow conditions, the pulsating pressure signals of the traditional stilling basin and stilling basin with a negative step are measured. At the same time, the Hilbert Huang transformation method is introduced to describe the energy distribution of different body types, different flow rates, and different characteristic points. Through the EMD decomposition of the pulsating pressure signal, several eigenmode functions are obtained, and the energy eigenvalues of each order of IMFs are extracted, and the relationship between the energy distribution, conversion and flow regime of IMFs is discussed, so as to reveal the pulsation characteristics of the stilling basin with a negative step.

2 Physical Modeling and Experimental Setup

The physical model test system is composed of water supply system, rectangular weir, upstream reservoir, drainage channel, WES weir and stilling pool. The model is made of organic glass. The length of the stilling pool is 2 m, the width is 0.5 m, the height of the tail sill is 0.2625 m, the height of the falling sill is 0.1 m, and the incident angle is 10°. The straight line segment is a steep slope drainage chute with a rectangular section with a slope of 53°. The flow conditions in the experiment were: 30 L/s, 50 L/s, 70 L/s, and 90 L/s. The flow of the model is controlled by a rectangular thin-walled weir with an accuracy of 0.1 mm. The SDA1000 high-performance digital sensor system is used to collect and process the pulsating pressure. The sampling frequency is $f=100$ Hz, the sampling time is $t=240$s, and the total number of samples is $N=24000$.

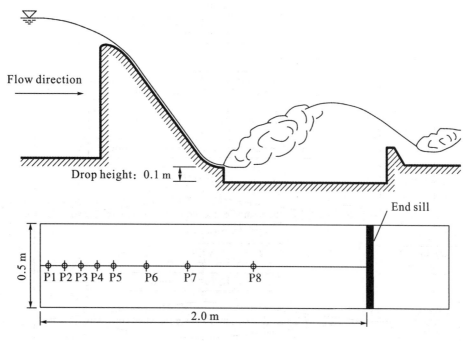

Figure 2. Experiment setup

In order to describe the position of the pressure measurement point, a one-dimensional coordinate system with the central axis of the stilling pool floor as the x-axis is established. The coordinate origin is set at the intersection of the front end of the stilling pool floor and the central axis, and the same is arranged along the central axis of the stilling pool. 8 measuring points.

Table 1. The measuring point position table on the central axis of the bottom plate of the stilling pool

Measuring point number	x/m	Measuring point number	x/m
P1	0.025	P5	0.225
P2	0.075	P6	0.325
P3	0.125	P7	0.45
P4	0.175	P8	0.64

3 HHT Method

The HHT method is a time-frequency analysis method proposed by Huang E et al. which is suitable for processing nonlinear and non-stationary signals. The method consists of two main parts, one part is empirical mode decomposition(EMD), the other part is Hilbert transform(HT) and its spectrum analysis. Compared with signal processing methods such as fast Fourier transform and wavelet transform, HHT transform does not require pre-analysis and research, nor does it need to set any basis functions(Huang et al., 1998). Instead, it is adaptively decomposed according to the characteristics of the original signal to obtain several IMF components. Each IMF component represents each frequency component in the original signal, and is decomposed in sequence from high frequency to low frequency. The following figure shows the flow chart of EDM decomposition(Jing and Li, 2016).

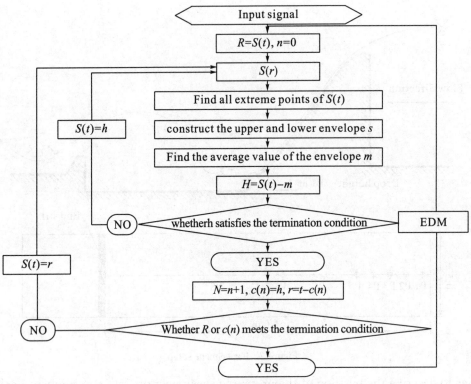

Figure 3. **EDM decomposition flow chart**

Any signal through the above process will be decomposed into a finite number of intrinsic mode functions and a residual term(Jing and Li, 2016, Tan et al., 2018). Several intrinsic mode functions $c_j(t)$ represent the intrinsic components of the signal, and the residual term represents the trend of the entire signal. Any signal can be represented by the following formula.

$$X(t) = \sum_{j=1}^{n} c_j(t) + r_n(t) \tag{1}$$

4 Experimental Resuils

4.1 Stationary test of water flow pulsating pressure

The pressure pulsation on the stilling basin with a negative step is a random process. The pulsating pressure signal measured by the model test is a random fluctuation graph of the pulsating pressure at each characteristic point. The waveform of pulsating pressure is an irregular and irregular time series. It is almost impossible for us to obtain the time-frequency characteristics of pulsating pressure from the time-frequency curve(Tang et al., Armenio et al., 2000). Random data needs to be processed and the time series converted to frequency domain analysis. At present, the most commonly used random data processing method is the spectrum analysis method. There is an important assumption in the use of spectrum analysis to analyze random signals. It is assumed that the measurement data is long enough and that the pressure pulsation process of the water flow is a stable process experienced by various states(Bowers and Tsai, 1969). Then the time average value can be used to replace the overall average value, which is irregular. The pulsating pressure waveform can be understood as the result of a series of orthogonal simple harmonics superimposed. On this basis, the random function theory intermediate frequency harmonic analysis method can be used to process the pulsating pressure signal(Abdul Khader and Elango, 1974).

When encountering actual engineering problems, it is often difficult to prove the validity of this hypothesis. Although the mathematical theory is very mature, it is still impossible to prove the assumption in actual engineering. In model tests, we usually think that as long as the measurement time is long enough, the pulsating pressure process is a smooth and random process. In fact, it is not rigorous without argumentation. The assumption is established under strict mathematical conditions. We usually think that a stable physical random process has two notable features: (1) The measured time series fluctuates randomly around a certain fixed value, and the fluctuation frequency is relatively uniform, without drastic mutations and deviations. (2) The mathematical statistical parameters of random data in adjacent time intervals are independent, and the characteristic value of the random fluctuation curve is constant. (Tan et al., 2018)

In this paper, the empirical mode decomposition method will be used to test the stability of the pulsating pressure of the water flow, and on this basis, the pulsating pressure sequence will be processed by the frequency spectrum analysis method. First, perform empirical modal decomposition of the pulsating pressure signal to obtain waveforms of different scales in the signal. Each waveform is an inherent modal function. The decomposition process is completed based on the inherent characteristics of the pulsating pressure signal, which can truly reflect the pulsating pressure. Inherent characteristics. When the experimental conditions are selected at $Q=90$ L/s, the pulsating pressure signal at $x/L=0.025$ on the central axis of the measuring point is decomposed by EMD. Through EMD decomposition, 10 IMFs components and 1 residual term are obtained. Each IMFs component and its corresponding Fourier spectrum are shown in Fig. 4.

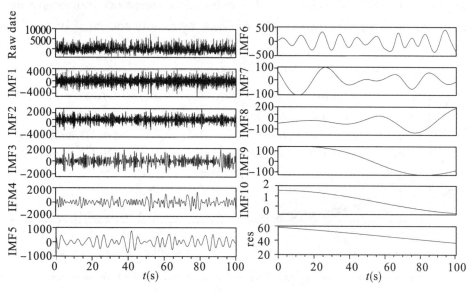

Figure 4. IMFs component and its Fourier spectrum(unit: Pa)

Through EMD decomposition, 10 IMFs and 1 residual term are obtained. From the spectrogram of each component, it can be seen that the spectral properties of each component are stable. The residual term represents the trend of the entire time series data. The time series of the pulsating pressure signal is Extremely stable, the residual curve changes gently. That is, under the working condition of $Q=90$ m³/s, the pulsating pressure acting on the bottom of the stilling basin with a negative step can be regarded as the steady random process of each state. For such various states undergoing a stable random process, the characteristic quantities acting on the bottom plate of the stilling basin with a negative step can be obtained through spectrum analysis, which is an important prerequisite for the spectrum analysis of random signals.

The HHT method can efficiently process nonlinear and non-stationary random signals. The EMD decomposition process is a reversible process, that is to say, the IMFs components and residuals obtained by the decomposition can be reconstructed to obtain the random signal again, which shows Irregular pulsating pressure waveform can be seen as the result of a series of orthogonal simple harmonics superimposed. At the same time, this method can be used to evaluate the smoothness of random signals, which cannot be done by fast Fourier transform and wavelet analysis.

4.2 Energy distribution and flow pattern analysis of IMFs components

The EMD decomposition separates the pulsating pressure signal in order from high frequency to low frequency, and the pulsating pressure signal is adaptively decomposed into several eigenmode functions. In order to describe the energy distribution of the IMFs components of the stilling basin with a negative step and the conventional stilling basin under different working conditions, this paper uses the root mean square as the statistical characteristic value of the pulsating pressure amplitude, which characterizes the intensity of the pulsating pressure. The model experiment measured the pulsating pressure signals of the characteristic points on the central axis of the stilling basin with a negative step and the bottom of the conventional stilling basin. The figure below shows the stilling basin with a negative step and the bottom of the conventional stilling basin when $Q=90$ m³/s. The result of the pulsating pressure signal decomposition of each feature point on the axis. The abscissa in the figure represents the IMFs of each order. The order of the eigenmode components decomposed by different feature points is not the same. The ordinate represents the intensity σ of the pulsating pressure.

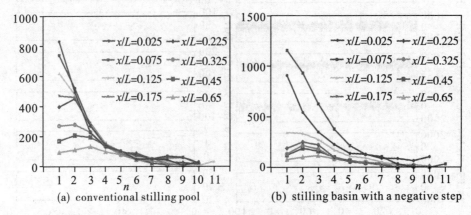

Figure 5. The intensity change of the pulsating pressure of the IMF component

The pulsating pressure signal intensity of each characteristic point on the bottom of the conventional stilling pool gradually decreases with the increase of the decomposition order of IMFs. The pulsating pressure intensity of the low-order IMFs component is much smaller than that of the high-order IMFs. The intensity of the separated IMFs components first If it is larger, the pulsating pressure intensity of the separated IMFs component is smaller, and the pulsating pressure intensity of the final separated IMFs component tends to zero. The intensity of pulsating pressure is mainly concentrated in the first 6 components. Judging from the characteristic points on the central axis of the bottom of the stilling pool, the pulsating pressure intensity at the head of the stilling pool is greater than that at the middle and rear, and the pulsating pressure intensity gradually decreases along the way.

The pulsating pressure intensity of the characteristic points of the stilling basin with a negative step is mainly concentrated in the first 6 components. On the whole, the pulsating pressure intensity of the IMFs decomposed first is greater than the last decomposed IMFs. The intensity of the pulsating pressure at $x/L = 0.025$ is much greater than the other positions, and the changing trend of the pulsating pressure intensity of each order of IMFs at the other positions is consistent. This is obviously different from the conventional still pool.

(a) flow pattern in the conventional stilling pool

(b) flow pattern in the stilling basin with a negative step

Figure 6. Flow pattern in the stilling pool

The reason why the pulsating pressure intensity distribution law of the stilling basin with a negative step and the conventional stilling basin IMFs component is different is due to the different shape of the hydraulic jump in the stilling basin. When $Q=90$ L/s, in the conventional stilling pool, the surface swirling and strong turbulent energy-dissipating jet produced by the hydraulic jump are close to the bottom, the bottom velocity is higher, and the water surface has large-scale swirling

(Nonaka et al., 2014). In the stilling basin with a negative step, the water flow rushes through the slum to form a powerful jet with huge energy, and a very obvious reverse swirl is formed below the main flow, forming a submerged impinging jet and a mixed flow state of hydraulic jump(Deng et al., 2008). There is a jet impact zone in the stilling basin with a negative step, and the pulsating pressure of the bottom of the stilling basin in this zone is much greater than other positions(Armenio et al., 2000).

Compared with the conventional stilling basin, the pulsating pressure at the attachment of the jet impact point is outstanding, and the pulsating pressure at other positions is greatly reduced(Riasi et al., 2010). The reason is that the submerged jet is formed in the stilling basin with a negative step. Due to the existence of the slum, the main flow can spread longitudinally, and strong shear and turbulent eddies are formed around the jet, and the pulsation in the jet impact area is very intense (Toso and Bowers, 1985).

In order to facilitate the discussion of the change law of the energy distribution of IMFs with the flow and relative position, this paper divides the decomposed IMFs into high frequency band, medium frequency band and low frequency band, and uses the percentage of the energy of each frequency band of the IMF components to the total energy of the signal as the characteristic value, that is, the energy characteristic The value P_i is

$$P_i = \frac{E_i}{E} \quad (i=H, M, L) \tag{2}$$

In the formula, H_E represents high-frequency energy, M_E represents mid-frequency energy, and L_E represents low-frequency energy.

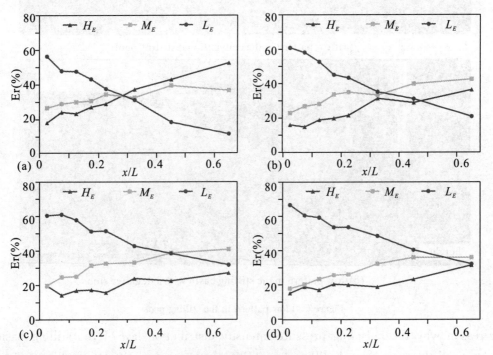

Figure 7. The proportion of high, medium and low frequency band energy at the characteristic points of the conventional stilling pool floor

Note: (a) $Q=30$ m³/s; (b) $Q=50$ m³/s; (c) $Q=70$ m³/s; (d) $Q=90$ m³/s.

As shown in the figure, under different flow rates, the proportion of pulsating pressure on the central axis of the conventional stilling pool gradually decreases with the increase of x/L, and the proportion of the middle and low frequency bands increases with the increase of x/L. The main reason for this energy conversion process in the stilling pool is that the water jump is formed in the stilling pool, and the water jump produces a strong turbulent energy dissipation jet close to the bottom, and the surface of the water flow produces large-scale swirling, as x/L The energy in the stilling pool is fully converted and the fluid energy dissipation effect is better.

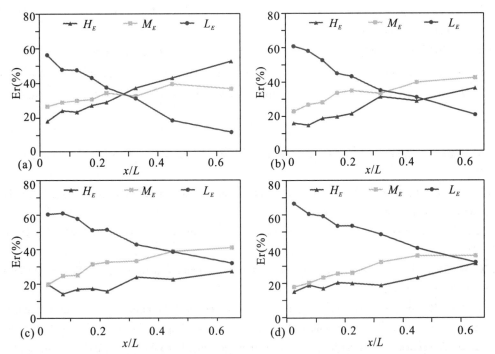

Figure 8. The proportion of high, medium and low frequency band energy at the characteristic points of the stilling basin with a negative step

Note: (a) $Q=30$ m³/s; (b) $Q=50$ m³/s; (c) $Q=70$ m³/s; (d) $Q=90$ m³/s.

As shown in the figure, under different flow rates, the proportion of pulsating pressure on the central axis of the bottom plate of the stilling basin with a negative step decreases gradually with the increase of x/L, and the proportion of the middle and low frequency bands increases with x/L. The increase gradually increases. The law of energy conversion in the stilling basin with a negative step is basically the same as that in the conventional stilling basin, but the gradient of energy change in the conventional stilling basin is larger than that in the stilling basin with a negative step. The high-frequency energy at the rear of the stilling basin is much lower than the mid-range. The gradient of energy change in the stilling basin with a negative step is relatively slow, especially in the middle of the stilling basin, where the degree of energy conversion is low.

5 Conclusions

In this study, the HHT method was used to analyze the pulsating pressure signal, and the pulsating pressure signal on the bottom plate of the stilling basin with a negative step was studied from a different perspective from the traditional one, and the following conclusions were obtained:

(a) By observing the IMFs and residual terms, it is believed that the pulsating pressure acting on the bottom of the stilling basin with a negative step can be regarded as a steady random process in each state.

(b) The results of EMD decomposition show that the pulsating pressure intensity of low-order IMFs is much smaller than that of high-order IMFs. EDM decomposition is gradually separated from high frequency to low frequency; conventional stilling basin and stilling basin with a negative step The energy of the pulsating pressure is mainly concentrated in the front of the stilling basin; the pulsating pressure of the stilling basin with a negative step is prominent near the jet impact point. The reason is that the submerged jet is formed in the stilling basin with a negative step, and the impact area The hydrodynamic pressure fluctuates violently.

(c) Under different flow rates, the proportion of the high frequency band of the pulsating pressure on the central axis of the bottom plate of the stilling pool gradually decreases with the increase of x/L, and the proportion of the middle and low frequency bands gradually increases.

References

[1] Abdul K M, Elango K. Turbulent pressure field beneath a hydraulic jump[J]. Journal of Hydraulic Research, 1974, 12:469−489.

[2] Armenio V, Toscano P, Fiorotto V. On the effects of a negative step in pressure fluctuations at the bottom of a hydraulic jump[J]. Journal of Hydraulic Research, 2000: 359−368.

[3] Bowers C E, Tsai F Y. Fluctuating Pressure in Spillway Stilling Basins[J]. Journal of the Hydraulics Division, 1969, 95: 2071−2080.

[4] Deng J, Xu W, Zhang J. A new type of plunge pool—Multi-horizontal submerged jets[J]. Science in China Series E: Technological Sciences, 2008, 51: 2128−2141.

[5] Huang N E, Shen Z, Long S R. The empirical mode decomposition and the Hilbert spectrum for nonlinear and non-stationary time series analysis[J]. Proceedings Mathematical Physical Engineering Sciences, 1998, 454: 903−995.

[6] Jing X, Li Q. A nonlinear decomposition and regulation method for nonlinearity characterization[J]. Nonlinear Dynamics, 2016, 83: 1−23.

[7] Kazemi F, Khodashenas S R, Sarkardeh H. Experimental Study of Pressure Fluctuation in Stilling Basins[J]. International Journal of Civil Engineering, 2016, 14: 13−21.

[8] Lu Y, Yin J, Yang Z. Numerical Study of Fluctuating Pressure on Stilling Basin Slab with Sudden Lateral Enlargement and Bottom Drop[J]. Water, 2021, 13: 238.

[9] Nonaka A J, Sun Y, Bell J B. Low Mach Number Fluctuating Hydrodynamics of Binary Liquid Mixtures[J]. Communications in Applied Mathematics Computational Science, 2014, 10: 1085−1105.

[10] Riasi A, Nourbakhsh A, Raisee M. Numerical modeling for hydraulic resonance in hydropower systems using impulse response[J]. Journal of Hydraulic Engineering, 2010, 136: 929−934.

[11] Tan X M, Yang Z G, Tan X M. Vortex structures and aeroacoustic performance of the flow field of the pantograph[J]. Journal of Sound Vibration, 2018, 432: 17−32.

[12] Tang N B, Tian Z. Power Experimental Study on Hydrodynamic Model of Stilling Basin with Lateral Flow[J]. Water Resources and Power, 2017, 35(6): 85−87.

[13] Toso J W, Bowers C E. Data acquisition and analysis of pressure fluctuations in hydraulic jumps[J]. Hydraulics and Hydrology in the Small Computer Age, 1985, ASCE: 1184−1189.

[14] Yan Z M, Zhou C T, Lu S Q. Pressure fluctuations beneath spatial hydraulic jumps[J]. Journal of Hydrodynamics, 2006, 18(6): 723−726.

[15] Zhivoderov V N, Tupikov N I. Effective fastening of the bottom of the stilling basin of a high-head dam by prestressed anchors[J]. Hydrotechnical Construction, 1994, 28: 243−252.

Effects of Weir Height on Water Level and Bed Profile in a Degrading Channel

Weiming Wu, Lu Wang, Ruihua Nie, Xudong Ma, Xingnian Liu

State Key Laboratory of Hydraulics and Mountain River Engineering,
Sichuan University, Chengdu 610065, China

Abstract

Weirs are usually built in degrading rivers for grade control, as they can raise the water level and create a non-erodible point to resist the upstream bed degradation. Up to the present, the impacts of weir height on the bed morpho-dynamic process in a degrading river are still unknown. This study conducted 7 experiments using a rectangular weir installed in a uniform sediment bed (median grain size $d_{50}=2.5$ mm), investigating the impacts of weir height z on the bed profile and water level in a degrading channel. The experimental results indicate that the initial upstream water level generally increases with increasing weir height; the backwater level and area of $z/h_0 \geqslant 0.43$ is much greater than those of $z/h_0 \leqslant 0.29$ (h_0 is the initial approach flow depth without weir). The equilibrium bed elevation upstream of the weir increases with increasing weir height. For $z/h_0 \geqslant 0.29$, the weir creates a non-degrading reach and its length increases with increasing weir height. The equilibrium bed slope upstream of the weir increases and decreases with increasing weir height for $z/h_0 \leqslant 0.29$ and $0.29 < z/h_0 \leqslant 0.71$, respectively.

Keywords: Weir height; Degrading channel; Water level; Bed profile; Equilibrium slope

1 Introduction

Bed degradation is a natural river process when the upstream sediment supply rate is less than the bed erosion rate (Galay 1983; Breusers et al. 1991). The bed degradation can destabilize river banks, instream structures, leading to downstream bed deposition and ecological damages. To protect the river bed from been excessively degraded, grade control structures have been widely constructed in the past decades (Rice et al. 1998; Radspinner et al. 2010).

Weirs are common grade control structures and have been extensively built in degrading rivers As the flow over the weir can cause local scour that may lead to structural damages or failure, many studies have investigated the local scour at weirs and proposed a series of scour predictors.

For grade control, it is important to understand the impacts of weir on the bed morpho-dynamic process in a degrading channel. Porto et al. (1999) conducted a field study in southern Italy, investigating the impacts of check dams on the equilibrium bed slope in Calabrian streams. Galia et al. (2016, 2017) surveyed several streams in Czech, finding that the check dams disrupt the sediment transport continuity, reduce the sediment transport rate and gradually increase the downstream grain size. Piton et al. (2016) conducted flume tests to study the impacts of check dams on the sediment transport. They found that, for a steady flow and upstream sediment supply, the check dam does not affect the sediment transport rate when it is silted up. Martín-Vide et al. (2006) experimentally studied the impacts of sequential bed sills on the equilibrium bed profile of a degrading channel. They found that the bed sill can create a non-erodible point, rotating the upstream bed profile around its crest. As a result, the upstream bed slope decreases gradually until reach the equilibrium bed profile. Based on the experimental data, the dependencies of equilibrium bed slope on initial slope and separation distance are evaluated; and an empirical predictor for the equilibrium bed slope is proposed (Martín-Vide et al. 2009).

However, the existing studies only focused on the impacts of check dams in a specific site or bed sills(with crests aligned with the initial bed level) on the bed equilibrium bed profile. The impacts of weir height on the flow characteristics and bed morpho-dynamic process in a degrading channel are still unknown. To fill this research gap, the present study conducted 7 flume tests to investigate the impacts of weir height on the water level and bed morphology in a degrading channel.

2 Experimental Setup

All experiments were conducted in a concrete flume (Fig. 1) at the State Key Laboratory of Hydraulics and Mountain River Engineering, Sichuan University, China. The flume is 30 m long, 1 m wide and 0.7 m deep. The water is supplied by a recirculating pump system. A uniform sediment with a median diameter $d_{50}=2.5$ mm and a geometric standard deviation $\sigma_g=1.23$ was used. The flow discharge was gauged by a rectangular thin-plate weir before the inlet of the flume. Three perforated tracery walls were installed near the flume inlet to ensure a unidirectional flow. A tailgate was used to control the water depth; and a 1 m wide, 1 m long and 1.5 m deep sediment trap was built before the tailgate to collect sediment. The flume had a 12 m long glass sidewall 10 m downstream of the inlet for observation. A 200 mm thick sediment layer greater than the maximum bed incision depth of each test was paved along the flume. A dismountable retention wall was installed at the end of the sediment layer to avoid unexpected bed erosion at the beginning of the test. Ten video cameras were used with transparent grid on the glass wall to measure the flow and bed profile. This measurement method has an accuracy of ±2 mm.

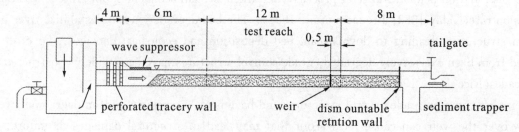

Figure 1. Sketch of the test flume

Six weir heights $z=0\sim50$ mm were used in the experiments. The weir was made of a 10 mm thick acrylic plate spanning the full width of the flume. The weir was installed 0.5 m before the end of the test reach. Seven tests(one without weir and six with weir) were carried out with no sediment feeding with a flow intensity $U_0/U_c=1.5$ for simulating considerable bed degradation. U_0 is the mean approach velocity; U_c is the critical velocity for sediment motion based on the logarithmic form of the velocity profile $U_c \mid u_{*c} = 5.75(5.53 h_0 \mid d_{50})$, in which the critical shear velocity of sediment motion u_{*c} is determined by the Shields curve(Melville 1997).

Before the formal test, a preliminary no-weir test is conducted to determine the opening degree of the tailgate to obtain a normal flow depth $h_0 = 70$ mm. For all tests, the only variable is the weir height, while the initial bed slope, flow depth and velocity were held constant. At the beginning of each test, the flume was filled slowly to avoid unexpected bed erosion. When the flume was filled up, the dismountable retention wall was removed and the test started. Each experiment stopped when the bed elevation variation is no more than 2 mm (less than the minimum grain size) in an hour. The experimental conditions of all tests are summarized in Table 1.

Table 1. Summary of experimental conditions

Test	U_0 (m/s)	U_c (m/s)	Q (L/s)	S_0	D_{50} (mm)	z (mm)	h_0 (mm)	T (h)
1	0.862	0.575	60.3	0.01	2.5	—	70	21
2	0.862	0.575	60.3	0.01	2.5	0	70	8
3	0.862	0.575	60.3	0.01	2.5	10	70	5.5
4	0.862	0.575	60.3	0.01	2.5	20	70	6
5	0.862	0.575	60.3	0.01	2.5	30	70	7
6	0.862	0.575	60.3	0.01	2.5	40	70	6.5
7	0.862	0.575	60.3	0.01	2.5	50	70	15

3 Results and Discussion

Fig. 2 shows the longitudinal water level at the beginning of each test. Fig. 2 indicates that, the water level upstream of the weir generally increases with increasing weir height. For $z/h_0 \geq 0.43$, the water level and are of the backwater zone is much greater than those of $z/h_0 \leq 0.29$.

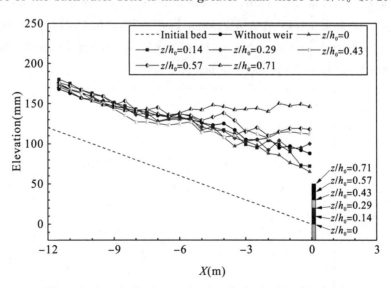

Figure 2. Longitudinal water level at the beginning of each test

Fig. 3 plots the equilibrium longitudinal bed profile of each test. Fig. 3 shows that the bed elevation immediately upstream of the weir is slightly lower (14~19 mm for all tests) than the weir crest, as the flow before weir generates a vortex eroding the sediment near the upstream weir face. Fig. 3 also shows that the upstream bed level generally increases with increasing weir height, as a higher weir can increase the water level and the erosion base level of the bed. In Fig. 3, the length of the non-degrading reach(i.e. the channel reach in which the equilibrium bed level is higher than the original bed level) upstream of the weir is defined as the protective length. For $z/h_0 \geqslant 0.14$, the weir starts to form a non-degrading reach and the protective length increases with increasing weir height.

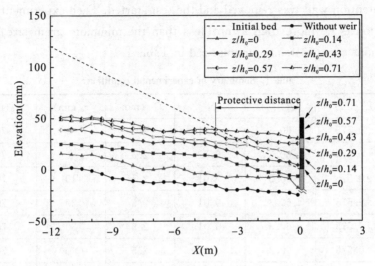

Figure 3. Equilibrium longitudinal bed profile of each test

Fig. 4 plots the dependence of equilibrium bed slope on the normalized weir height z/h_0. Fig. 4 shows that the equilibrium slope increases and decreases with increasing weir height for $z/h_0 \leqslant 0.29$ and $0.29 < z/h_0 \leqslant 0.71$, respectively. As mentioned previously, the weir can reduce the upstream bed degradation as they can increase the water level and the erosion base level. For $z/h_0 \leqslant 0.29$, the increase in upstream water level is not significant and the bed degradation was mainly inhibited by the increased erosion base level(Fig. 2). Thus, the upstream bed elevation increases and less sediment deposits downstream, leading to a greater bed slope. For $0.29 < z/h_0 \leqslant 0.71$, the water level and area of the backwater zone are much greater than those of $z/h_0 < 0.29$(Fig. 2), such that more sediment deposits in the downstream reach to reduce the bed slope.

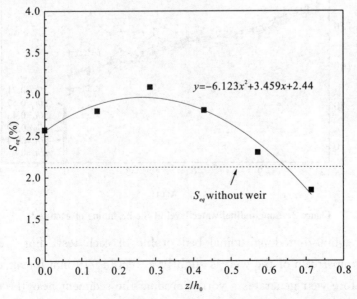

Figure 4. Dependence of equilibrium slope S_{eq} on the normalized weir height z/h_0

4 Conclusions

This study conducted 7 experiments investigating the effect of the weir height on the water level and bed profile in a degrading channel. The experimental results indicate that, at the beginning of each test, the upstream water level generally increases with increasing weir height; the backwater level and area of $z/h_0 \geqslant 0.43$ is much greater than those of $z/h_0 \leqslant 0.29$. The experimental results also indicate that the bed elevation immediately upstream of the weir is slightly lower (14~19 mm for all tests) than the weir crest; and the equilibrium bed elevation upstream of the weir increases with increasing weir height. For $z/h_0 \geqslant 0.14$, the weir creates a non-degrading reach and its length increases with increasing weir height. The equilibrium bed slope upstream of the weir increases and decreases with increasing weir height for $z/h_0 \leqslant 0.29$ and $0.29 < z/h_0 \leqslant 0.71$, respectively.

Acknowledgements

This work was funded by the National Key R&D Program of China (2019YFC1510701), the National Natural Science Foundation of China (U20A20319, 51909177) and the Fundamental Research Funds for the Central Universities (YJ201935).

References

[1] Bormann N E, Julien P Y. Scour downstream of grade-control structures [J]. Journal of Hydraulic Engineering, 1991, 117(5): 579−594.

[2] Breusers H, Raudkivi A J. Scouring [M]. AA Balkema, 1991.

[3] D'Agostino V, Ferro V. Scour on alluvial bed downstream of grade-control structures [J]. Journal of Hydraulic Engineering, 2004, 130(1): 24−37.

[4] Galay V J. Causes of river bed degradation [J]. Water Resources Research, 1983, 19(5): 1057−1090.

[5] Galia T, Škarpich V. Response of bed sediments on the grade-control structure management of a small piedmont stream [J]. River Research and Applications, 2017, 33(4): 483−494.

[6] Galia T, Škarpich V, Hradecký J, et al. Effect of grade-control structures at various stages of their destruction on bed sediments and local channel parameters [J]. Geomorphology, 2016, 253: 305−317.

[7] Lenzi M A, Marion A, Comiti F. Local scouring at grade-control structures in alluvial mountain rivers [J]. Water Resources Research, 2003, 39(7).

[8] Lenzi M A, Marion A, Comiti F, et al. Local scouring in low and high gradient streams at bed sills [J]. Journal of Hydraulic Research, 2002, 40: 731−739.

[9] Li Y N, Han C N, Yang Y. Riverbed incision damagable to hydraulic structure and its countermeasures [J]. Hydo-Science and Engineering, 2014: 58−64. (in Chinese)

[10] Ma X, Wang L, Nie R, et al. Case study: model test on the effects of grade control datum drop on the upstream bed morphology in shiting river [J]. Water, 2019, 11(9): 1898.

[11] Martín-Vide J P, Andreatta A. Disturbance caused by bed sills on the slopes of steep streams [J]. Journal of Hydraulic Engineering, 2006, 132(11): 1186−1194.

[12] Martín-Vide J P, Andreatta A. Channel degradation and slope adjustment in steep streams controlled through bed sills [J]. Earth Surface Processes and Landforms, 2009, 34(1): 38−47.

[13] Melville B W. Pier and abutment scour: integrated approach [J]. Journal of Hydraulic Engineering, 1997, 123(2): 125−136.

[14] Piton G, Recking A. Effects of check dams on bed-load transport and steep-slope stream morphodynamics [J]. Geomorphology, 2017, 291: 94−105.

[15] Porto P, Gessler J. Ultimate bed slope in calabrian streams upstream of check dams: field study[J]. Journal of Hydraulic Engineering-Asce, 1999,125(12): 1231−1242.

[16] Radspinner R R, Diplas P, Lightbody A F, et al. River training and ecological enhancement potential using in-stream structures[J]. Journal of Hydraulic Engineering, 2010,136(12), 967−980.

[17] Rice C E, Kadavy K C. Low-drop grade-control structure[J]. Transactions of the Asae, 1998, 41(5): 1337−1343.

[18] Wang L, Melville B W, Guan D. Effects of upstream weir slope on local scour at submerged weirs[J]. Journal of Hydraulic Engineering, 2018,144(3):04018002.

[19] Wang L, Melville B W, Guan D, et al. Local scour at downstream sloped submerged weirs[J]. Journal of Hydraulic Engineering, 2018,144(8).

[20] Wang L, Melville B W, Shamseldin A Y, et al. Impacts of bridge piers on scour at downstream river training structures: submerged weir as an example[J]. Water Resources Research, 2020,56(4): 026720.

[21] Wang L, Melville B W, Whittaker C, et al. Scour estimation downstream of submerged weirs[J]. Journal of Hydraulic Engineering, 2019,145(12): 06019016.

[22] Yu G A, Wang Z Y, Zhang K, et al. Effect of artificial step-pools on improving aquatic habitats and stream ecological in incised river channel[J]. Journal of Hydroelectric Engineering, 2008,39(2):162−167. (in Chinese)

[23] Zhang Y Y, He Z J, Huang B J, et al. Study on influence of riverbed undercutting on site selection of water intake project[J]. Yangtze River. 2013,44: 52−55,116. (in Chinese)

Experimental Study on Clear-water Scour Evolution Downstream of Rock Weirs

Wen Zhang[1], Lu Wang[2], Xingnian Liu[3], Ruihua Nie[4], Gang Xie[5]

[1,2,3,4] State Key Laboratory of Hydraulics and Mountain River Engineering,
Sichuan University, Chengdu 610065, China

[5] China Institute of Water Resources and Hydropower Research, Beijing 100036, China

Abstract

Rock weirs are common river restoration structures for grade-control, water level rise and instream habitat enhancement. The flow over a rock weir can cause local scour, leading to structural instability. This study conducted 13 flume tests using uniform sediment (median grain size $d_{50}=1.27$ mm) to investigate the temporal evolution of clear-water scour downstream of rock weirs. Various combinations of discharge, water depth, weir height were adopted to study the impacts of densimetric Froude number, submergence, and structural void ratio on the development of clear-water scour depth and length. The results show that the scour depth and length increase with increasing densimetric Froude number, but decrease with increasing submergence. The results also show that the scour depth and length decrease and increase with increasing void ratio, respectively.

Keywords: Clear-water scour; River restoration; Rock weir; Temporal evolution

1 Introduction

Rock weirs are common permeable river restoration structures used to grade control, water level rise and instream habitat enhancement (Radspinner et al., 2010). Typically, a rock weir is composed by loose rocks and spans the full width of the channel (USBR, 2016), and should be completely overtopped during flood events (NRCS, 2001). The rock weir can control bed degradation during high-flow events (NRCS, 2007; Nichols & Polyakov, 2019) and enhance instream habitat during low-flow events in rivers (Martens & Connolly, 2010; Crispell & Endreny, 2009; Gordon et al., 2013). However, the existence of a rock weir can alter the local flow and induce local scour, leading to structural instability or failure. Therefore, it is important to understand the local scour at rock weirs for safe design.

In the past decades, many studies have investigated the scour downstream of impermeable weirs (Ben Meftah & Mossa, 2006, 2019; Gaudio et al., 2000; Lenzi et al., 2002; D'Agostino & Ferro, 2004; Bhuiyan et al., 2009; Tregnaghi et al., 2009; Wang et al., 2019). However, the conclusions based on impermeable weirs are not applicable to rock weirs. Thus, to fill this research gap, many studies have investigated the clear-water scour downstream of rock weirs in the last ten years (Scurlock et al., 2012; Pagliara et al., 2014, 2015, 2016; Pagliara & Kurdistani, 2013, 2015). For clear-water scour, it normally takes a very long time to reach the equilibrium scour depth (Melville & Coleman, 2000). However, as the peak flood flows may last only a few hours or days in the field, the duration of a single flood event is insufficient for a scour hole to reach the equilibrium stage. Thus, the existing predictors of clear-water scour at rock weirs associated with the peak flow rate of a flood hydrograph may be overly conservative. For improving the accuracy of scour estimation, it is important to study the clear-water scour evolution downstream of rock weirs.

This study conducted 13 flume experiments to investigate the clear-water scour evolution downstream of rock weirs. The impact of densimetric Froude number, submergence, and structural void ratio on the evolution of scour depth and length were analyzed and discussed.

2 Experimental Set-up

Experiments were conducted in a 12 m long, 0.5 m deep, and 0.5 m wide glass-walled flume [shown in Fig. 1(a)]. A flow straightener is set at the inlet of the flume to ensure a unidirectional flow. A tail-water gate is set at the downstream end of the flume to control the tail-water depth. Transparent grids are pasted on both sides of the flume to measure the water and bed profile using cameras. The accuracy of the measurement method is ±2 mm.

Fig. 1(b) shows the main variables and the coordinate system. A uniform coarse sand (median diameter, d_{50}=1.27 mm, relative submerged particle density, =1.65) were used. The rock weir was constructed by uniform loose crushed rocks (rock size, D=2.4 cm) and located 6 m downstream from the flume inlet, and had the same width as the flume. The upstream and downstream weir slopes were equal to the angle of repose of rock (φ, taken as 40 degree). To avoid weir settlement on the sediment bed, a foundation (build of the same rock material as the weir, with a depth of about 3~3.5 times the weir height) was set below the weir (USBR, 2016).

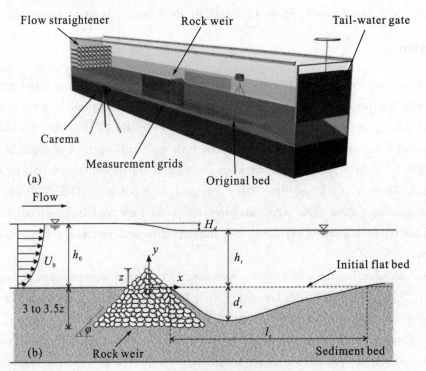

Figure 1. Experimental setup

Note: (a) experimental flume and measurement devices; (b) definition of variables.

Table 1 summarizes the experimental conditions in this study. Various combinations of discharge (Q), weir height(z), and tailwater depth(h_t) were tested in a total of 13 runs. Each test run started with a flat-bed and lasted about 9 to 12 hours. During each test run, the discharge was steady; the upstream approach flow was quasi-uniform; and the tail-water level was constant. The development of the scour hole was recorded by the cameras placed on both sides of flume; and the scour depth(d_s) and length(l_s) at a particular time were extracted from these image records. It is worth mentioning that the test duration is not enough to reach the equilibrium stage of clear-water scour. Because the equilibrium clear-water scour stage normally takes an extremely long time to reach, perhaps infinite time, the final equilibrium scour is not studied in this study.

Table 1. **Summary of experimental conditions**

No.	Q(m³/s)	z(mm)	H_d(mm)	h_t(mm)	h_0(mm)	U_0(m/s)	S_0(‰)	t(min)
1	0.016	30	5	100	105	0.296	0.79	720
2	0.018	30	7	100	107	0.332	1.03	660
3	0.020	30	10	100	110	0.363	1.31	660
4	0.025	30	2	150	152	0.327	0.63	720
5	0.028	30	4	150	154	0.369	0.82	720
6	0.032	30	6	150	156	0.41	1.03	720
7	0.018	40	9	100	109	0.326	1.03	600
8	0.022	40	7	120	127	0.346	0.92	720
9	0.029	40	6	150	157	0.364	0.82	690
10	0.018	50	14	100	114	0.312	1.03	720
11	0.028	50	7.5	150	157.5	0.361	0.82	660
12	0.032	50	9	150	159	0.402	1.03	570
13	0.034	50	12	150	162	0.417	1.15	540

Note: H_d = the water level difference across the weir; h_0 = average approach flow depth; U_0 = average approach flow velocity; S_0 = flume slope; t = time.

3 Results

Fig. 2 shows the temporal evolution of the clear-water scour depth and length downstream of rock weirs for all experiments. Generally, the scour depth and length grow fast for $t<300$ min(initial fast stage) slow down for $t>300$ min(progressing stage). This is consistent to the clear-water scour at concrete weir-like structures. However, as mentioned previously, all tests of the present study have not reached the equilibrium stage and the equilibrium clear-water scour dimension is not investigated in this paper.

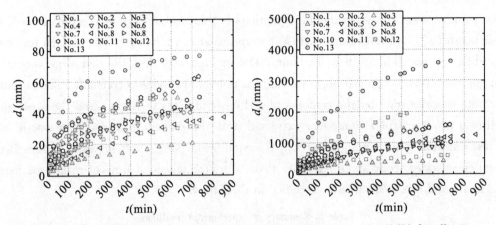

Figure 2. Temporal evolution of clear-water scour depth(a) and scour length(b) for all tests

4 Discussion

For a rock weir under steady flow conditions, the scour depth and length (d_s and l_s) at a particular time can be expressed as:

$$d_s/l_s = f(b, z, D, \mu, \rho, g, h_t, h_0, Q, \rho_s, d_{50}, t, t_e, d_{se}/l_{se}) \tag{1}$$

where ρ and ρ_s = water and sediment mass densities, respectively; μ = fluid dynamic viscosity; g = gravity acceleration; b = weir width; t_e = time to reach equilibrium scour stage; and d_{se} and l_{se} = equilibrium scour depth and length, respectively. For a constant h_t, h_0 is not independent but is determined by Q and z. Besides, uniform sediments are used in this study. Thus, assuming constant fluid viscosity, the normalized downstream clear-water scour depth and length for rock weirs at a particular time can be expressed as:

$$\frac{d_s}{d_{se}}, \frac{l_s}{l_{se}} = f\left[\frac{Q}{b z \sqrt{g\left(\frac{\rho_s}{\rho}-1\right)d_{50}}}, \frac{z}{h_t}, \frac{D}{z}, \frac{t}{t_e}\right] \tag{2}$$

where the first parameter in the right side of Eq. (2) is termed as densimetric Froude number (F_d) which has been widely used in the study on local scour at low-head rock structures (Pagliara et al., 2014); z/h_t is the normalized weir height representing the submergence of the weir; D/z is the normalized rock size reflecting the structural void ratio of the rock weir placed (i.e., the larger the D/z, the larger the void ratio of the rock weir). As the equilibrium scour stage is not reach in this study, the time scale T_z (characteristic time at $d_s=z$) is used in this study to replace the equilibrium time scale t_e. Accordingly, the equilibrium length scale of the scour depth and length are replaced by d_{sz} and l_{sz} (scour depth and length at T_z), respectively. Similar treatment can be found in Guan et al. (2019). Hence, Eq. (2) is modified as:

$$\frac{d_s}{d_{sz}}, \frac{l_s}{l_{sz}} = f\left[\frac{Q}{b z \sqrt{g\left(\frac{\rho_s}{\rho}-1\right)d_{50}}}, \frac{z}{h_t}, \frac{D}{z}, \frac{t}{T_z}\right] \tag{3}$$

This section will evaluate the impacts of F_d, D/z, and h_t/z on the temporal evolution of clear-water scour downstream of rock weirs.

4.1 Impacts of densimetric froude number

Fig. 3 shows the effect of F_d on the clear-water scour evolution downstream of a rock weir. Fig. 3 indicates that, at a particular normalized time t/T_z and with z/ht and D/z held constant, both d_s/d_{sz} and l_s/l_{sz} are higher with a greater F_d. This trend is consistent to that of equilibrium clear-water scour depth or length downstream of rock weirs(Scurlock et al., 2012). For clear-water scour, there is no sediment supply from the upstream of the rock weir and the scour dimension is only affected by the strength of the overflow for a given submergence and structural void. Therefore, the increase in F_d leads to a stronger overflow, resulting in a larger clear-water scour hole downstream of the rock weir.

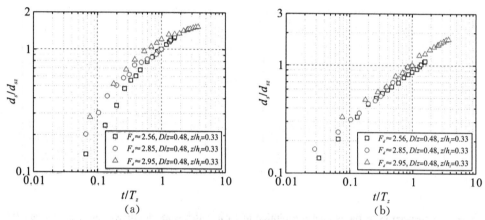

Figure 3. Dependences of the temporal evolution of scour depth(a) and length(b) on F_d

4.2 Impacts of submergence

Fig. 4 shows the effects of z/h_t on the clear-water scour evolution downstream of a rock weir. In Fig. 4, at a particular normalized time, the increase in z/h_t increases both d_s/d_{sz} and l_s/l_{sz}. For clear-water scour, the upstream sediment does not move. The flow over the weir accelerates at the weir crest and forms a jet to generate a scour hole. For a small z/h_t(high submergence), the overflow jet is observed to be a surface jet; and the scour hole downstream of the rock weir is caused by the increasing jet thickness and turbulence mixing with the tailwater. For a high z/h_t(low submergence), the jet appears to be an impinge jet directly impacting on the bed, leading to a larger scour hole.

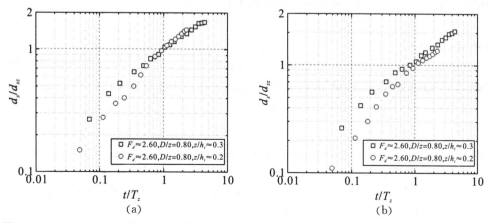

Figure 4. Dependences of the temporal evolution of clear-water scour depth(a) and length(b) on z/h_t

4.3 Impacts of structural void ratio

Fig. 5 shows the effect of D/z on the clear-water scour evolution downstream of a rock weir. In Fig. 5, d_s/d_{sz} and l_s/l_{sz} slightly decrease and increase with increasing D/z, respectively. The increase in D/z increases the structural void ration of the rock weir, reducing the intensity of downstream turbulence structures but promotes the turbulence structures to evolve far downstream (Leu et al., 2008). As the scour hole dimension downstream of a weir is dependent on the turbulence structures in the scour hole(Guan et al., 2014), the weakened downstream turbulence intensity due to the increased D/z reduces the scour depth; and the widened turbulence area entrains the sediment to further downstream and increases the scour length.

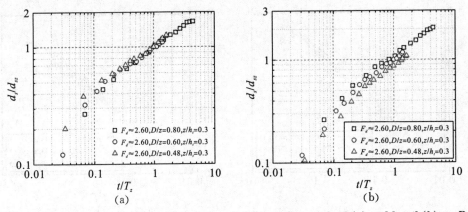

Figure 5. Dependences of the temporal evolution of clear-water scour depth(a) and length(b) on D/z

5 Conclusions

This study conducted flume tests to investigate the temporal evolution of the clear-water scour hole downstream of rock weirs. The impact of densimetric Froude number(F_d), submergence(z/h_t), and structural void ratio(D/z) are analyzed and discussed.

The experimental results show that the scour hole develops rapidly and slowly in the initial fast stage and progressing stage, respectively. The normalized downstream clear-water scour depth (d_s/d_{sz}) and scour length(l_s/l_{sz}) increase with increasing with F_d and z/h_t. The experimental also indicates that d_s/d_{sz} and l_s/l_{sz} decrease and increase with increasing D/z, respectively. The findings of this study could be useful for the safe design of rock weirs.

Acknowledgements

This work was funded by the National Key R&D Program of China(2019YFC1510701), the National Natural Science Foundation of China(U20A20319, 51909177) and the Fundamental Research Funds for the Central Universities(YJ201935).

References

[1] Ben M M, Mossa M. New approach to predicting local scour downstream of grade-control structure[J]. Journal of Hydraulic Engineering, 2020, 146(2): 04019058.

[2] Ben Meftah M, Mossa M. Scour holes downstream of bed sills in low-gradient channels[J]. Journal of Hydraulic Research, 2006, 44(4): 497−509.

[3] Bhuiyan F, Hey R D, Wormleaton P R. Effects of vanes and w-weir on sediment transport in meandering channels

[J]. Journal of Hydraulic Engineering, 2009, 135(5): 339−349.

[4] Crispell J K, Endreny T A. Hyporheic exchange flow around constructed in-channel structures and implications for restoration design[J]. Hydrological Processes, 2009, 23(8): 1158−1168.

[5] D'Agostino V, Ferro V. Scour on alluvial bed downstream of grade-control structures [J]. Journal of Hydraulic Engineering, 2004, 130(1): 24−37.

[6] Gaudio R, Marion A, Bovolin V. Morphological effects of bed sills in degrading rivers [J]. Journal of Hydrualic Research, 2000, 38(2): 89−96.

[7] Gordon R P, Lautz L K, Daniluk T. Spatial patterns of hyporheic exchange and biogeochemical cycling around cross-vane restoration structures: Implications for stream restoration design[J]. Water Resources Research, 2013, 49(4): 2040−2055.

[8] Guan D, Liu J, Chiew Y M, et al. Scour evolution downstream of submerged weirs in clear water scour conditions[J]. Water, 2019, 11(9): 1746.

[9] Lenzi M A, Marion A, Comiti F, & Gaudio R. Local scouring in low and high gradient streams at bed sills [J]. Journal of Hydraulic Research, 2002, 40(6): 731−739.

[10] Leu J M, Chan H C, Chu M S. Comparison of turbulent flow over solid and porous structures mounted on the bottom of a rectangular channel[J]. Flow Measurement and Instrumentation, 2008, 19(6): 331−337.

[11] Martens K D, Connolly P J. Effectiveness of a redesigned water diversion using rock vortex weirs to enhance longitudinal connectivity for small salmonids [J]. North American Journal of Fisheries Management, 2010, 30(6): 1544−1552.

[12] Melville B W, Coleman S E. Bridge scour[M]. Water Resources Publication, 2000.

[13] Nichols M H, Polyakov V O. The impacts of porous rock check dams on a semiarid alluvial fan[J]. Science of The Total Environment, 2019, 664: 576−582.

[14] NRCS(Natural Resources Conservations Service). Design of rock weirs. Washington. DC, US Dept. of Agriculture, NRCS, 2001.

[15] NRCS(Natural Resources Conservations Service). Grade Stabilization Techniques. Washington. DC, US Dept. of Agriculture, NRCS, 2007.

[16] Pagliara S, Kurdistani S M. Scour downstream of cross-vane structures [J]. Journal of Hydro-environment Research, 2013, 7(4): 236−242.

[17] Pagliara S, Kurdistani S M. Clear water scour at J-hook vanes in channel bends for stream restorations[J]. Ecological Engineering, 2015, 83: 386−393.

[18] Pagliara S, Kurdistani S M, Cammarata L. Scour of clear water rock W-weirs in straight rivers [J]. Journal of Hydraulic Engineering, 2014, 140(4): 06014002.

[19] Pagliara S, Kurdistani S M, Palermo M, et al. Scour due to rock sills in straight and curved horizontal channels[J]. Journal of Hydro-environment Research, 2016, 10: 12−20.

[20] Pagliara S, Palermo M, Kurdistani S M, et al. Erosive and hydrodynamic processes downstream of low-head control structures[J]. Journal of Applied Water Engineering & Research, 2015, 3(2): 122−131.

[21] Radspinner R R, Diplas P, Lightbody A F, et al. River training and ecological enhancement potential using in-stream structures[J]. Journal of Hydraulic Engineering, 2010, 136(12): 967−980.

[22] Scurlock S M, Thornton C I, Abt S R. Equilibrium scour downstream of three-dimensional grade-control structures [J]. Journal of Hydraulic Engineering, 2012, 138(2): 167−176.

[23] Tregnaghi M, Marion A, Coleman S. Scouring at bed sills as a response to flash floods[J]. Journal of Hydraulic Engineering, 2009, 135(6): 466−475.

[24] USBR(United States Bureau of Reclamation). Rock weir design guidance[M]. Denver, Colorado, USBR Technical Service Center, Sedimentation and River Hydraulics Group, 2016.

[25] Wang L, Melville B W, Whittaker C N, et al. Scour estimation downstream of submerged weirs [J]. Journal of Hydraulic Engineering, 2019, 145(12): 06019016.

Experimental Study on Flow Characteristics in Curve Open Channel with Suspended Vegetation

Qiaoling Zhang[1], Hefang Jing[2], Chunguang Li[3]

[1, 3] School of Mathematics and Information Science, North Minzu University, Yinchuan 750021, China

[2] School of Civil Engineering, North Minzu University, Yinchuan 750021, China

Abstract
The suspended vegetation in natural rivers has large impact on the flow structure. In this study, suspended vegetation, which was replaced by glass rods, was placed in an area of the bend in a U-shaped flume to investigate the water flow characteristics of the bend with suspended vegetation. According to the depth in the water and the arrangement of vegetation, six typical cases were designed and carried out. The flow velocity was measured with a particle imaging velocimetry(PIV), and the water level was measured with an ultrasonic water level detector. According to the measured results, hydraulic elements, such as the distribution of water, the distribution of velocity, the lateral water level gradient and the head loss under different cases were calculated, compared and analyzed. The research results show that the distribution of suspended rigid vegetation in a bend will raise the water level of the vegetation area, increase the loss of the water head and the hydraulic slope compared with the no vegetation case. Furthermore, the distributions of the two velocity gradients near the concave bank and the convex bank were obviously different under different vegetation arrangements. In addition, the depth and the position of the vegetation in the water have greater impact on the flow velocities of the vegetation area.

Keywords: Suspended vegetation; U-shaped open channel; PIV; Water flow

1 Introduction

Almost all natural rivers are curved. Suspended vegetation such as water hyacinth, water soft-shelled turtle, big frog and Sophora japonicus grows widely in rivers, lakes and offshore environments (Han, 2017). Their developed root system absorbs nutrients such as nitrogen, phosphorus and heavy metals in the water, so they play a great role in purifying the water body(Dierssen et al., 2015). At the same time, the flow character of curved channel with suspended vegetation cannot be simply described by the hydraulic factor of the open channel flow under the action of the centrifugal force of the bend, the river bank and the vegetation. Therefore, the study of the characteristics of curved rivers with suspended vegetation is of great significance.

In recent years, there have been many research results on physical experiments and numerical simulations of river flows with aquatic vegetation. In terms of experiments, Yue(2007) et al. used Particle Image Velocimetry(PIV) to conduct a series of measurements on vegetation canopy water flow in open channels; Huai(2009) et al. measured the time averaged velocity and turbulence behaviour of a steady uniform flow with fully submerged artificial rigid vegetation using a three dimensional Micro Acoustic Doppler Velocimetry (ADV); Shucksmith(2010) et al. conducted a flume experiments on open channel flow with natural vegetation; Han(2018) et al. used rigid cylindrical glass rods instead of vegetation, investigated the vertical distribution of longitudinal velocity and shear stress of a suspended vegetation-covered flow. Jing(2020) et al. used three dimensional laser Doppler Velocimetry (3D-LDV) to measure and analyze the flow characteristics of a single flume with unsubmerged rigid vegetation.

In terms of numerical simulation, Zhang(2013) et al. developed a depth-averaged model using the finite volume method on a staggered curvilinear grid to simulating the flow in a 180° curved open channel partly covered by emerged vegetation; Rao (2014) et al. built a moveable two-layer combination model based on the continuous porous media model to describe the floating bed; Jing (2016) et al. modified the momentum equations by adding an additional source term, and developed A 2-D depth averaged RNG k-ε model to simulate the flow in a typical reach of the Upper Yellow River with non-monotonic banks. Cai (2017) et al. used the lattice Boltzmann method to numerically simulate the river flow with emergent vegetation;

Judging from the current research situation, many scholars have focused on the straight and emergent vegetation open channel flow. In this study, a series of experiments on the water flow characteristics of suspended vegetation with developed roots replaced by glass rods were conducted in the curve section of the U-shaped flume using PIV and other instruments, and the effects of suspended vegetation on water flow characteristics under different arrangements and different depths of water near the top of the bend were investigated.

2 Experimental Facilities and Design of Case

This experiment was carried out in a U-shaped folding circulating water flume in the hydraulics laboratory of the author's unit. The flume is 38 meters in total, of which the length of the straight section water inlet and outlet area are both 16 meter, the inner bend radius is 1.6 meter, and the flow section dimensions of the trough is 0.8 meters by 0.8 meters, the bottom of the flume slope is fixed at 1‰. A double-layer glass frame with 10 mm diameter drills evenly arranged on the top of the flume was designed to fix the suspended rigid vegetation, which replaced by cylindrical glass rods.

The main experimental instrument is the PIV system includes high-speed CMOS camera, tracer particles and laser system. The water level is measured by an ultrasonic real-time measurement probes arranged at a distance of 6cm from the shore. The laser arranged under the bottom of the water flume is emitted upward from the bottom of the water flume through the glass plate. The camera shoots the displacement of the tracer particles in the water from the outside of side wall of the flume, and the flow velocity distribution is obtained through cross-correlation and auto-correlation algorithms. The plane of laboratory flume is shown in Fig. 1.

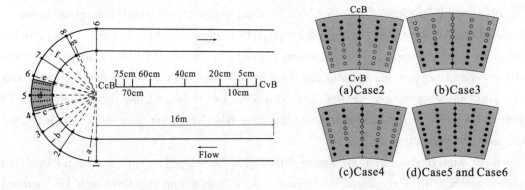

Figure 1. Position of velocity and water level observation and Vegetation Arrangements
Notes: CcB=Concave bank; CvB=Convex Bank.

The vegetation area was arranged in the range of 79°~101° with the entrance of the bend, the lateral distance between two adjacent glass rods was 0.09 m, and the central angle of the two adjacent glass rods in the same arc was 5°. A total of six cases named Case1 to Case6 were set up in this experiment. The characteristic parameters of various cases are shown in Table 1, the different vegetation arrangements are shown in Fig. 1.

Table 1. Characteristic parameters of various cases

Case	Q(m³/s)	H(m)	V(m/s)	Re	Fr	H_1(m)	Vegetation Arrangements
Case1						—	No vegetation
Case2							Vegetation only near the concave bank
Case3							Vegetation only near the convex bank
Case4	0.052	0.14	0.464	42000	0.36	0.07	Vegetation near both banks
Case5							Vegetation evenly distributed across the entire section
Case6						0.035	Vegetation evenly distributed across the entire section

Notes: $Q=$ the quantity of flow; $H=$ the tail gate water level; $V=$ the average velocity; $H_1=$ the depth of vegetation into water; $Fr=$ the Froude number; $Re=$ the Reynolds number.

Seven velocity measurement sections are set along the way, and the angles with the entrance section of the curve are 10°, 45°, 78°, 90°, 102°, 135° and 155°(The degrees mentioned below are the angles between the entrance section), which denoted as a, b, c, d, e, f, and g, each section has five measuring lines along the horizontal direction, and the distance from the convex bank are 5cm, 20cm, 60cm and 75cm respectively. There were seven sections for water level observation, and the angles are 0°, 30°, 60°, 78°, 90°, 102°, 120°, 150° and 180°, which are numbered 1~9. Each observation section is shown in Fig. 1.

3 Experimental Results and Analysis

3.1 Water level measurement results and analysis

Fig. 4 shows the longitudinal distribution of the water level under different cases.

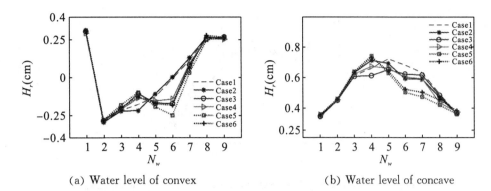

(a) Water level of convex (b) Water level of concave

Figure 4. The distribution of water level along the bank under different vegetation arrangement cases

Notes: N_w = the water level number; $H_r (=z-H)$ = the relative water level; z = the water level at that point.

It can be seen from Fig. 4 that he water level of concave bank was higher than that of convex bank. Generally speaking, when the depth of vegetation into water was 0.07 m, the upstream water level before the vegetation area, which located from section 1 to section 4, was the highest when vegetation was evenly arranged across the entire section in Case5, The water level on opposite side is lower than that without vegetation when vegetation is arranged on one side. For the rest, when vegetation is arranged on both banks and only on the opposite side, the water level rises sequentially. In the vegetation area (sections 4~6), the water level drops to less than that in the case of no vegetation in case1. The water level of the convex bank drops to the lowest in the last row of the vegetation area. After leaving the vegetation area, the water level of the convex bank began to rise rapidly, but the water level is still lower than that in case1, while the water level of the concave bank still keep a downward trend, but the water level gradient slows down, and the water level was close to the value of case1 until the exit of the bend. This is because the resistance to the current side is increased when vegetation is arranged on one side, the water level on the local bank appears to be high, and the water level on the opposite bank decreases; The resistance of the water flow by the vegetation is relatively large when the number of vegetation doubles and relative water entry depth increased, resulting in a higher water level.

3.2 The characteristics and analysis of the lateral gradient along the way

Formula(1) provides the calculation formula for the horizontal water gradient:

$$J_y = \frac{H_a - H_b}{B} \quad (1)$$

where J_y = the lateral water gradient; H_a, H_b = the water level at the concave and convex banks, in cm; B = the width of the flume.

The distributionof horizontal water gradient along the way under different cases is shown in Fig. 5.

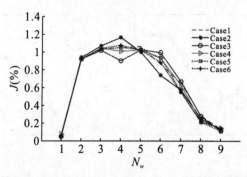

Figure 5. The horizontal gradient along the way under different cases

It can be seen from Fig. 5 that the horizontal water gradient approximately shows a trend of "increasing first and then decreasing". When suspended vegetation was placed near the top of bend, it had the greatest impact on the cross-section 4 and section 6. For section 4, the water level of the convex bank was raised when vegetation was only arranged on the convex bank in Case3, reducing the horizontal gradient of the water; the water of concave bank is further raised on the basis of centrifugal action when vegetation was only arranged on the concave bank in Case2, reducing the water level of the convex bank, and the water horizontal gradient was larger than other conditions, the maximum profile drop occurs at section 4(78°) and reached 1.2%; in the rest cases include cases of vegetation is placed on both sides bank and arranged across the entire section, the horizontal gradient of the water is close to that in the case of no vegetation which between Case2 and Case3. The horizontal gradient of the water at section 6 is located between Case2 and Case3.

4 Measurment Results of Flow and Analysis

4.1 Distribution of longitudinal velocity along the water depth

At section d, under different working conditions, the longitudinal velocity distribution at different positions along the vertical direction is shown in Fig. 6.

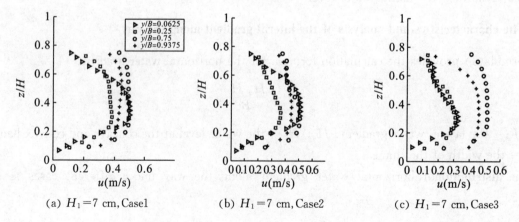

(a) $H_1=7$ cm, Case1　　(b) $H_1=7$ cm, Case2　　(c) $H_1=7$ cm, Case3

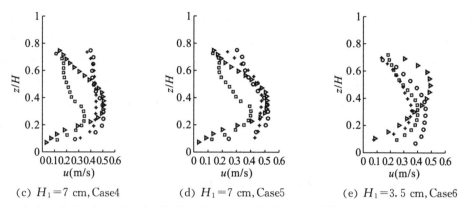

(c) $H_1 = 7$ cm, Case4   (d) $H_1 = 7$ cm, Case5   (e) $H_1 = 3.5$ cm, Case6

Figure 6. Longitudinal velocity distribution along water depth under different vegetation arrangements

Notes: y = the distance from the convex bank.

It can be seen from Fig. 6 that the longitudinal velocity distribution in the curve along the water depth shows a trend of "large in the middle and small at the two ends", and the velocity on the concave bank is greater than that on convex bank. The water flow is affected by the dual effects of vegetation and bends when suspended vegetation is arranged in the bend, and the flow velocity of the vegetation area is significantly reduced. By comparing several cases, it can be found that the suspended vegetation has a smaller resistance to the water flow below the roots of the vegetation. In areas above the roots of vegetation, different cases have a greater impact on the longitudinal velocity distribution. When vegetation is arranged only on the concave bank as Fig. 6(b), the two velocity gradients near the concave bank and the convex bank generally only exist in the upper water area. When vegetation is arranged only on the convex bank as Fig. 6(c), there are two more obvious velocity gradients, which extend to the bottom of the water. The velocity changes relatively drastically when vegetation is arranged on both banks as Fig. 6(d), since y2 and y4 are located at the junction of vegetation and non-vegetation. There is no obvious flow velocity gradient when vegetation is evenly arranged across the entire section as Fig. 6(d) and Fig. 6(e). When there is no vegetation, the maximum vertical velocity is mostly distributed near the bottom area ($z/H < 0.25$). When vegetation is arranged, the maximum velocity is distributed near the roots of the vegetation.

4.2 Longitudinal velocity distribution along the way

When horizontal plane is 7.2 cm, and the two measurement lines are 20 cm from the shore, the distribution of flow velocity along the route under different cases are shown in Fig. 7.

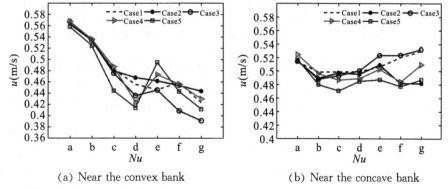

(a) Near the convex bank   (b) Near the concave bank

Figure 7. Changes of velocity along the way under different vegetation arrangements

Notes: Nu = the velocity number; u = the longitudinal velocity.

It can be seen from Fig. 7 that the flow velocity of the convex bank decreases along the curve due to the centrifugal force of the curve, and the concave bank velocity increases gradually. After vegetation is placed in the bend, the flow velocity of vegetation area and the downstream drastic changed relatively. The velocity of the vegetation area has a significant decrease, which is most obvious in the Case5, the distribution of the longitudinal velocity along the way under the condition of Case4 is relatively close to that under the condition of no vegetation in Case1, and there is no obvious distribution law of the minimum velocity of concave and convex banks.

5 Head Loss and Hydraulic Gradient in Vegetation Area

According to the energy equation of the liquid movement:

$$Z_1 + \frac{P_1}{\rho g} + \frac{V_1^2}{2g} = Z_2 + \frac{P_2}{\rho g} + \frac{V_2^2}{2g} + h_w \tag{2}$$

Since the atmospheric pressures of the upstream and downstream sections are the same, the head loss is

$$h_w = Z_1 - Z_2 + \frac{V_1^2}{2g} - \frac{V_2^2}{2g} \tag{3}$$

where h_w=head loss, in m; Z_1 and Z_2=the water level, in m; V_1 and V_2=the flow velocity, in m/s.

The hydraulic gradient is the head loss per unit length. The formula is

$$J = \frac{h_w}{L} \tag{4}$$

where J=the hydraulic gradient, in %; L = the length of the vegetation area, in m.

The head loss and the hydraulic gradient of calculation results are shown in Table 2.

Table 2. Section flow rate, water level, head loss and hydraulic gradient

Case	H_1(m)	V_1(m/s)	V_2(m/s)	h_w(cm)	J(%)
Case1	—	0.4565	0.4566	0.004	0.005
Case2		0.4563	0.4571	0.021	0.025
Case3	0.07	0.4561	0.4570	0.023	0.028
Case4		0.4556	0.4568	0.033	0.039
Case5		0.4539	0.4602	0.166	0.198
Case6	0.035	0.4545	0.4588	0.113	0.135

Notes: V_1 and V_2= the average velocity(is equal to the flow quantity divided by the flowing area).

It can be seen from Table 2 that the head loss is minimal due to the centrifugal force and side wall friction when there is no vegetation. There is the largest head loss when the vegetation evenly distributed across the entire section in Case5. As the number of vegetation and the depth of vegetation entry water increase, the resistance to water flow increases, and the head loss increases accordingly. The head loss under vegetation on both banks is greater than that in the case of vegetation on one side, this is because the flow velocity of the concave bank is larger, and the resistance of the water flow by the vegetation is greater, and the greater the head loss. The change trend of hydraulic gradient is exactly the same as the head loss because the fixed length of the vegetation area.

6 Conclusions

This study uses PIV and other instruments to measure the water flow velocity and water level when suspended vegetation is arranged in the local area of the bend in the laboratory, and the lateral water gradient and water head loss under different cases are comparative analyzed. After suspended vegetation is arranged in the bend, the main conclusions are as follows:

(1) The overall water level of the vegetation area will be changed, but the distribution law of the water level along the bend will not be changed. The water level in the upper reaches of the vegetation area is generally high, but when vegetation is arranged on the opposite bank, the upstream water level does not rise but falls; the water level near the concave bank first rises and then falls, and the water level near the convex bank first falls and then rises.

(2) It will locally change the lateral water gradient of the bend, but the distribution law of the lateral water gradient is basically unchanged. The water lateral gradient is close to zero at the entrance section of the bend, and then shows a gradually increasing trend, reaching a maximum near the top of the bend, and then gradually decreasing, but at the exit section of the bend, it is still greater than zero under the effect of centrifugal force.

(3) The "upward decrease and downward increase" feature is more obvious, and the velocity along the vertical direction is "the two ends are small and the middle is large"; the longitudinal velocity in the vegetation area is significantly reduced.

(4) The water flow resistance will be increased, which will lead to the increase of the head loss and the hydraulic gradient. In comparison, when the depth of entry water is the same, the head loss is the smallest when vegetation only arranged near one side bank, and it is the largest when the vegetation evenly distributed across the entire section; when the vegetation layout is the same, the greater the depth of vegetation entry, the greater the head loss.

Acknowledgements

The research reported herein is funded by the Project of Key Research and Development Planned by Science and Technology Department of Ningxia (2019BEG03048) and the Western First Class Discipline Construction for Mathematics in Ningxia (NXYLXK2017B09). The author would like to thank all those who participated in the project.

References

[1] Cai Y J, Jing H F, Li C G. Numerical simulation of vegetation flow based on MRT- LBE model[J]. People's Yellow River, 2017, 39(10): 15−21. (in Chinese)

[2] Dierssen H M, Chlus A, Russell B. Hyperspectral discrimination of floating mats of seagrass wrack and the microalgae Sargassum in coastal waters of Greater Florida Bay using airborne remote sensing[J]. Remote Sensing of Environment, 2015, 167: 247−158.

[3] Han L J. Research on the hydraulic characteristics of floating vegetation flow[J]. Wuhan University, 2017: 1−6. (in Chinese)

[4] Han L J, Zeng Y H, Chen L, et al. Modeling streamwise velocity and boundary shear stress of vegetation-covered flow[J]. Ecological Indicators, 2018, 92.

[5] Huai W X, Zeng Y H, Xu Z G. Three-layer model for vertical velocity distribution in open channel flow with submerge rigid vegetation[J]. Advances in Water Resources, 2009, 32(4): 487−492.

[6] Jing H F, Cai Y J, Wang W H. Investigation of open channel flow with unsubmerged rigid vegetation by the lattice Boltzmann method[J]. Journal of Hydrodynamics, 2020, 32(4): 771−783.

[7] Jing H F, Li Y T, Li C G. Numerical study of flow in the Yellow River with non-monotonous banks[J]. Journal of Hydrodynamics, 2016, 28(1): 142−152.

[8] Rao L, Qian J, Ao Y. Influence of artificial ecological floating beds on river hydraulic characteristics[J]. Journal of Hydrodynamics, 2016, 26(3): 474−481.

[9] Shucksmith J D, Boxall J B, Guymer I. Effects of emergent and submerged natural vegetation on longitudinal mixing in open channel flow[J]. Water Resources Research, 2010, 46: W04504.

[10] Yue W, Meneveau C, Parlange M B. A comparative quadrant analysis of turbulence in a plant canopy[J]. Water Resources Research, 2007, 43: W054225.

[11] Yang Y Y, Ma Y S, Zhan Z R. Refined numerical simulation of three-dimensional hydrodynamics in vegetation areas under the action of submerged and suspended vegetation[J]. Journal of hydrodynamics, 2020, 1−7. (in Chinese)

[12] Zhang M, Li C W, Shen X. Depth-averaged modeling of free surface flows in open channels with emerged and submerged vegetation[J]. Applied Mathematical Modelling, 2013, 37(1−2): 540−553.

Experimental Study on Responses of Bed Morphology and Water Level to Unsteady Sediment Supply

Qihang Zhou[1], Lu Wang[2], Ruihua Nie[3], Xudong Ma[4], Xingnian Liu[5], Gang Xie[6]

[1,2,3,4,5] State Key Laboratory of Hydraulics and Mountain River Engineering,
Sichuan University, Chengdu 610065, China

[6] China Institute of Water Resources and Hydropower Research, Beijing 100048, China

Abstract

Flash flood, landslide or debris flow can carry a lot of sediment into rivers, causing rapid adjustment of river morphology. The study conducts three flume tests using uniform sediment (median grain size $d_{50} = 3.6$ mm), investigating the responses of bed morphology and water level to three sediment supply modes (constant, trapezoidal and triangular) at the same time average sediment supply rate. The experimental results show that for the constant sediment supply mode, the riverbed elevation, water level and bed slope increase firstly with time and then decrease with the end of sediment feeding. For the trapezoidal sediment supply mode and the triangle periodic sediment supply mode, the responses of bed elevation, water level and bed slope are similar to that of the constant sediment supply mode, but the change of channel morphology is gentler than that of the constant sediment supply mode. Generally, the more unsteady the sediment supply, the more unstable the riverbed and water level. In addition, the river morphology resopond strongly to the excessive sediment supply in hydrography boundary layer (located near the sediment supply position).

Keywords: Bed morphology; Sediment supply; Water level

1 Introduction

Flow regimes and sediment transport are two main control factors of channel morphology and river channel bars (Venditti et al., 2012). The unbalance of sediment supply and river transport capacity may cause a river to significantly change its form through eroding or depositing (Yager et al., 2012). Generally, channels with higher sediment supply tend to have more complex bed morphology. In mountain rivers, due to excessive sediment supply, sedimentation and riverbed elevation make the water level exceed the flood control standard, causing river flood disasters (Recking, 2014; Zheng et al., 2019). Therefore, it is important to understand the responses of bed morphology and water level to sediment supply for engineering safety and flood control.

Sediment supply in mountain rivers is often caused by landslides, debris flows, flash floods and other geological hazards (An et al., 2017a; Bel et al., 2017; Bordoni et al., 2020; Borga et al., 2014; Cama et al., 2017; Destro et al., 2018). The sediment supply in those processes varies in space and time, expressing unsteady characteristic. The unsteady sediment supply can induce more complex bed morphology than the steady sediment supply. Especially in mountain flood hazards, a mount of unsteady sediment transport can affect inhabitants by changing flow resistance or the flow section

(Fig. 1). In mountain streams, these flood risks depend not only on flow intensity, but also sediment transport(Gan et al., 2018; Recking, 2014; Zheng et al., 2019). Thus, the risk assessment of mountain areas should not only consider the hydrological conditions, but also the sediment supply conditions.

Figure 1. (a) The sediment transport during flash floods in Caopo River Basin, China; (b) Structural damage due to excessive and unsteady sediment supply during mountain flood hazards in Caopo town, China

In the past decades, the morphodynamic response of mountainous rivers to the change of sediment supply has been widely concerned(Dean et al., 2016; Kammerlander et al., 2017; Nelson et al., 2015; Nelson and Morgan, 2018; Pfeiffer et al., 2017; Recking, 2012). Sediment often enters rivers in the form of pulses with flash floods, landslides and debris flows rather than a more continuous supply in mountainous rivers(An et al., 2017a; Cui et al., 2003a; Cui et al., 2003b). Cui et al. (2003a, b) thought the sediment pulses can be eliminated by translation or dispersion in mountain streams. The results of their experiments and numerical predictions both proved that those two mechanisms can operate in any given river. Nelson et al. (2015) conducted a series of flume experiments to investigate the morphodynamic response of variable-width channel to the changes in sediment supply, finding variable channel width and riffle-pool topography may have enhanced dispersion of sediment pulses. An et al. (2017a) explored the evolution of gravel bed river to cyclic hydrographs and repeated sediment pulses with a one-dimensional morphodynamic model, indicating the bed elevation and surface grain size distribution respond strongly in a specific region(hydrography boundary layer) near the sediment feed point(Wong and Parker, 2006).

However, the above researches only considered the influence of sediment pulses on riverbed morphology. Due to the sediment supply during flood events is usually not a continuous process, this paper reports three sediment supply modes(constant, trapezoidal and triangular) to fully understand the influence of excessive sediment supply on river morphology in flood events. A series of flume experiments were carried out to investigate the responses of bed morphology and water level to unsteady sediment supply.

2 Experimental Setup

All experiments were conducted in a 9.0 m long, 0.6 m wide and 0.8 m deep glass-walled recirculating flume with a slope of 0.005 at the State Key Laboratory of Hydraulics and Mountain River Engineering, Sichuan University(Fig. 2). The sediment used in experiments was well-sorted gravel, with median particle size d_{50}=3.6 mm, geometric standard deviation σ_g=1.15 and density ρ=2650 kg/m³. Uniform sediment was laid in the bed of the flume, where the layer thickness was 0.2 m. Flow passed though the PVC pipes to dissipate energy. The gravel was supplied by a funnel sediment feeder at the upstream of the flume, and the feeding rate was controlled by the speed of the sediment feeder motor. The sediment washed away from the upstream was deposited into a sediment recess in the downstream. A tailgate was installed at the end of the flume to control the flow depth. The flow was considered to be uniform when the water surface gradient approaches the river bed slope.

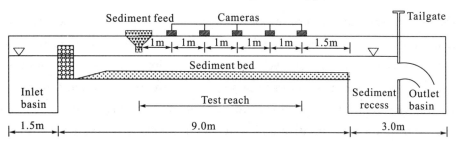

Figure 2. **Plan view of flume**

The flow discharges were measured by a rectangular weir before the inlet of the flume. The bed elevation and water level were measured using webcams with transparent grid papers on one side of the flume, which has an accuracy of ±2 mm. The bed elevation and water level at 5 cross sections were measured during each experimental test (Fig. 2), distance from sediment supply position 1 m, 2 m, 3 m, 4 m and 5 m, respectively. Table 1 summarizes the hydraulic conditions of all experiments in this study. In Table 1, d_{50} is the median particle size; Q is the flow discharge; h_0 is the mean approach flow depth; U_0 is the mean approach flow velocity; U_c is the critical mean approach flow velocity; S is the bed slope; Re is the Reynolds number; Fr is the Froude number.

Table 1. **Summary of experimental conditions**

Run number	d_{50} (mm)	Q (L/s)	h_0 (m)	U_0 (m/s)	U_c (m/s)	S	Re (10^6)	Fr	Feeding mode
1	3.6	27.23	0.06	0.756	0.633	0.005	4.54	0.99	constant
2	3.6	27.23	0.06	0.756	0.633	0.005	4.54	0.99	trapezoidal
3	3.6	27.23	0.06	0.756	0.633	0.005	4.54	0.99	triangular

Three sediment supply modes were applied in experiments (Fig. 3). These hydrographs were characterized by (a) constant sediment supply, the feeding rate per unit width $g_b = 76$ g·m^{-1}·s^{-1}; (b) trapezoidal sediment supply, g_b ranges from 38 g·m^{-1}·s^{-1} to 114 g·m^{-1}·s^{-1}; (c) triangular sediment supply, g_b also ranges from 38 g·m^{-1}·s^{-1} to 114 g·m^{-1}·s^{-1} and the feeding time is 10 min for each triangular cycle. For those three sediment supply modes, the total feeding time and total amount of sediment recharge is 100 min and 273.6 kg, respectively. It should be emphasized that the sediment transport for each experimental test is saturated sediment transport. After the sediment supply ended, consecutive identical floods were continued until the system adjusted itself to its initial equilibrium.

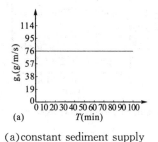 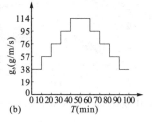 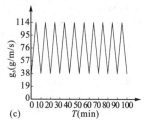

(a) constant sediment supply (b) trapezoidal sediment supply (c) triangle sediment supply

Figure 3. **Sediment supply modes**

3 Results and Discussion

3.1 Evolution of riverbed elevation and water level

Fig. 4 shows the variation of bed elevation and water level with time at three streamwise positions. Before 5 minutes of sediment supply, the flow is in quasi-uniform flow condition. At the moment of sediment feeding, the water level tends to decline firstly due to the compressing flow, and then rises sharply. For the constant sediment supply mode, the bed elevation and water level in each position reach the maximum values with continuous sediment feeding at about 70~80 min. After sediment supply ended, the bed elevation and water level gradually decrease with time. In addition, the change of upstream channel morphology is stronger than that of downstream channel. Specifically, the bed elevation and water level rise to 0.267 m and 0.307 m at 1 m distance from sediment supply position, 0.251 m and 0.296 m at 2 m distance from sediment supply position, 0.231 m and 0.285 m at 3 m distance from sediment supply position. For the trapezoidal sediment supply mode, the responses of bed elevation and water level to the sediment supply show an increasing firstly and then decreasing trend, which is similar to those of the constant sediment supply mode. They have a maximum values at about 100 minutes, while the maximum is smaller than that of the constant sediment supply. For the triangle sediment supply mode, the bed elevation increases with time continuously, while the water level shows a triangle fluctuation slightly. Generally, the more unsteady the sediment supply, the more unstable the riverbed and water level response.

The numerical and flume experimental results of An et al. (2017b), Parker et al. (2007) and Wong and Parker(2006) shown that there is a specially region near the sediment feed point, called "Hydrography Boundary Layer" (HBL), where the bed elevation and river morphology respond strongly to the hydrography. However, the riverbed elevation and slope do not respond to the hydrography outside of the hydrography boundary layer. In this study, the HBL can easily explain why the riverbed elevation and water level rise greatly at the upstream than that at the downstream. In addition, the response of river morphology is a complex process to excessive sediment supply. Once entering river channels, the excessive sediment can rapidly change the river topography and texture. With the increase of sediment supply, the sediment load reaches saturated transport, causing sediment deposition and bed rising. Meanwhile, the high-intensity sediment transport can increase the flow flux and flow resistance, and then cause flood risks.

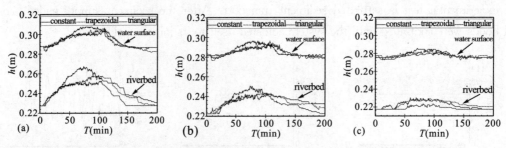

Figure 4. Evolution of riverbed elevation and water level with time at difference positions

Note: (a) 1 m distance from sediment supply position; (b) 2 m distance from sediment supply position; (c) 3 m distance from sediment supply position.

3.2 Bed and water surface slope adjustment

Fig. 5 shows the adjustment process of mean riverbed and water surface slope with time for different sediment supply mode. The initial river bed and water surface slope is about 0.5% at 0 min. Due to the impacts of excessive sediment supply of upstream, the bed slope increases rapidly with time and then decreases with the end of sediment supply. However, the slope of water surface increases gentler than the riverbed. For the constant sediment supply mode, the bed has a maximum slope 1.95% at about 70 min, while the water surface has a maximum value 1.25% at about 65 min. As time passed, the bed and water surface slope gradually decrease, returning to the initial slope. For the trapezoidal sediment supply mode and triangle sediment supply mode, the change of slope is gentler than that of the constant sediment supply mode, showing more unstable regimes.

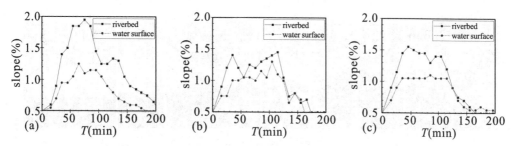

Figure 5. The change of bed and water surface slope with time

Note: (a) constant sediment supply; (b) trapezoidal sediment supply; (c) triangle sediment supply.

3.3 Temporal patterns of riverbed and water level evolution

In experiments, the sediment transport regime can be accurately characterized as lower-regime plane bed without dunes and ripples. The responses of river morphology to excessive sediment supply are dramatic (Fig. 6). In the process of sediment supply, the bed elevation rises continuously and the rising water level is maintained at 1 m distance from sediment supply position. With the end of sediment supply, the bed and water surface fall back to the initial level. For different sediment supply mode, the bed elevation and water depth increases significantly due to the strong sediment transport. Among them, the response of river bed and water level is dramatic for the constant sediment supply [Fig. 6(a)]. The riverbed deposition and water level are 0.236 m and 0.28 m at 80 min for the constant sediment supply mode, respectively. While the riverbed deposition and water level is gentler for the trapezoidal sediment supply mode and triangle sediment supply mode.

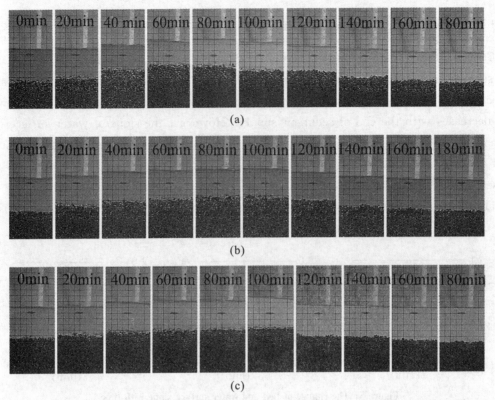

Figure 6. Photographs of bed elevation and water level at 1 m distance from sediment supply position
Note:(a) constant sediment supply;(b) trapezoidal sediment supply;(c) triangle sediment supply.

4 Conclusions

This paper presents an experimental study on the responses of bed morphology and water level to unsteady sediment supply during flood events. Based on the experimental results, the following conclusions are obtained:

(a) For the constant sediment supply, the response of channel morphology is strongest. The riverbed elevation, water level and bed slope increase firstly and then decrease with time.

(b) For the trapezoidal sediment supply and triangle sediment supply, the responses of bed elevation, water level and bed slope are similar to those of the constant sediment supply mode, but the change of channel morphology is gentler than that of the constant sediment supply mode.

(c) Due to the "hydrography boundary layer" (HBL), the river morphology resopond strongerly to the excessive sediment supply near the feeding position than that at the position outside of HBL.

Acknowledgements

This study was supported by the National Key Research and Development Program of China (2019YFC1510701) and the National Natural Science Foundation of China(U20A20319).

Notation

The following symbols are used in this paper:
d_{50} = median particle size(mm);
Fr = Froude number;

g_b = sediment supply rate per unit width(g · m^{-1} · s^{-1});

h_0 = mean approach flow depth(m);

Q = flowrate(L/s);

Re = Reynolds number;

S = bed slope;

U_0 = mean approach flow velocity(m/s);

U_c = critical mean approach flow velocity(m/s);

σ_g = geometric standard deviation;

ρ = sediment density(kg/m^3);

References

[1] An C, Cui Y, Fu X. Gravel-bed river evolution in earthquake-prone regions subject to cycled hydrographs and repeated sediment pulses[J]. Earth Surface Processes and Landforms, 2017, 2042: 2426−2438.

[2] An C, Fu X, Wang G. Effect of grain sorting on gravel bed river evolution subject to cycled hydrographs: Bed load sheets and breakdown of the hydrograph boundary layer[J]. Journal of Geophysical Research: Earth Surface, 2017, 122: 1513−1533.

[3] Bel C, Liébault F, Navratil O. Rainfall control of debris-flow triggering in the Réal Torrent, Southern French Prealps [J]. Geomorphology, 2017, 291: 17−32.

[4] Bordoni M, Galanti Y, Bartelletti C. The influence of the inventory on the determination of the rainfall-induced shallow landslides susceptibility using generalized additive models[J]. Catena, 2020, 193: 104630.

[5] Borga M, Stoffel M, Marchi L. Hydrogeomorphic response to extreme rainfall in headwater systems: Flash floods and debris flows[J]. Journal of Hydrology, 2014, 518: 194−205.

[6] Cama M, Lombardo L, Conoscenti C. Improving transferability strategies for debris flow susceptibility assessment: Application to the Saponara and Itala catchments(Messina, Italy)[J]. Geomorphology, 2017, 288: 52−65.

[7] Cui Y, Parker G, Lisle T. Sediment pulses in mountain rivers: 1. Experiments[J]. Water Resources Research, 2003, 39(9): 1239.

[8] Cui Y, Parker G, Pizzuto J. Sediment pulses in mountain rivers: 2. Comparison between experiments and numerical predictions[J]. Water Resources Research, 203, 39(9): 1240.

[9] Dean D J, Topping D J, Schmidt J C. Sediment supply versus local hydraulic controls on sediment transport and storage in a river with large sediment loads[J]. Journal of Geophysical Research: Earth Surface, 2016, 121(1): 82−110.

[10] Destro E, Amponsah W, Nikolopoulos E I. Coupled prediction of flash flood response and debris flow occurrence: Application on an alpine extreme flood event[J]. Journal of Hydrology, 2013, 558: 225−237.

[11] Gan B, Liu X, Yang X. The impact of human activities on the occurrence of mountain flood hazards: Lessons from the 17 August 2015 flash flood/debris flow event in Xuyong County, South-Western China[J]. Geomatics, Natural Hazards and Risk, 2018, 9: 816−840.

[12] Kammerlander J, Gems B, Kößler D. Effect of bed load supply on sediment transport in mountain streams[J]. International Journal of Sediment Research, 2017, 32(2): 240−252.

[13] Nelson P A, Brew A K, Morgan J A. Morphodynamic response of a variable-width channel to changes in sediment supply[J]. Water Resources Research, 2015, 51(7): 5717−5734.

[14] Nelson P A, Morgan J A. Flume experiments on flow and sediment supply controls on gravel bedform dynamics[J]. Geomorphology, 2018, 323: 98−105.

[15] Parker G, Hassan M, Wilcock P. Adjustment of the bed surface size distribution of gravel-bed rivers in response to cycled hydrographs, in Gravel-Bed Rivers Ⅵ: From Process Understanding to River Restoration[M]. New York: Elsevier, 2007.

[16] Pfeiffer A M, Finnegan N J, Willenbring J K. Sediment supply controls equilibrium channel geometry in gravel rivers[C]. Proceedings of the National Academy of Sciences of the United States of America, 2017, 114(13): 3346-3351.

[17] Recking A. Influence of sediment supply on mountain streams bedload transport[J]. Geomorphology, 2012, 175-176: 139-150.

[18] Recking A. Relations between bed recharge and magnitude of mountain streams erosions[J]. Journal of Hydro-environment Research, 2014, 8(2): 143-152.

[19] Venditti J G, Nelson P A, Minear J T. Alternate bar response to sediment supply termination[J]. Journal of Geophysical Research: Earth Surface, 2012, 117(F2).

[20] Wong M, Parker G. One-dimensional modeling of bed evolution in a gravel bed river subject to a cycled flood hydrograph[J]. Journal of Geophysical Research, 2006, 111, F03018.

[21] Yager E M, Turowski J M, Rickenmann D. Sediment supply, grain protrusion, and bedload transport in mountain streams[J]. Geophysical Research Letters, 2012, 39(10).

[22] Zheng X, Chen R, Luo M. Dynamic hydraulic jump and retrograde sedimentation in an open channel induced by sediment supply: experimental study and SPH simulation[J]. Journal of Mountain Science, 2019, 16: 1913-1927.

Experiments on the Deterministic and Non-deterministic Effects of Water Inflow on Channelized Configurations During the Lacustrine Delta Evolution

Xiaolong Song[1,2], Haijue Xu[1,2], Yuchuan Bai[1,2]

[1] State Key Laboratory of Hydraulic Engineering Simulation and Safety,
Tianjin University, Tianjin 300350, China

[2] Institute for Sedimentation on River and Coastal Engineering,
Tianjin University, Tianjin 300350, China

Abstract

The stochastic characteristics of river-delta systems pose large uncertainties for the environmental management on deltas. In this study, two sets of each-repeated physical modelling experiments were conducted, to tentatively explore the deterministic and non-deterministic effects of the upstream water-inflow on lacustrine deltaic channels. Results indicated that the single-main channel length increased positively with discharge, and was inevitably accompanied with the non-positive changing of both the fluctuation characteristics of the relative position of bifurcating node for different-order branching channels, and the average-period of all the geometrical indicators of higher-order branching channels in the middle delta; the major non-deterministic highlights were the positive and negative correlation between average period with average value, on the relative position of bifurcating node and the branching channel length respectively in the upper delta under the low-discharge conditions. These findings refresh our understanding of bifurcation instability and lay the foundation for stochastic modelling of lacustrine-delta morphology.

Keywords: Lacustrine deltaic channels; Water inflow; Deterministic and non-deterministic effects; Bifurcation instability; Stochastic modelling

1 Introduction

A The lacustrine delta area is abundant in defocusing channelized and sedimentary configurations as the result of a complex interplay of upstream river inflow and intake basin. In China, the formation and development process of lacustrine delta is an im-portant part of alluvial river systems. The famous ones include Dongting Lake Delta and Poyang Lake Delta (Guo et al., 2020). In 2020, facing the severe flood situation there, a higher level of understanding of systematic evolution of lacustrine deltaic channels becomes urgent. In a scientific sense, the self-organizing channels in delta have received con-siderable deterministic attention, but still lack the scientific observation of some ran-dom characteristics, especially considering this shortage has been highlighted in the actual environmental protection and flood prevention works.

The important roles of external forcing on the evolution of natural alluvial delta were emphasized in lots of literature(Whipple and Trayler, 1996; Hartley et al., 2005), but were not accurate enough to interpret the inner mechanical details of delta morphology(Clarke et al., 2010). Numerical studies on the influence of autogenic mechanisms found some complex driving modes on the cyclic aggradation-incision(Nicholas and Quine, 2007a, b) and avulsion processes(Jerolmack and Paola, 2007) over large time-scales in delta, while inevitably ignored some representative factors of nature. Also, due to some measuring technical reasons, such as, the difficulty of isolating the effects of internal forcing from external forcing, and the overlong time periods for real-time observation of the changes of delta in the field, the experimental physical modelling is considered as a much more realistic and valuable technique for studying the autogenic processes in an artificial controlled water-sediment environment herein. Many successful physical models have provided various dynamic information of delta morphology, and discussed the deterministic relationships between the external water-sediment supply and the channelized-sedimentary configurations(Ashworth et al., 2004; Davies and Korup, 2007; Kim and Jerolmack, 2008; Van Dijk et al., 2009; Clarke et al., 2010) under calm environmental conditions. For example, the instability of such configurations during switching of flow patterns from sheetflow to channelized flow were effectively identified, the dependent effects of the upstream input conditions for initiating the autocyclic erosion-aggradation process were confirmed in the form of "critical gradient", the observed cyclic sedimentation packages were suggested to be regarded as a signature of the autogenic "pulse" of deltas, etc.

However, very few investigations have been conducted on the response of alluvial delta to stochastic disturbance. The bifurcated channels are actually highly unstable and sensitive to internal-external disturbances, a small difference could be enough to initiate the redistribution of channelized configurations in delta. Over the years, due to the construction of the large-scale reservoirs and soil conservation engineering, and the instability of hydrology-weather-human conditions, something profound are happening inside the rivers and lakes in China, in particular, the abnormal water-sediment flux. The Ganjiang River Delta of Poyang Lake experienced the heaviest ever rainfalls in the summer of 2020; yet before that, the water levels in dry season have always been noticeably dropping, the enormous deepening of riverbed and the cutoff of distributary channels resulted in some new features in this river-delta system. Considerable issues of water engineering and environmental conservation on a relatively short-time scale, such as, the development pattern of interior multi-channels, the connectivity of upstream river and downstream delta system, need to be evaluated from a new perspective.

Many scientists have focused on modern systems theory to seek thorough comprehension of dynamic processes of a nonlinear organic system. The transfer function, a mathematical model that describes the relation between the input and output based on the differential equations was commonly used in the relevant analysis. In particular, for the highly regular and also interactive complexity systems that are dominated by non-deterministic equations of motion, such as river system, the random nature of the discharge or rainfall forcing have been interpreted as a fundamental additive noise in the implement of Stochastic Differential Equation(SDEs) for identifying river processes, e.g., the forecasting of river flows generated by rainfall-runoff processes(Mondal and Wasimi, 2005), the probabilistic description of river characteristics including water quality constituents(Boano et al., 2006), soil water balance dynamics(Manfreda and Fiorentino, 2008), hydraulic geometry relationship (Song et al., 2019; Song et al., 2020). Understandably, this stochastic modeling tool also has great application potential in the probabilistic management of delta on the basis of mastering some stochastic characteristics of morphology evolution. Thus, such basic observation data would play a fundamental and vital role.

Discharge controls have a notable relationship with the delta morphology. In this paper, the objective is to investigate both the deterministic and non-deterministic effects of water-inflow on channelized configurations under the steady-averaged discharge conditions based on a physical modeling method in a lacustrine shallow-water delta. By comparing the repeated experiments in low and high discharge cases, it is possible to preliminarily distinguish the primary and secondary disturbance effects from flow-instability and sediment-environment and to further advance the understanding of the nature of the stochastic changing deltaic channels.

2 Experimental Design

2.1 Experiment facilities

The experiments were carried out at Tianjin University, China, in a self-developed alluvial delta-simulator basin [this facility has also been described in previous article (Xin et al., 2019)]. This working space (3.6 m long, 1.7 m wide, and 0.2 m high) is consisted of three functional areas along a longitudinal x direction: Inlet region ($x=-0.5\sim 0$ m), Deltaic region ($x=0\sim 3.0$ m), and Tailgate region ($x=3.0\sim 3.1$ m). A complete water circulation system with controlled discharge and sediment supply into the inlet channel is a guarantee for the sustainable development of artificial river-delta. An image recording system was fixed above the basin for collecting flow-fields and topographic information. Figure 1 shows the schematic diagram of the experimental facility.

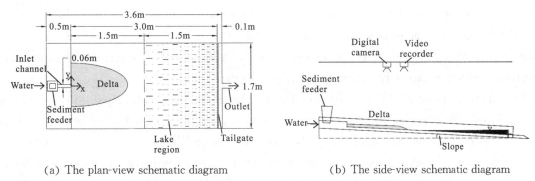

(a) The plan-view schematic diagram   (b) The side-view schematic diagram

Figure 1. The schematic diagram of the experimental facility

2.2 Experiment methods

The experiments During the preparatory phase, the Inlet and Deltaic region were firstly sanded up to thickness of 10 cm and 5 cm respectively. Non-cohesive sediment (median grain size $d_{50}=0.62$ mm, non-uniformity coefficient $C_u=d_{60}/d_{10}=2.387$, density $\rho_s=2650$ kg/m³) was used in the experiment for preventing flocculation and facilitating the natural formation and manual observation of deltaic channels (Hoyal and Sheets, 2009; Clarke et al., 2010). Then, the inlet channel was molded with a set of glass plates in the middle of Inlet region (the cross-section dimension was 6 cm in width and 5 cm in depth), the height of the basin bottom bracket was adjusted to reach the specified slope of 1‰

(here, this gentle value was chosen with care through performance analysis of preliminary experiments in order to ensure the clear and complete development of delta). The downstream Lake region was formed by delivering water into the basin gradually up to the designed lake level through the Tailgate. Fig. 1 shows this initial stable Lake region($x=1.5\sim3.0$ m). After above, starting the water pumps and the sediment feeder system, keeping the designed discharge and sediment feed rate constant, running the experiment for 30 h.

During the recording phase, the images of the delta morphology were collected at an interval of 5 min; the flow fields were measured by use of the particle tracking velocimetry(PTV) system(Bai and Xu, 2007) every one hour; the delta topography were measured every 10 cm longitudinally and 2 cm transversely using a laser range finder with the precision of 1mm after every two to three hours of running.

2.3 Experiment runs

Major external factors influencing the lacustrine shallow-water delta evolution include water inflow, sediment supply and lake level. Our group has investigated the roles of the latter two and revealed some stochastic characteristic information of delta morphology under steady conditions in previous studies(Xin et al., 2019; Bai et al., 2018; Xu et al., 2019). Hence, the present work aims to study the effects of different steady discharge values on the autogenic deltaic behavior. As shown by the literatures(Clarke et al., 2010; Van Dijk et al., 2009; Whipple et al., 1998), absolute water discharge typically affects the sediment transport capacities and the degree of development of channelized configurations in the delta plain. And also, bifurcations are generally unstable: a very small disturbance may rechannel the flow into one of the bifurcations, and initiate the other channels closure(Van Dijk et al., 2012). Alluvial deltas are thus a kind of stochastic changing organic system. Therefore, we conducted four runs with two stable water supplies and two repeats for each to tentatively explore the deterministic and non-deterministic effects of water inflow on channelized configurations, considering there is fundamental need to take the initial limited-runs experiments for accumulating effective experience prior to starting considerable tests.

This experimental study follows the 'similarity of processes' concept(Hooke, 1968). The perfect producibility of a lacustrine shallow-water delta evolution requires the low Froude number(it is around 0.1 herein) and maintains the dominance of bedload movement(Xin et al., 2019). To make sure that the upstream flow velocity exceeds the sediment-moving incipient velocity and the deltaic landscapes are remarkable during the experiments, the discharges for Runs 1 and Runs 2 were determined as 50 cm^3/s and 100 cm^3/s(the adjustment range of flowmeter is 0$\sim$100 cm^3/s) respectively based on a comprehensive consideration of the relevant experiments and field measurements. Also, the ratios of sediment supply(Q_s) to water supply(Q), S_v, were set to be not exceeding 0.2‰(Clarke et al., 2010; Zhang et al., 2016). Table 1 shows the detailed experimental parameters(necessarily, the repeated runs under the same conditions are also fully presented).

Table 1. Details of the experimental scenarios

Run	$Q(\text{cm}^3/\text{s})$	$Q_s(\text{g/s})$	S_v	Basin slope	Duration(h)
1-a	50	0.26	0.2‰	1‰	30
1-b	50	0.26	0.2‰	1‰	30
2-a	100	0.26	0.1‰	1‰	30
2-b	100	0.26	0.1‰	1‰	30

3 Results and Analysis

3.1 Evolution processes of lacustrine deltaic channels

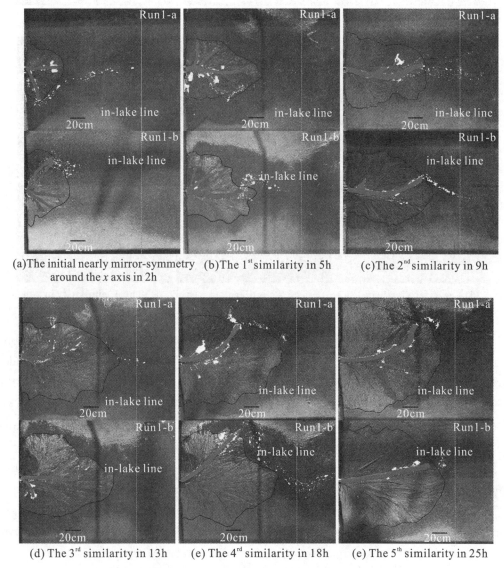

(a) The initial nearly mirror-symmetry around the x axis in 2h
(b) The 1st similarity in 5h
(c) The 2nd similarity in 9h
(d) The 3rd similarity in 13h
(e) The 4rd similarity in 18h
(e) The 5th similarity in 25h

Figure 2. The quasi-periodic similarity of deltaic channels in Runs 1

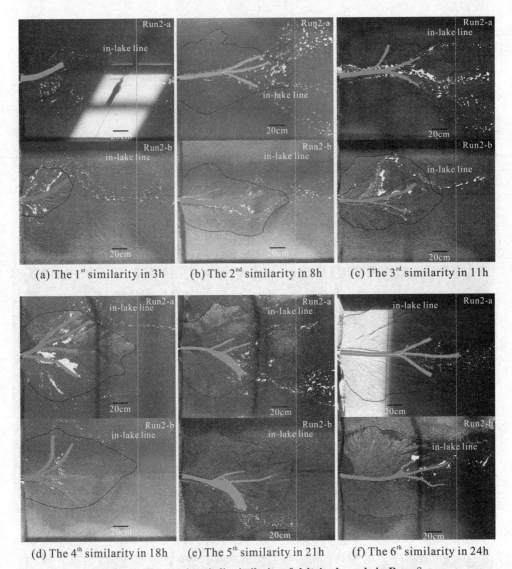

(a) The 1st similarity in 3h (b) The 2nd similarity in 8h (c) The 3rd similarity in 11h

(d) The 4th similarity in 18h (e) The 5th similarity in 21h (f) The 6th similarity in 24h

Figure 3. The quasi-periodic similarity of deltaic channels in Runs 2

The delta morphology was always characterized by complex channelized configurations in the plane as the system moved forward. Under the two discharge conditions, it can be found that the channels from the repeated experiments evolved similar features at intervals, in particular, the typical formations of a single-main channel with a few avulsion channels. However, most of the experiment time, the flow adjusted itself in a seemingly irregular manner. Taken as a whole, the channels in the delta behaved like a stable self-adaptive control system.

Fig. 2 shows the channel evolution process under the condition of low-flows in Runs 1. The development of water flow obviously was subjected to stochastic factors in the non-longitudinal direction in the initial experiment stage: the channel geometries in the two repeated groups (Run1-a and Run1-b) have the trend of mirror-symmetry around the longitudinal x axis, as shown in the subfigure in 2 h. As the delta continued to grow, starting from 5 h, both the two group of channelized configurations were similar and also distinctly different in an approximately regular cycle (every 4.5 h on average) without change in the extrinsic environments. Fig. 2 shows that the commonality of performances of the quasi-periodic similarity (this phrase means both follow a quasi-period cycle to keep the similar) of deltaic channels was the left-center development of main channel. While in their respective transition periods as recorded in our video, the change of channels was a multiple-mode convergence of fairly common avulsion, lateral migration, and abandonment, and full of intricate detail. Overall, a cyclical phenomenon of plain aggradation and incision gradually gave way to the dominated lateral migration of channels with the mature development of delta. And many times, the channels of Run1-a tended to backfill more earlier and migrate mainly around the left area; Run1-b, on the other hand, was the opposite.

Fig. 3 shows the multiple channels under the condition of high-flows in Runs 2. Comparing with Runs 1, the non-longitudinal sediment-environmental impacts were less apparent at the beginning of the experiment, from our video observation of no trend of mirror-symmetry around the longitudinal x axis for the two repeated runs. And, the 1st similarity of channelized configurations in 3 h has not yet show avulsion characteristics [Fig. 3(a)]. While after this moment, the flow became more unstable and the whole plain was accompanied by repetitive processes of aggradation towards the toe with weak avulsion and backfilling from the toe to the apex with stronger avulsion at intervals of about 4 h. The channelized configurations here had few lateral moving parts in the cycles in comparison with Runs 1. In the transition phases as recorded in our video, both Run2-a and Run2-b shared some deep and intricate universals. According to the continuous observation, mainly, the channels of Run2-b tended to trench and backfill around the mid delta relatively smoothly; yet those of Run2-a maintained strongly mid-trenching for long in the lower delta and the occasional backfilling process was very rapid. A few visible lateral migrations of the main channel only occurred in the later stage of this experiment.

3.2 Geometrical characteristics of lacustrine deltaic channels

The qualitative analysis on distributary channels of delta shows that only the single-main channels are quasi-periodically trustworthy enough, the avulsion and lateral migration of the channels have much spatial-temporal complexity and uncertainty under the different discharge conditions. In this section, the channel characteristics were analyzed by mainly using the following four quantitative indicators, including the main channel length (S), the relative position of bifurcation node (I), the bifurcation angle (A), and the branching channel length (s^*). Fig. 4 shows such a schematic of geometrical characteristics of channelized configurations.

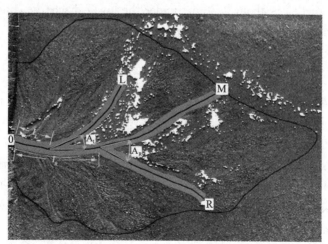

Figure 4. A schematic of geometrical characteristics of channelized configurations

Note: "L" represents the left branching channel, "R" represents the right branching channel, "M" represents the main channel.

The primary deterministic effects of discharge(Q) mainly contain:

(1) The relations of the fluctuation characteristics of the single-main channel length(S) and discharge(Q) were basically positive, except the stable performance of average amplitude under the high-discharge conditions.

(2) The absolutely identical relation-patterns between the indicators-(I, A and s^*) with discharge(Q) during the development of different order of branching channels, were the positive correlations on the parameters of average value and average amplitude in the "1st branch" cases, and the negative correlation on the parameter of average period in the "2nd+ branches" cases, respectively. The outliers were the relation between the average period of indicator-(I) with discharge(Q) in the "1st branch" cases, and the relations between both the average value and average amplitude of indicator-(I) with discharge(Q) in the "2nd+ branches" cases, respectively.

On the other hand, to those secondary non-deterministic effects of discharge(Q), based on the deterministic aspects above, the multiple comparisons between the left and right 1st branches came to the following findings:

(1) The characteristic indicator-(A) remained in full positive relationship with discharge(Q). Non-deterministic dependences were not observed from whole to part.

(2) The key outliers identified were, under the low-discharge conditions, the positive correlation between average period with average value on the indicator-(I), and, the negative correlations between both the average amplitude and average period with average value on the indicator-(s^*).

In other words, the most notable thing is that, accompanying the positive change of the single-main channel length(S) with the increase of discharge(Q), were the three non-positive fluctuating parameters of the indicator-(I) around the different order of branching channels, and the negatively changing of average-period of the indicators-(I, A, and s^*) in the middle delta(here the average value of indicator-(I) is about 0.44 m relative to the average longitudinal length of mature delta(1.0 m) under the significant influence of higher discharge, so we say "middle delta"), which highlight the impacts of flow instability; meanwhile, during the avulsion processes in delta, the non-deterministic effects of discharge could be clearly observed from the performances of the fluctuation characteristics (especially, average period versus average value) of the indicators-(I, s^*) in the upper delta[here the average value of indicator-(I) is about 0.28 m] under the lower discharge conditions, which indicated the highly stochastic impacts of sediment motion and sorting.

4 Conclusions

In the absence of direct measurements, channel geometry is worthwhile to collect and is reliable enough for exploring the continuous redistribution of water-sediment in delta plain as the result of channel bifurcations (Dong et al., 2020). But few experimental studies directly measured the geometrical characteristics of deltaic channels like us. As some literature documented, flow partitioning on the delta is a function of river water discharge (Ilyicheva et al., 2015; Ilyicheva, 2008), and most natural and simulated bifurcations possess asymmetrical flow partitioning because of the instability of symmetrical bifurcations (Kleinhans et al., 2008; Edmonds and Slingerland, 2008; Bolla Pittaluga et al., 2015). Our research findings on the deterministic effects of discharge on the overall development of channels and non-deterministic effects of discharge on the left-right branching channels are not only consisted with such studies above, but also provide some fresh information of bifurcation instability without involving excessive hydraulic parameters (such as channel width, sinuosity, depth, cross-sectional area, etc.). Although many researchers have applied one-dimensional models to evaluate bifurcation stability by coupling downstream branches to an upstream branching at a bifurcation node(Wang et al., 1995; Bolla Pittaluga et al., 2003; Kleinhans et al., 2008; Dong et al., 2020), obviously there were the lack of morphodynamic feedback therein: as flow varies, the channel network will naturally undergo changes including avulsions and channel migration, which in turn will subvert the original channel network. From our perspective, the accurate predictions of flow partitioning in lacustrine delta plain would be well improved by deterministically including the position indicator-(I) of a bifurcation node and the factor of flow turbulence in the middle delta, and by non-deterministically adding the noise factors of sediment properties in the upper delta. This study helps to establish the basic database and accumulate effective experience for the future probabilistic analysis on delta morphology with more testing and the application of stochastic models in delta management practice.

Acknowledgements

This work was supported by the (National Natural Science Foundation of China) under Grant (No. 51879182).

References

[1] Ashworth P J, Best J L, Jones M. Relationship between sediment supply and avulsion frequency in braided rivers[J]. Geology, 2004, 32(1): 21−24.

[2] Bai Y. Experimental analysis of the formation process of lacustrine shallow-water delta[J]. Journal of Hydraulic Engineering, 2018, 49(5): 549−560.

[3] Bai Y, Xu D. Study on particle tracking velocimetry in complex surface flow field[J]. Proceedings of the 20th National Conference on Hydrodynamics, 2007, 430−438.

[4] Boano F, Revelli R, Ridolfi L. Stochastic modelling of DO and BOD components in a stream with random inputs[J]. Advances in Water Resources, 2006, 29(9): 1341−1350.

[5] Bolla P M, Coco G. And, Kleinhans M G. A unified framework for stability of channel bifurcations in gravel and sand fluvial systems[J]. Geophysical Research Letters, 2015, 42(18): 7521−7536.

[6] Bolla P M, Repetto R, Tubino M. Channel bifurcation in braided rivers: Equilibrium configurations and stability[J]. Water Resources Research, 2003, 39(3).

[7] Clarke L, Quine T A, Nicholas A. An experimental investigation of autogenic behaviour during alluvial fan evolution [J]. Geomorphology, 2010, 115(3−4): 278−285.

[8] Davies T R, Korup O. Persistent alluvial fanhead trenching resulting from large, infrequent sediment inputs[J]. Earth Surface Processes and Landforms, 2007,32(5): 725−742.

[9] Dong T Y. Predicting Water and Sediment Partitioning in a Delta Channel Network Under Varying Discharge Conditions[J]. Water Resources Research, 2020, 56(11).

[10] Edmonds D A, Slingerland R L. Stability of delta distributary networks and their bifurcations[J]. Water Resources Research, 2008, 44(9).

[11] Guo R, Zhu Y, Liu Y. A Comparison Study of Precipitation in the Poyang and the Dongting Lake Basins from 1960—2015[J]. Scientific Reports, 2020, 10(1): 1−12.

[12] Hartley A J. Climatic controls on alluvial-fan activity, Coastal Cordillera, northern Chile[J]. Geological Society, 2005, 251(1): 95−116.

[13] Hooke R L. Model geology: prototype and laboratory streams: discussion[J]. Geological Society of America

Bulletin, 1968, 79(3): 391-394.

[14] Hoyal D, Sheets B. Morphodynamic evolution of experimental cohesive deltas[J]. Journal of Geophysical Research: Earth Surface, 2009, 114(F2).

[15] Ilyicheva E A. Dynamics of the Selenga river network and delta structure[J]. Geography and Natural Resources, 2008(4): 57-63.

[16] Ilyicheva E I, Gagarinova O V, Pavlov M V. Hydrologo-Geomorphological Analysis of Landscape Formation Within the Selenga River Delta[J]. Geography and Natural Resources, 2015, 36(3): 263-270.

[17] Jerolmack D J, Paola C. Complexity in a cellular model of river avulsion[J]. Geomorphology, 2007, 91(3-4): 259-270.

[18] Kim W, Jerolmack D J. The pulse of calm fan deltas[J]. The Journal of Geology, 2008, 116(4): 315-330.

[19] Kleinhans M G. Bifurcation dynamics and avulsion duration in meandering rivers by one-dimensional and three-dimensional models[J]. Water Resources Research, 2008, 44(8).

[20] Manfreda S, Fiorentino M. A stochastic approach for the description of the water balance dynamics in a river basin[J]. 2008.

[21] Mondal M S, Wasimi S A. Periodic transfer function-noise model for forecasting [J]. Journal of Hydrologic Engineering, 2005, 10(5): 353-362.

[22] Nicholas A, Quine T. Crossing the divide: Representation of channels and processes in reduced-complexity river models at reach and landscape scales[J]. Geomorphology, 2007a, 90(3-4): 318-339.

[23] Nicholas A P, Quine T A. Modeling alluvial landform change in the absence of external environmental forcing[J]. Geology, 2007b, 35(6): 527-530.

[24] Song X, Zhong D, Wang G. A study of the stochastic evolution of hydraulic geometry relationships[J]. River Research and Applications, 2019, 35(7): 867-880.

[25] Song X. Stochastic evolution of hydraulic geometry relations in the lower Yellow River under environmental uncertainties[J]. International Journal of Sediment Research, 2020.

[26] Van Dijk M. Contrasting morphodynamics in alluvial fans and fan deltas: effect of the downstream boundary[J]. Sedimentology, 2012, 59(7): 2125-2145.

[27] Van Dijk M, Postma G, Kleinhans M G. Autocyclic behaviour of fan deltas: an analogue experimental study[J]. Sedimentology, 2009, 56(5): 1569-1589.

[28] Wang Z. Stability of river bifurcations in ID morphodynamic models[J]. Journal of Hydraulic research, 1995, 33(6): 739-750.

[29] Whipple K X. Channel dynamics, sediment transport, and the slope of alluvial fans: experimental study[J]. The Journal of Geology, 1998, 106(6): 677-694.

[30] Whipple K X, Trayler C R. Tectonic control of fan size: the importance of spatially variable subsidence rates[J]. Basin Research, 1996, 8(3): 351-366.

[31] Xin W, Bai Y, Xu H. Experimental study on evolution of lacustrine shallow-water delta [J]. Catena, 2019, 182: 104125.

[32] Xu H. Experimental study on formation and morphologic evolution of alluvial lake deltas [J]. Journal of Hydroelectric Engineering, 2019, 38(1): 52-62.

[33] Zhang X. The development of a laterally confined laboratory fan delta under sediment supply reduction [J]. Geomorphology, 2016, 257: 120-133.

Free-Surface Characteristics of Undular Hydraulic Jumps in a Shallow Channel

Hong Hu, Ailing Cai, Ruidi Bai, Hang Wang

State Key Laboratory of Hydraulics and Mountain River Engineering,
Sichuan University, Chengdu 610065, China

Abstract

This paper focuses on the free-surface characteristics of undular hydraulic jumps in shallow rectangular channel with an aspect ratio between 0.035 and 0.058. Compared with full hydraulic jumps with breaking free surface and air entrainment, undular hydraulic jumps consist of several standing waves with smooth free surface and low to null flow aeration. The geometry of the first three waves is studied in this experiment. Comparing the present flow pattern observations with the previous experiments under similar flow conditions, relationships between several characteristic geometry parameters and the Froude number are investigated. Detailed flow depth measurements are performed with acoustic displacement meters(ADMs). The three-dimensional free-surface profile and depth fluctuations are recorded along both longitudinal and transverse directions. The wave amplitude and its streamwise attenuation are analyzed for different Froude numbers. Both Froude number and Reynolds number of the undular hydraulic jump affect its time-averaged free-surface geometry, instantaneous free-surface fluctuations, wave amplitude and its longitudinal attenuation to different degrees.

Keywords: Undular hydraulic jump; Free surface; ADM; Free-surface fluctuation; Wave amplitude

1 Introduction

Hydraulic jumps are a special hydraulic phenomenon which is often encountered at downstream of water discharge structures such as dams, sluice gates and weirs. It occurs at the transition from upstream supercritical flow to downstream subcritical flow, typically characterized by big sounds, spray and splashing. In some textbooks, hydraulic jumps are classified by considering the upstream Froude number(Fr), where $Fr = v/\sqrt{gh_1}$, h_1 is the upstream flow depth, v is the upstream flow velocity and g is the gravity acceleration. Undular hydraulic jumps are defined with $1 < Fr < 1.7$ while full hydraulic jumps are defined for $Fr > 1.7$. Based on the equations of conservation of mass and momentum in a horizontal rectangular channel, the sequent depth h_2 derived from the pre-jump depth and inflow Froude number is

$$h_2 = h_1 \cdot \frac{1}{2}(\sqrt{1+8\,Fr^2}-1) \tag{1}$$

Compared with full hydraulic jumps characterized by breaking free surface, remarkable roller and large amount of air entrainment, undular hydraulic jumps are characterized by standing waves and smooth wavy-shaped free surface (Fig. 1). It often occurs in natural rivers and artificial water conveyance channels, attracting attention due to its special flow patterns. The channel sidewalls must be constructed with sufficient height, because, once the undular jump occurs, the first crest depth will be greater than the sequent depth(Chanson and Montes 1995).

Experimental study can obtain information about the development of the free-surface shapes of undular hydraulic jumps and provide physical reference data for the validation of relevant numerical simulation study. Lateral shock waves appear near both sidewalls at the toe of undular jumps when the inflow Froude number is larger than 1.2 (Ohtsu et al. 2003). According to whether the lateral shock waves cross at the first wave crest or not, undular hydraulic jumps are divided into two types: Case I for no crossing and Case II for crossing. According to whether the first wave crest is breaking or not, each type of undular jump encompasses non-breaking and breaking undular jumps (Chanson and Montes 1995, Ohtsu 2003). Free-surface characteristics of undular jumps consist of the depths of wave crest and trough, the horizontal length between the jump toe and the first wave crest, the length between the first and second wave crests, which have been experimentally studied in detail by Ohtsu et al. (2003). The free-surface fluctuations of full hydraulic jumps (not undular jumps) were investigated by Murzyn et al. (2009) and Chachereau and Chanson(2011) using non-intrusive acoustic displacement meters(ADMs).

This paper focuses on Case I of undular hydraulic jumps (Fig. 1). The free-surface shapes, geometric characteristics and depth fluctuations of the first three waves are investigated in a horizontal rectangular shallow channel. The present flow pattern observations are compared with the previous experiments under similar flow conditions, with additional description of the second and third waves. ADMs are used to record the water depth of the undular jumps along and across the channel, aiming to describe the mean shape and the fluctuations of the free surface at different locations with different Froude numbers. The effects of the Froude number on the standing wave amplitude and its attenuation in the streamwise direction are also investigated.

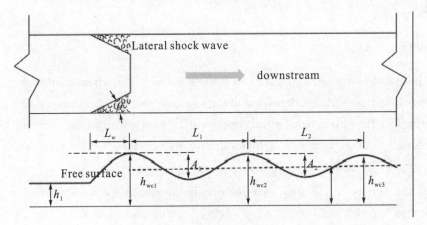

Figure 1. Schematic characteristics of free-surface geometry of undular hydraulic jumps

2 Experimental Facility and Instrumentation

The experimental facility consists of the test section and the water supply system. The test section is a horizontal rectangular channel made of glass. The channel is 7.20 m long and 1.40 m wide. The water discharge in the channel is controlled by an upstream full-width sluice gate(Fig. 2). The maximum discharge capacity of the recirculation water supply system is greater than 150 L/s, pumping water from an underground reservoir to the channel. The flow rate Q is controlled by a valve and measured by a rectangular weir as

$$Q = m_0 b \sqrt{2g} H^{\frac{3}{2}} \tag{2}$$

where H is the water head on the weir, b is the width of the weir, m_0 is the discharge coefficient given by $m_0 = 0.403 + 0.053 \frac{H}{P_1} + \frac{0.0007}{H}$ with P_1 the height of the weir.

Six identical acoustic displacement meters(ADMs, model: MIC+25/IU/TC) are used to measure the instantaneous free-surface elevations. The measurement range is between 25 and 350 mm from the sensor head. A calibration experiment was conducted to obtain the linear relationship between the voltage(V) and the distance(mm) for each sensor. Considering the symmetry of the experimental flume, the measurements are carried out on the right side of the flume centerline. The six ADMs are mounted on a plastic board spanning over the flume and perpendicular to the sidewalls, moving freely from upstream to downstream. A schematic diagram is shown in Fig. 2. The x direction is along the centerline, with x being the distance from the upstream sluice gate to the measurement spot, and the y direction is perpendicular to the centerline, with y_i being the distance from the right wall to the No. i^{th} sensor, where y_1 to y_6 is 65 mm, 165 mm, 265 mm, 410 mm, 555 mm and 700 mm, respectively. Each sensor is scanned at 100 Hz for 1 min upstream the lateral shock wave and for 3 min downstream the starting point of the lateral shock wave.

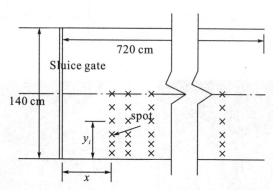

Figure 2. Experimental flume(top view) and the measurement spot arrangement

3 Experimental Conditions

The present experiments focus on undular hydraulic jumps in the shallow rectangular horizontal channel, namely, the aspect ratio of the channel is smaller than 0.06 and the slope equals 0. The inflow depth is controlled by the sluice gate opening and the downstream is free outflow. The present experiments include two parts. In the first part, the geometric characteristics of the undular hydraulic jumps are observed and recorded under different Froude and Reynolds numbers. The range of the Froude and Reynolds numbers of this part of experiments is shown in Table 1(a). In the second part, the water depths of three non-breaking undular hydraulic jumps are recorded utilizing ADMs. The overall free-surface shapes and fluctuations are demonstrated and analyzed. The flow conditions of the tested undular hydraulic jumps are shown in Table 1(b).

Table 1(a). Experimental conditions of the first part

Gate opening(cm)	x_1(cm)	h_1(mm)	Q(L/s)	v(m/s)	Fr	$Re(\times 10^4)$
3~13	82~420	19.8~92.2	35.7~185.6	1.22~1.55	1.51~2.91	2.46~11.63

Table 1(b). Experimental conditions of the second part

Gate opening (cm)	x_1(cm)	h_1(mm)	Q(L/s)	v(m/s)	Fr	$Re(\times 10^4)$	Number of wave crests
8	206	53.93	104.0	1.38	1.89	6.85	8
10	249	67.00	130.2	1.39	1.71	8.43	6
12	251	83.35	163.8	1.40	1.55	10.39	5

4 Experimental Results

A widely accepted upper limit of the inflow Froude number of undular hydraulic jump is $Fr=1.7$, but many studies indicated that the upper limit can reach a higher value depending on the aspect ratio and Reynolds number. Fig. 3(a) shows an undular hydraulic jump with $Fr=2.9$ and $Re=2.46\times 10^4$. It is characterized by non-breaking free surface, no aeration, only one complete standing wave, and the free surface remains horizontal after the second wave crest. It is observed in present experiment that, when the Reynolds number is greater than 4.3×10^4, the undular hydraulic jumps exhibit similar shapes, consisting of several standing waves with clear crests and troughs, as shown in Fig. 3(b). While the first wave is non-breaking and formed across the flume width, two local minor breakings are visible at the second wave crest with slight "white water", from about 1/4 B from each sidewall, respectively, where B is the flume width.

(a, right) $Fr=2.91$, $Re=2.4\times 10^4$ (b, left) $Fr=1.71$, $Re=8.4\times 10^4$

Figure 3. Typical free-surface shapes of undular hydraulic jumps with different relatively high Froude and Reynolds numbers

The characteristic geometric parameters are quantified and compared with the experimental data of Ohtsu et al. (2003) in Fig. 4. The solid triangulars represent for the present data, where the black triangulars represent for the data with $Re>6.5\times 10^4$, the same as the data of Ohtsu et al. (2003) (the hollow rectangulars), and the orange triangulars represent for the data with $4.3\times 10^4 < Re < 6.5\times 10^4$. With similar experimental conditions ($Re>6.5\times 10^4$, Case I), the present data agree well with the data of Ohtsu et al. (2003) in terms of the ratio of the first crest depth to the inflow depth h_{wc1}/h_1 and the ratio of the first-to-second-crest horizontal distance to the sequent depth L_1/h_2 (see Fig. 1) as functions of Fr, as shown in Figs. 4(a) and 4(b). However, the present data of the ratio of the toe-to-first-crest horizontal distance to the inflow depth L_w/h_1 is more scattered than the data of Ohtsu et al. (2003), due to the fluctuations of the toe position of the lateral shock wave which may have increased the measurement error of L_w in present experiment. The relationship between the angle θ of the lateral shock wave from the sidewall and Fr in present experiment biases significantly from the correlation provided by Ohtsu et al. (2003). The average lateral shock wave angle is found about 38° and doesn't change over Fr, as can be seen in Fig. 4(d).

The free-surface shape development of the first three waves of the undular hydraulic jumps is investigated in terms of the ratio of the second-to-third-crest horizontal distance to the sequent depth L_2/h_2, the ratio of the second crest depth to the inflow depth h_{wc2}/h_1 and the ratio of the third crest depth to the inflow depth h_{wc3}/h_1, as shown in Fig. 5. The trends of the increase in h_{wc2}/h_1, h_{wc3}/h_1 and h_{wc1}/h_1 with increasing Fr almost overlap, so do the decreasing trends of L_2/h_2 and L_1/h_2 with Fr. It is demonstrated that, for a given Fr, the shapes of the first three waves are similar in terms of the wave-length and crest-depth. These shapes varies when Fr changes.

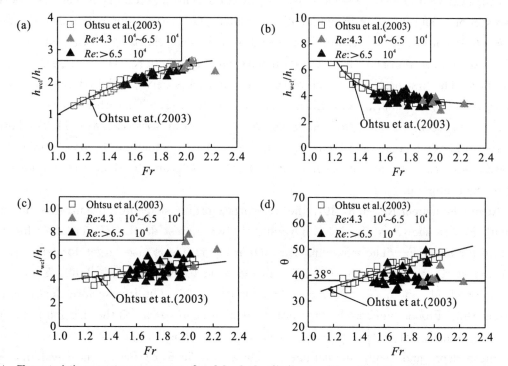

Figure 4. **Characteristic geometry parameters of undular hydraulic jumps with comparation to previous experimental data**

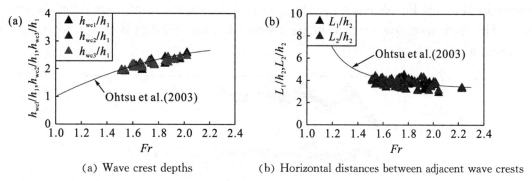

(a) Wave crest depths   (b) Horizontal distances between adjacent wave crests

Figure 5. **Characteristics of the first three waves**

A simultaneous scanning of the six ADMs along the flume allows for the measurement of mean water depth(h/h_1) of the undular hydraulic jumps at six longitudinal cross-sections(y_1 to y_6) from the right sidewall to the centerline. The six longitudinal cross-sections can be divided into three groups in the near-wall area(y_1) where the water surface profile is smooth and wave crests and troughs are hardly seen, the transition area(y_2) where the first wave crest is observed clearly and the downstream surface profile is more fluctuating than that in the near-wall area, and the central area(y_3, y_4, y_5, y_6) where several wave crests and troughs are obvious. Three sections in different areas with the same Fr and three sections in the same areas with different Fr are presented below in terms of both mean depth(h/h_1) and depth fluctuations(h'/h_1).

Fig. 6 shows similar inflow free-surface fluctuation levels of $h'/h_1=0.029$, independent of Fr or the distance to the sidewall. Fig. 6(a) compares the dimensionless mean depth(h/h_1) and depth fluctuations(h'/h_1) for $Fr=1.71$ in the near-wall area(y_1), the transition area(y_2) and the central area(y_3). The rise and fall of the free surface increase from the sidewall to the centerline. For instance, there are 5 wave crests and 5 wave troughs in the central area while zero wave crest and trough in the near-wall region. The wave crest depth in the central section is deeper than those in the other two sections, and the wave trough depth is smaller. The surface fluctuation data show only one peak($h'/h_1=0.14$) at $(x-x_0)/h_1=1.8$ in the near-wall section, one peak($h'/h_1=0.19$) at $(x-x_0)/h_1=3.0$ in the transition section and three peaks(the maximum $h'/h_1=0.25$) at $(x-x_0)/h_1=3.0$ in the central section. The first peak fluctuation of each section occurs between the lateral shock wave and the first wave crest, at $0.33L_w$, $0.72L_w$ and $0.53L_w$ respectively. The other peaks of the central section($h'/h_1=0.23$ at $(x-x_0)/h_1=6.57$ and $h'/h_1=0.23$ at $(x-x_0)/h_1=11.34$) both occur between the local wave crests and troughs. The free-surface fluctuations become stable at further downstream, with the average values of $h'/h_1=0.082$, 0.098 and 0.13 in each section from the sidewall to the centerline.

For Reynolds number $Re>4.3\times10^4$, the free-surface profile shape in the central section does not change with Fr, as shown in Fig. 6(b). Overall, the wave crest depth h/h_1 is larger for large Fr compared to for small Fr. The maximum fluctuation magnitude is also larger for large Fr, which occurs at almost the same location, i.e., $h'/h_1=0.30$ at $0.56L_w$ for $Fr=1.89$, $h'/h_1=0.25$ at $0.56L_w$ for $Fr=1.71$, and $h'/h_1=0.22$ at $0.6L_w$ for $Fr=1.56$. The free-surface fluctuations under the different three Froude numbers become nearly equal at downstream as the rise and fall of the free surface level gradually flatten out, reaching a constant value about 0.13.

The mean depth and depth fluctuations are shown in Fig. 6(c) for the near-wall region. The downstream water depth is larger for a large Fr than for a small Fr, while the fluctuations are nearly the same for different Froude numbers. Only one peak fluctuation value is shown at the same location $1/3L_w$. The fluctuation downstream the peak location is hardly affected by Fr, showing a constant value about $h'/h_1=0.082$.

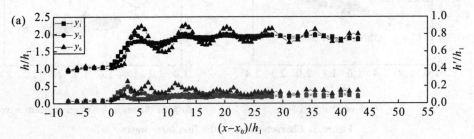

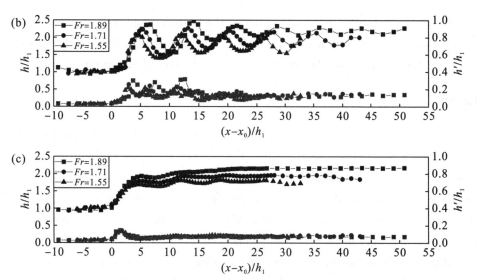

Figure 6. Mean depth and depth fluctuations of undular hydraulic jumps(a) in different transverse sections for $Fr=1.71$ and in(b) central and(c) near-wall sections for three different Froude numbers

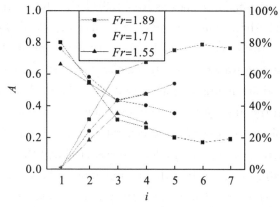

Figure 7. Crest-to-trough amplitude and the degree of amplitude attenuation in central section

Fig. 6(b) shows that, in the central region, the wave crest water depth(h_{wc1}/h_1, h_{wc2}/h_1, ⋯) decreases slightly along the flume while the rise and fall of the free surface flatten out obviously. Define the ratio A_i of the depth difference between the adjacent crest(h_{wci}) and trough(h_{wti}) to the inflow depth(h_1):

$$A_i = \frac{h_{wci} - h_{wti}}{h_1} \tag{3}$$

to describe the magnitude of rise and fall of the free surface. Fig. 7 shows that, with the streamwise development of the jumps, this wave amplitude keeps decreasing and is finally stabilized. For a larger Froude number, the first wave amplitude appears to be larger and the amplitude at the end of jump is smaller.

$$\eta = \left(1 - \frac{A_i}{A_1}\right) \times 100\% \tag{4}$$

Eq. (4) is used to indicate the degree of the amplitude attenuation, as also plotted in Fig. 7. The attenuation degree of the first three crest-to-trough amplitudes increases linearly. A larger Fr leads to a faster amplitude attenuation. At the end of the jumps, attenuation degree under different Fr approaches to different values, i.e., 80% for $Fr=1.89$, 50% for $Fr=1.71$, and 30% for $Fr=1.55$, respectively. It can be inferred that, downstream the fourth amplitude, the degree of amplitude attenuation is stable around a certain value, which is related to Fr. The attenuation degree increases slowly along the jump, and it must be close to 100% at the end of the jump.

5 Conclusion

Characteristic free-surface geometry and fluctuations of undular hydraulic jumps in a shallow channel is studied in this paper. The effects of Fr on the free-surface characteristics are compared with previous experiments of Ohtsu et al. (2003) under similar flow conditions. The relationships of h_{wc1}/h_1 and L_1/h_2 with Fr in the present experiments are in good agreement with the previous experiments. Compared with previous experimental data, the present observations of the relationship between L_w/h_1 and Fr are more scattering. The relationship between θ and Fr in the present experiment is different from that in the previous experiments, with θ fluctuating around an angle of 38°. The number of the standing waves is found to be related to the Reynolds number. There is only one complete standing wave in undular hydraulic jumps for $Re=2.46\times10^4$, while several complete standing waves are shown with similar shapes as $Re>4.3\times10^4$. The relationships between Fr and the crest water depths of the first three standing waves, h_{wc1}/h_1, h_{wc2}/h_1 and h_{wc3}/h_1, and the horizontal distance between the wave crests, L_1/h_2 and L_2/h_2, are also almost the same, indicating similar shapes of the first three standing waves. Acoustic displacement meters are used to measure the water depth and fluctuations of undular hydraulic jumps with $Fr=1.56$, 1.71 and 1.89 along the flume, revealing sidewall effects on the jump shapes. The longitudinal cross-sections can be divided into the side-wall section, the transition section and the central section. From the near-wall to the central sections, the crest depth increases, the trough depth decreases, and the surface fluctuation magnitude increases. The maximum value of the surface fluctuation appears between the starting point of lateral shock wave and the first wave crest. The surface fluctuation tends to be stable at downstream, of which the value increases toward the central section and is basically independent of Fr. The free-surface fluctuation values have three peaks in the central section and only one peak in the side-wall section. This paper studies the wave amplitude of the undular hydraulic jumps. The first wave amplitude increases with the increase of Fr, and the amplitude decreases with the streamwise development of the jump. The degree of amplitude attenuation increases with the increase of Fr, and the attenuation degree for different Fr approaches to different values at downstream of the jump.

References

[1] Iwao O, Youichi Y, Hiroshi G. Flow Conditions of Undular Hydraulic Jumps in Horizontal Rectangular Channels[J]. Journal of Hydraulic Engineering, 2003, 129(12): 948−955.

[2] Montes J S, Chanson H. Characteristics of Undular Hydraulic Jumps: Experiments and Analysis[J]. Journal of Hydraulic Engineering, 1998, 124(2): 192−205.

[3] Chanson H. Characteristics of Undular Hydraulic Jumps: Experimental Apparatus and Flow Patterns[J]. Journal of Hydraulic Engineering, 1995, 121(2): 129−144.

[4] Zhang G. On the estimation of free-surface turbulence using ultrasonic sensors [J]. Flow Measurement and Instrumentation, 2018, 60: 171−184.

[5] Murzyn F, Chanson H. Free-surface fluctuations in hydraulic jumps: Experimental observations[J]. Experimental Thermal and Fluid Science, 2009, 33(7): 1055−1064.

[6] Chachereau Y, Chanson H. Free-surface fluctuations and turbulence in hydraulic jumps[J]. Experimental Thermal

and Fluid Science, 2011, 35(6): 896−909.

[7] Valero D, Chanson H, Bung D B. Robust estimators for free surface turbulence characterization: A stepped spillway application[J]. Flow Measurement and Instrumentation, 2020,

[8] Chanson H. Current knowledge in hydraulic jumps and related phenomena. A survey of experimental results[J]. European Journal of Mechanics / B Fluids, 2008, 28(2): 191−210.

[9] Ohtsu I, Yasuda Y, Gotoh H. Non-Breaking Undular Hydraulic Jump[J]. 2010, 34(4): 567−573.

[10] Reinauer R, Hager W H. Non-breaking undular hydraulic jump[J]. Journal of Hydraulic Research, 2010, 33(5): 683−698.

Influence of Asymmetry on the Performance of a Breakwater Type Point Absorber Wave Energy Converter

Qi Zhang[1], Binzhen Zhou[2], Chaohe Chen[3], Peng Jin[4]

[1,4] College of Shipbuilding Engineering, Harbin Engineering University, Harbin 150001, China

[2,3] School of Civil Engineering and Transportation, South China University of Technology, Guangzhou 510006, China

Abstract

A hybrid system that integrating a point absorber wave energy converter into a floating breakwater is capable of both shoreline protection and wave energy conversion. The structural asymmetry of the hybrid system greatly influences its performance but the mechanism is not well understood. In this paper, a semi-analytical method is proposed to model the interactions between waves and asymmetric structures. Symmetric and asymmetric breakwater type wave energy converters are used in comparative studies to study the effect of structural asymmetry on the transmission coefficient, reflection coefficient, and energy conversion efficiency. Results show that increasing the bottom slope and letting the shorter wall face the incident waves can improve wave attenuation and energy conversion performance.

Keyword: Breakwater; Wave energy converter; Asymmetry; Wave attenuation; Energy conversion efficiency

1 Introduction

Although wave energy has the characteristics of high energy density and wide utilization range (Falnes J et al., 2007), its development is limited by high cost(Astariz S et al., 2015). It is believed that combining wave energy converters(WECs) with an off-the-shelf breakwater could effectively reduce construction, installation, and maintenance costs(Mustapa M A et al., 2017). Besides the hybrid system is capable of both shoreline protection and wave energy extraction. Recently, the concept of integrating a point absorber WEC into a floating breakwater(breakwater-PA) becomes a hot spot.

For a two-dimensional(2D) heaving PAWEC, the energy conversion efficiency is limited by the well-known maximum of 50%. Therefore many devices are designed to have an asymmetric structure to break the limitation. Berkeley Wedge, an asymmetric breakwater-PA applying a sharp needle-like bottom(Madhi F et al., 2014) could improve the energy conversion efficiency to 96.34%. Its wave attenuation was also improved significantly. Zhang et al.(Zhang H M et al., 2020) modify the not easily manufactured Berkeley wedge bottom to a much simpler triangular baffle bottom, achieving similar wave attenuation and energy conversion performance.

Despite its wide application, the influence of structural asymmetry on the wave attenuation and energy conversion performance of a breakwater-PA is not well understood. This paper presents a parametric study to investigate this problem. A semi-analytical method is established to calculate wave interactions of asymmetric floating structures. Three kinds of models with symmetric or asymmetric triangular and rectangular bottoms are used to study the influence of the type of bottom and the bottom slope on the performance of a breakwater-PA. Results show that structural asymmetry slightly influences the wave attenuation but greatly influences the energy conversion efficiency.

2 Mathematical Model

As indicated in Fig. 1, a float with asymmetric bottom is discretized into several steps using the boundary discretization method (BDM). The cartesian coordinate system is used with its origin on the calm water surface. The float is divided into a left part and a right part according to the vertical line passing the lowest point of the bottom. The left part is divided into P_1 steps on the right part P_2 steps. The coordinates of the left and right corner points are (xl_{P_1}, zl_{P_1}) and (xl_{P_2}, zl_{P_2}), respectively. The fluid domain is therefore divided into (P_1+P_2+1) subdomains—i.e. $I^{l,0} x \leqslant xl_{P_1}$, $I^{l,p} xl_p \leqslant x \leqslant xl_{p-1}$, $I^1 xl_1 \leqslant x \leqslant xr_1$, $I^{r,p} xr_{p-1} \leqslant x \leqslant xr_p$ and $I^{r,0} x \geqslant xr_{P_2}$. The float is subjected to a train of regular waves traveling in the positive x-direction. The water depth is d.

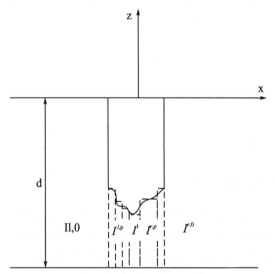

Figure 1. A float with arbitrary bottom shape and the discretization of the fluid domain

the velocity potential φ in the frequency domain, which satisfies Laplace's equation, can be decomposed into the expression as

$$\varphi = \varphi_{(0)} + \sum_{L=1}^{3} \xi_{(L)} \varphi_{(L)} \qquad (1)$$

where $\varphi_{(0)}$ is the sum of incident velocity potential φ_I and diffraction potential φ_D. when $L=1, 2$ and 3, $\varphi_{(L)}$ is the radiation potential due to surge, heave, and pitch motion. $\xi_{(L)}$ is the response amplitude in Lth motion mode. Besides, φ also satisfies the boundary conditions on the seabed, free surface, the solid boundary of the float, and in far-field (Li Y C et al., 1990).

2.1 The expressions of the velocity potential in each subdomain

Based on the solution of diffraction and radiation problems of a rectangular buoy (Li Y C et al., 1990), the spatial velocity potential in each subdomain i.e. $I^{l,0} x \leqslant xl_{P_1}$, $I^{l,p} xl_p \leqslant x \leqslant xl_{p-1}$, $I^1 xl_1 \leqslant$

$x \leqslant xr_1$, $I^{r,p} xr_{p-1} \leqslant x \leqslant xr_p$ and $I^{r,0} x \geqslant xr_{P_2}$. can be expressed as follows:

$$\varphi_{(L)}^{l,0} = \left(-\frac{igA}{\omega}\right)^{\delta_{0,L}} \left\{ \delta_{0,L} e^{ik_0(x-xl_{P_1})} Z_0(k_0 z) + O_{0,(L)} e^{-ik_0(x-xl_{P_1})} Z_0(k_0 z) + \sum_{m=1}^{\infty} O_{m,(L)} e^{k_m(x-xl_{P_1})} Z_m(k_m z) \right\}$$

$$\varphi_{(L)}^{l,p} = \left(-\frac{igA}{\omega}\right)^{\delta_{0,L}} \left\{ \varphi_{(L)}^{p,l,p} + (F_{0,(L)}^{l,p} + R_{0,(L)}^{l,p} x) Y_0(\lambda_0^{l,p} z) + \sum_{n=1}^{\infty} (F_{n,(L)}^{l,p} e^{\lambda_n^{l,p} x} + R_{n,(L)}^{l,p} e^{-\lambda_n^{l,p} x}) Y_n(\lambda_n^{l,p} z) \right\}$$

$$\varphi_{(L)}^{1} = \left(-\frac{igA}{\omega}\right)^{\delta_{0,L}} \left\{ \varphi_{(L)}^{p,1} + (F_{0,(L)}^{1} + R_{0,(L)}^{1} x) Y_0(\lambda_0^{1} z) + \sum_{n=1}^{\infty} (F_{n,(L)}^{1} e^{\lambda_n^{1} x} + R_{n,(L)}^{1} e^{-\lambda_n^{1} x}) Y_n(\lambda_n^{1} z) \right\} \quad (2)$$

$$\varphi_{(L)}^{r,p} = \left(-\frac{igA}{\omega}\right)^{\delta_{0,L}} \left\{ \varphi_{(L)}^{p,r,p} + (F_{0,(L)}^{r,p} + R_{0,(L)}^{r,p} x) Y_0(\lambda_0^{r,p} z) + \sum_{n=1}^{\infty} (F_{n,(L)}^{r,p} e^{\lambda_n^{r,p} x} + R_{n,(L)}^{r,p} e^{-\lambda_n^{r,p} x}) Y_n(\lambda_n^{r,p} z) \right\}$$

$$\varphi_{(L)}^{r,0} = \left(-\frac{igA}{\omega}\right)^{\delta_{0,L}} \left\{ T_{0,(L)} e^{ik_0(x-xr_{P_2})} Z_0(k_0 z) + \sum_{m=1}^{\infty} T_{m,(L)} e^{-k_m(x-xr_{P_2})} Z_m(k_m z) \right\}$$

where ω represents the angular frequency, g donates the acceleration of gravity, A is the incident wave amplitude, $\delta_{i,L}$ is the Kronecker delta function, $O_{m,(L)}$, $F_{n,(L)}^{l,p}$, $R_{n,(L)}^{l,p}$, $F_{n,(L)}^{1}$, $R_{n,(L)}^{1}$, $F_{n,(L)}^{r,p}$, $R_{n,(L)}^{r,p}$ and $T_{m,(L)}$ are the unknown coefficients to be determined in. k_m, $\lambda_n^{l,p}$, λ_n^{1} and $\lambda_n^{r,p}$ are the eigenvalues and $Z_m(k_m z)$, $Y_n(\lambda_n^{l,p} z)$, $Y_n(\lambda_n^{1} z)$ and $Y_n(\lambda_n^{r,p} z)$ are the corresponding eigenfunctions in each subdomain (Li Y C et al., 1990). $\varphi_{(L)}^{p,l,p}$, $\varphi_{(L)}^{p,1}$ and $\varphi_{(L)}^{p,r,p}$ is solutions related to the radiation conditions, their expressions are as follows:

$$\varphi_{(L)}^{p,l,p} = -\frac{i\omega}{2sl_p}\left[(z+d)^2 - x^2\right]\delta_{2,L} + \frac{i\omega}{2sl_p}\left[(z+d)^2(x-x_0) - \frac{(x-x_0)^3}{3}\right]\delta_{3,L}$$

$$\varphi_{(L)}^{p,1} = -\frac{i\omega}{2sl_1}\left[(z+d)^2 - x^2\right]\delta_{2,L} + \frac{i\omega}{2sl_1}\left[(z+d)^2(x-x_0) - \frac{(x-x_0)^3}{3}\right]\delta_{3,L} \quad (3)$$

$$\varphi_{(L)}^{p,r,p} = -\frac{i\omega}{2sr_p}\left[(z+d)^2 - x^2\right]\delta_{2,L} + \frac{i\omega}{2sr_p}\left[(z+d)^2(x-x_0) - \frac{(x-x_0)^3}{3}\right]\delta_{3,L}$$

2.2 Solution of the unknown coefficients

Continuity of pressure and normal velocity at the interfaces between each pair of adjacent subdomains should also be satisfied (Zhang W C et al., 2016). Combining these governing equations and boundary conditions, taking the first $(M+1)$ terms for $O_{m,(L)}$, $T_{m,(L)}$ and $(N+1)$ terms for $F_{n,(L)}^{l,p}$, $R_{n,(L)}^{l,p}$, $F_{n,(L)}^{1}$, $R_{n,(L)}^{1}$, $F_{n,(L)}^{r,p}$, $R_{n,(L)}^{r,p}$ in the infinite series, making some arrangements yield four sets of linear systems of $2(M-N)+2(P_1+P_2)(N+1)$ complex equations for each L.

$$EX_{(L)} = B_{(L)} \quad (4)$$

where $X_{(L)} = [O_{m,(L)}, F_{n,(L)}^{l,p}, R_{n,(L)}^{l,p}, F_{n,(L)}^{1}, R_{n,(L)}^{1}, F_{n,(L)}^{r,p}, R_{n,(L)}^{r,p}, T_{m,(L)}]^T$. E is the coefficient matrix decided by the shape of the float. Substituting the solved unknown coefficients into Eq. (2), the expression of velocity potential could be obtained.

2.3 Transmitted, reflected, and absorbed wave energy

As the velocity potential is solved, the optimal power take-off (PTO) damping b^{opt} and the corresponding optimal energy conversion efficiency η could be easily expressed according to (Zhang H M et al., 2020). And the expression of reflection coefficient K_r and transmission coefficient K_t are

$$K_r = \left|\frac{\varphi_D + \varphi_R}{\varphi_I}\right|_{x=-\infty} = A_{0,(0)} + \frac{\omega O_{0,(2)}\xi_{(2)}}{-igA}$$
$$K_t = \left|\frac{\varphi_I + \varphi_D + \varphi_R}{\varphi_I}\right|_{x=+\infty} = T_{0,(0)} + \frac{\omega T_{0,(2)}\xi_{(2)}}{-igA} \quad (5)$$

2.4 Validation

To validate the semi-analytical model, an energy conservation test is conducted. Case 3 in Table 1 with $h_1=0.6$ m, $h_2=6$ m, and $B=1.8$ m is applied. The result is shown in Fig. 2. The relation $K_r^2 + K_t^2 + \eta = 1$ is satisfied and the model is validated.

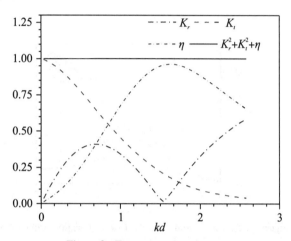

Figure 2. Energy conservation test

Table 1. Description of the bottom shape of the floaters

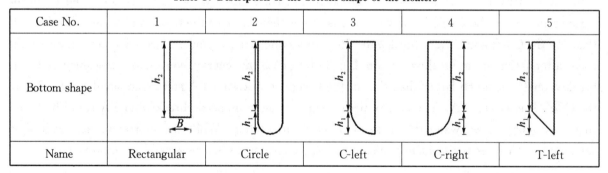

Case No.	1	2	3	4	5
Bottom shape					
Name	Rectangular	Circle	C-left	C-right	T-left

3 Influence of Asymmetry on the Performance of a Breakwater-PA

3.1 Performance comparison between the C-left, C-right, and Circle breakwater-PAs

In this section, we investigate the transmission coefficient, reflection coefficient, and energy conversion efficiency of the C-left, C-right, and Circle breakwater-PAs. Incident waves from $T=4$ s to 7 s are applied. The geometric parameters are $h_1=0.6$ m and $h_2=6$ m.

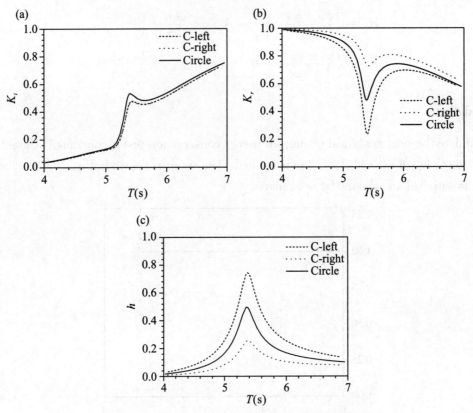

Figure 3. K_t, K_r, and η of the C-left, C-right, and Circle

In Fig. 3(a), the K_t curves of the C-left and C-right breakwater-Pas are identical. No matter which wall faces the incident wave, the proportion of transmitted wave energy is not changed. In the period range K_t of the Circle breakwater-PA is larger than those of the C-left and C-right breakwater-Pas. As the C-left and C-right breakwater-PAs are asymmetric, asymmetric could slightly improve the wave attenuation of a breakwater. In Fig. 3(b)(c), the energy conversion efficiency η Circle breakwater-PA is much larger than that of the C-right breakwater-PA and much smaller than that of the C-left breakwater-PA. The energy performance of an asymmetric breakwater-PA could be better than a symmetric breakwater-PA, depending on its orientation. While the shorter vertical wall of an asymmetric PA faces the incident wave, its energy conversion efficiency could be improved.

3.2 Influence of bottom slope on the performance of a breakwater-PA

In this section, we investigate the influence of the bottom slope on the transmission coefficient, reflection coefficient, and energy conversion efficiency of the T-left and Rectangular breakwater-PAs. Incident waves from $T=4$ s to 7 s are applied. The geometric parameters are $h_2=6.6$ m for the Rectangular breakwater-PA and $h_1=0.6$ m and $h_2=6$ m, $h_1=1.2$ m and $h_2=5.4$ m, and $h_1=1.8$ m and $h_2=4.8$ m for three T-left cases.

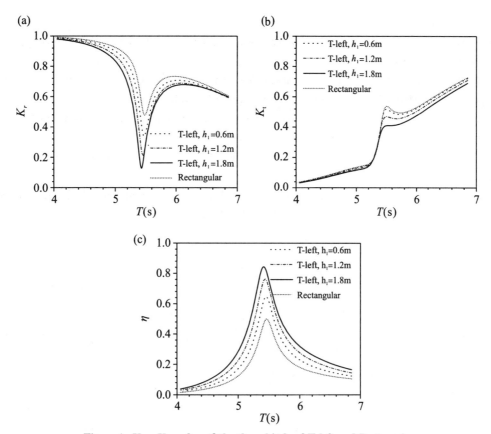

Figure 4. K_r, K_t and η of the three kinds of T-left and Rectangular

In Fig. 4(a), as $h_1 =$ increases, the bottom slope of breakwater-PA increases, and K_t slightly decreases. One can consider that as the bottom slope increases, the asymmetry of the breakwater-PA increases. Therefore, as the asymmetry of breakwater-PA increases, its wave attenuation is slightly improved. In Fig. 4(b) and (c), the energy conversion efficiency η of a breakwater-PA with a larger bottom slope is higher. An increase in the asymmetry of the breakwater greatly improves its energy performance.

4 Conclusion

This paper proposes a semi-analytical method to study the influence of asymmetry on the wave attenuation and energy conversion performance of a breakwater type point absorber wave energy converter(breakwater-PA). Asymmetry slightly influences the proportion of the transmitted energy of a breakwater-PA but greatly influences the proportions of the reflected wave energy and absorbed wave energy. Important remarks are concluded as follows.

(1) An increase in the asymmetry of a breakwater-PA could slightly improve its wave attenuation, as long as the shorter wall faces the incident waves.

(2) An increase in the asymmetry of a breakwater-PA greatly enhances its energy conversion efficiency.

References

[1] Falnes J. A review of wave energy extraction[J]. Marine Structures, 2007, 20(4): 185−201.

[2] Astariz S, Iglesias G. The economics of wave energy: a review[J]. Renewable and Sustainable Energy Reviews,

2015, 45(45): 397−408.

[3] Mustapa M A, Yaakob O B, Ahmed Y M. Wave energy device and breakwater integration: a review[J]. Renewable and Sustainable Energy Reviews, 2017, 77: 43−58.

[4] Madhi F, Sinclair M E, Yeung R W. The "berkeley wedge": an asymmetrical energy-capturing floating breakwater of high performance[J]. Marine Systems & Ocean Technology, 2014, 9(1): 5−16.

[5] Zhang H M, Zhou B Z, Vogel C. Hydrodynamic performance of a floating breakwater as an oscillating-buoy type wave energy converter[J]. Applied Energy, 2020, 257: 113996.

[6] Li Y C, Teng B. Wave action on maritime structures [M]. Dalian: The Press of Dalian University of Technology, 1990.

[7] Zhang W C, Liu H X, Zhang L. Hydrodynamic analysis and shape optimization for vertical axisymmetric wave energy converters[J]. China Ocean Engineering, 2016, 30(6): 954−966.

Influence of Sandspit Variability on Coastal Management Processes at the "Bouche du Roi" Inlet, Benin

Stephan Korblah Lawson[1], Hitoshi Tanaka[2], Keiko Udo[3],
Nguyen Trong Hiep[4], Xuan Tinh[5]

[1,2,4,5] Department of Civil and Environmental Engineering, Tohoku University,
Aobayama Campus, Sendai, Japan

[3] International Research Institute of Disaster Science, Tohoku University,
Aobayama Campus, Sendai, Japan

Abstract

The Bouche du Roi Inlet in Benin, West Africa, is among several morpho-dynamic coastal features along the West African coast. Discharge from the Mono River and Lake Aheme is drained through this inlet. A notable geological feature at the inlet are two distinct updrift and downdrift sandspits. The continuous accumulation of longshore sediment transport by the updrift sandspit results in rapid elongation of the sandspit. This leads to the constant migration of the inlet. Consequently, there is the near or complete closure of the inlet which exacerbates flooding events in surrounding communities during rainy seasons. To maintain the inlet opening, artificial breaching campaigns are usually carried out to minimize the risks of flooding around nearby communities. In this present study, satellite images were obtained from 1984—2020 to analyze the long-term sandspit morphology. These images were further used to examine how sandspit morphology affects coastal management processes at the Bouche du Roi inlet. The findings from this study indicate that, there is the need to implement permanent stabilization measures at the inlet to reduce the costs and man hours associated with the artificial breaching campaigns.

Keywords: Bouche du Roi Inlet; Breaching; Flooding; Sandspit; Satellite images

1 Introduction

A common coastal feature at river mouths and inlets is the existence of sandspits or barriers. These coastal features are controlled by complex forming and morphological processes (Dan et al., 2011). Growth of sandspits is predominantly sustained by the accretion of longshore sediment transport (Pradhan et al., 2015). Coastal areas with significantly high amount of longshore sediment transport greatly influences the variability of sandspits. As such, the relationship between longshore sediment transport and sandspit evolution is a critical area of interest to coastal researchers. Insufficient studies in this field results in poorly designed protection structures to prevent floods or beach erosion in coastal areas. Therefore, the aim of this study is to analyze the long-term sandspit morphology using satellite images and understand how sandspit morphology affects coastal management processes at the Bouche du Roi inlet.

2 Study Area

The study area is the Bouche du Roi Inlet, located in Benin, West Africa. The inlet drains discharge from the Mono River and Lake Aheme into the Gulf of Guinea [Figs. 1(a) and (b)]. The Mono River comprises an approximate 400 km river network and a catchment area of about 30,000 km². The river is an international river shared between Togo and Benin with the upstream section in Togo and the downstream section in Benin. Situated about 135 km upstream of the Bouche du Roi inlet is the Nangbeto Dam (located in Togo). The dam was constructed between 1984—1987 to serve as a hydro-electric generation dam. Lake Aheme on the other hand receives discharge from the upstream Couffo River which is subsequently released through the inlet.

The Bouche du Roi inlet possesses two prolonged sandspits which creates a lagoon on the landward side of the inlet. The inlet is constantly migrating due to the accumulation of longshore sediment transport on the updrift spit. This leads to the near or complete closure of the inlet and exacerbating flooding events in surrounding communities.

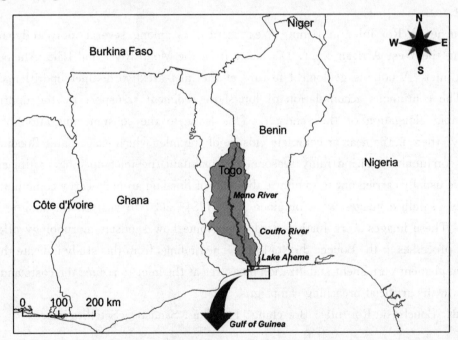

(a) Map of Mono River Basin, Benin, West Africa

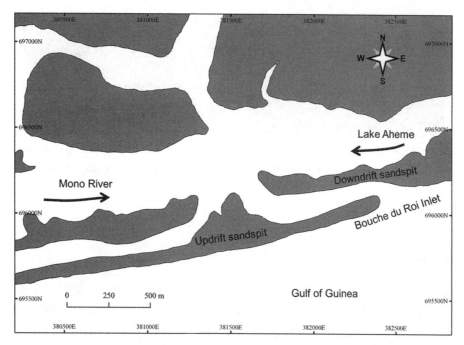

(b) Bouche du Roi Inlet, Benin, West Africa

Figure 1. Map of study area

Along the West African coast, chronic beach erosion has been a major issue battling countries on this coast. Studies by(Addo et al., 2008; Almar et al., 2014; Laïbi et al., 2014; Giardino et al., 2018; Guerrera et al., 2021; Lawson et al., 2021) have highlighted the problems of coastal erosion in West Africa. Coupled with insufficient funding to alleviate this problem, the lack of prioritization on coastal management by policy makers and planners has hindered a long-lasting solution to this predicament. However, coastal areas in this region have seen a significant rise in mitigation measures in recent years. Some examples of such projects are the jetty and groyne systems at Ada(Ghana) and Aneho(Togo) which are located on the updrift side of the Bouche du Roi inlet[Figs. 2(a)(b)]. On the downdrift side of the inlet are the breakwater and groyne systems at Cotonou(Benin) and Lekki (Nigeria) [Figs. 2 (c) (d)]. These projects give an indication of the improvement in coastal management along the coast by stakeholders. Nonetheless, an enormous gap still exists in the attempt to completely protect the West African coast against the forces driving beach erosion and its associated problems.

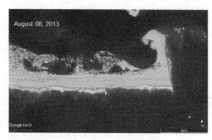 

(a) Jetty and Groyne Systems at Ada, Ghana   (b) Jetty and Groyne Systems at Aneho, Togo

(c) Breakwater and Groyne Systems at Cotonou, Benin　(d) Breakwater and Groyne Systems at Lekki, Nigeria

Figure 2. Beach protection structures on the updrift[(a) &(b)] and downdrift[(c) &(d)] of the Bouche du Roi Inlet

3　Materials and Methods

3.1　Data collection

In this study, the sandspit variations were analyzed by utilizing freely available satellite images from Landsat 5(TM), Landsat 7(EMT+), Landsat 8(OLI) and Sentinel-2(MSI) data catalogs. These were further supplemented with Google Earth images to fill the no-data gaps from the Landsat collections. Hence, the Landsat and Sentinel-2 images served as the main source of satellite images for the study. In total, 309 images were acquired from 1984—2020 which were used to perform the long-term morphological analysis.

3.2　Image processing

Preprocessing of the Landsat and Sentinel-2 images was one of the most crucial stages in the image processing stage. Here, the 30 m/pixel Landsat images were preprocessed using pan sharpening and down sampling techniques. The resulting Landsat images had a spatial resolution of 15 m/pixel. In the case of the Sentinel-2 images, the 20 m/pixel images were down-sampled to 10 m/pixel by bilinear interpolation. The use of these techniques helped to improve the accuracy of the detected shorelines(Vos et al., 2019). In addition, the resolution of the images acquired from Google Earth were 2 m/pixel and 20 m/pixel.

Two approaches were used in the detection of the shorelines after the preprocessing stage. For the Landsat and Sentinel-2 images, the shorelines were defined using the Modified Normalized Difference Index(MNDWI) method developed by Xu(2006). Images acquired from Google Earth were rectified into a single coordinate system(World Geodetic System-84) using an affine transformation with a baseline 90 degrees from the north. The differences in color intensity of the wet and dry sand served as the basis for shoreline detection. A summary on the details of the acquired satellite images for the study is presented in Table 1 below.

Table 1. Information on acquired and processed satellite images used in the study

Image source	No. of images	image resolution	Down-sampled resolution
Landsat(5, 7 & 8)	113	30 m/pixel	15 m/pixel
Senitnel-2	163	20 m/pixel	10 m/pixel
Google Earth	33	2 m/pixel & 20 m/pixel	—

3.3 Sandspit analysis

After detection of shorelines for the Bouche du Roi inlet, the next step involved detailed analysis of sandspit variability at the inlet. This was achieved by defining a coordinate system specific to this study. In this system, the alongshore coordinate of the updrift sandspit is represented by x_1 and the sandspit tip defined as x_1 tip. The area of the updrift sandspit is also denoted as A_1. Definition of these parameters aided in examining the morphological trends at the Bouche du Roi inlet for the study period. The sandspit parameters used in the analysis have been illustrated in Fig. 3 below.

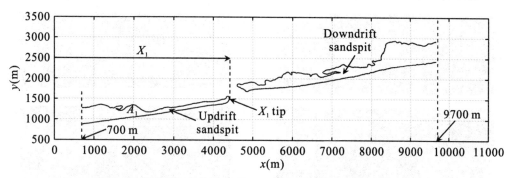

Figure 3. Definition of sandspit parameters used in sandspit analysis

4 Results and Discussions

4.1 Long-term sandspit variability at the Bouche du Roi inlet

Figures 4(a) to(h) show some prominent morphological changes at the Bouche du Roi inlet which were observed from the acquired satellite images. A typical sandspit behavior at the inlet is the eastward growth of the spit and the resulting migration of the inlet. [Figs. 4(a)(b)]. This direction of growth corresponds to the west to east transport of longshore sediment along the West African coast. The rapid elongation of the updrift sandspit leads to narrowing of the inlet which exacerbates flooding events in surrounding communities. Furthermore, there are instances where elongation of the updrift spit results in the complete closure of the inlet[Fig. 4(e)]. As a result, periodic artificial breaching of the sandspit(denoted as BR) is undertaken to mitigate the associated risks of flooding[Figs. 4(c)(f)]. After breaching events have been conducted, the cycle of sandspit elongation begins due to the accumulation of longshore sediment transport[Fig. 4(d)]. Another interesting observation at the inlet during updrift spit elongation is the intrusion of the spit into the estuary area[Fig. 4(g)]. This can be attributed to changes in water depth due to the recent dredging works executed in the study area to reduce the occurrence of floods.

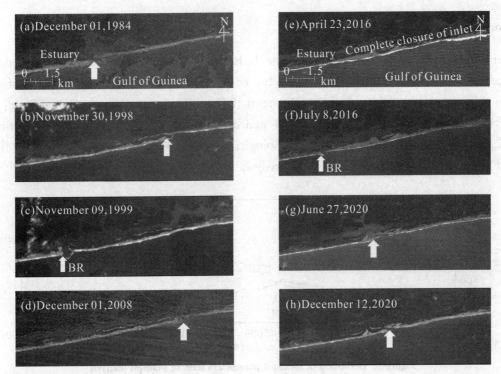

Figure 4. Selected satellite images depicting prominent morphological changes at the Bouche du Roi Inlet

4.2 Characteristics of sandspit morphological trends

Figs. 5 and 6 show the time variation of the updrift sandspit tip position (x_1) and sandspit area (A_1) for the Bouche du Roi inlet. Although the Landsat, Sentinel-2 and Google Earth Images have different spatial resolutions as indicated in Section 3.2, it was observed that the results obtained were in close agreement. The morphological characteristics (spit elongation and area changes) which were investigated showed similar trends due to the constant elongation of the updrift sandspit. Furthermore, results from the sandspit analysis and inspection of the satellite images revealed a total of five (5) breaches of the updrift sandspit during the study period. Most of these breaches were artificial breaches based on information gathered from dredging records within the area (Antoine et al., 2012; Ndour et al., 2018).

As shown in Fig. 5, the breaching campaigns (i.e. BR-02, BR-03 and BR-04) indicate that the critical sandspit tip location which would necessitate breaching is at an average distance of 8000 m based on the initial spit location defined for this study. Upon breaching of the updrift spit, the breached portion attaches to the downdrift sandspit before the elongation process begins. During the elongation process of the updrift spit from the west to east, the downdrift spit retreats accordingly.

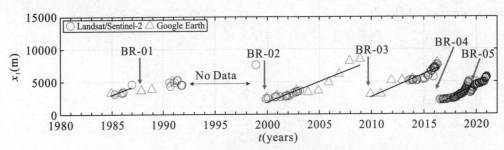

Figure 5. Time variation of sandspit x_1 coordinates from 1984—2020

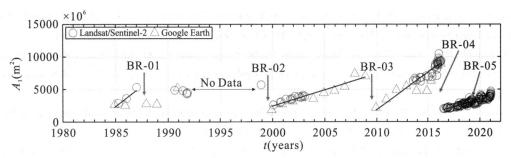

Figure 6. Time variation of sandspit area changes(A_1) from 1984—2020

5 Conclusions

The variability of the sandspit at the Bouche du Roi inlet and its impact on coastal management practices have been investigated in this study. This was achieved by acquiring remote sensing images from 1984—2020 to perform long-term sandspit analysis. Common morphological evolutions observed from the satellite images include sandspit elongation, frequent sandspit breaching and intrusion of the sandspit into the estuary area. Furthermore, results from the study revealed the constant migration of the updrift sandspit from the west to east direction of the Benin coast. In addition, the near or complete closure of the inlet due to the sandspit migration has led to numerous breaching campaigns at the Bouche du Roi inlet sandspit. Consequently, a permanent solution in the form of sandspit stabilization would be required at the inlet to attenuate the painstaking man hours required in artificial breaching campaigns.

Acknowledgements

The authors would like to express theirutmost gratitude to the Japanese Government Ministry of Education, Culture, Sports, Science and Technology (MEXT) Scholarship Program (MONBUKAGAKUSHO Scholarship) of which the first author of this paper is a beneficiary.

References

[1] Almar R, Hounkonnou N, Anthony E J, et al. The Grand Popo beach 2013 experiment, Benin, West Africa: from short timescale processes to their integrated impact over long-term coastal evolution[J]. Journal of Coastal Research, 2014, 70: 651—656.

[2] Antoine G, Anthony E, Lucien M O. Contribution of the Landsat Image Series to the Study of the Spatio-Temporal Dynamics of the Mouth of the Mono and Couffo Rivers Estuary in Benin, Before and After Construction of the Nangbéto Dam on the Mono[J]. Teledetection [Remote Sensing], 2012, 10(4): 179—198.(in French).

[3] Appeaning Addo K, Walkden M, Mills J P. Detection, measurement and prediction of shoreline recession in Accra, Ghana[J]. ISPRS Journal of Photogrammetry and Remote Sensing, 2008, 63(5): 543—558.

[4] Dan S, Walstra D J R, Stive M J F, et al. Processes controlling the development of a river mouth spit[J]. Marine Geology, 2011, 280(1—4): 116—129.

[5] Giardino A, Schrijvershof R, Nederhoff C M, et al. A quantitative assessment of human interventions and climate change on the West African sediment budget[J]. Ocean and Coastal Management, 2018, 156: 249—265.

[6] Guerrera F, Martín M, Tramontana M, et al. Shoreline changes and coastal erosion: The case study of the coast of togo(bight of benin, west africa margin)[J]. Geosciences(Switzerland), 2021, 11(2): 1—21.

[7] Laïbi R A, Anthony E J, Almar R, et al. Longshore drift cell development on the human-impacted Bight of Benin sand barrier coast, West Africa[J]. Journal of Coastal Research, 2014, 70(66): 78—83.

[8] Lawson S K, Tanaka H, Udo K, et al. Sandspit elongation and littoral drift at the Volta river mouth, Ghana[J]. Journal of Japan Society of Civil Engineers(JSCE) B2(Coast. Eng.), 2021, 77(2).(in Japanese)

[9] Ndour A, Laïbi R A, Sadio M, et al. Management strategies for coastal erosion problems in west Africa: Analysis, issues, and constraints drawn from the examples of Senegal and Benin[J]. Ocean and Coastal Management, 2018, 156: 92−106.

[10] Pradhan U, Mishra P, Mohanty P K, et al. Formation, growth and variability of sand spit at Rushikulya River Mouth, South Odisha Coast, India[J]. Procedia Engineering, 2015, 116(1): 963−970.

[11] Vos K, Splinter K D, Harley M D, et al. CoastSat: A Google Earth Engine-enabled Python toolkit to extract shorelines from publicly available satellite imagery[J]. Environmental Modelling and Software, 2019, 122.

[12] Xu H. Modification of normalised difference water index(NDWI) to enhance open water features in remotely sensed imagery[J]. International Journal of Remote Sensing, 2006, 27(14): 3025−3033.

Influence of Synchronization Time of Hybrid CPU/GPU 1D-2D Coupled Model for Urban Rainfall-runoff Process

Donglai Li[1], Jingming Hou[2], Yangwei Zhang[3], Yongde Kang[4]

[1,2,4] State Key Laboratory of Eco-hydraulics in Northwest Arid Region of China,
Xi'an University of Technology, Xi'an 710048, China

[3] Department of Civil Engineering, Technische Universität Berlin, Berlin 13355, Germany

Abstract

With the improvement of the methods and models, the 1D-2D coupled hydrodynamic model has increasingly become an essential tool to simulate and predict the rainfall-flood process and has been widely used in the field of urban area as well. The synchronization time is a vital parameter in the coupled model, now most of the coupled models apply the time step of 2D surface as the synchronization time. However, this method would affect the computational efficiency, especially when the data amount is large or the model is carried out on different platforms and in various type of codes. To evaluate the impact of synchronization time on the rainfall-runoff process, a new hydrodynamic model that couples the Graphics Processing Units(GPU) Accelerated Shallow Water Model(GSWM) and the Storm Water Management Model(SWMM) is applied to an ideal urban catchment to simulate rainfall-runoff-drainage processes with customized synchronization time after verification. The results show that the computational efficiency can be improved by 8.70%~27.37% in different scenarios compared with the method applying 2D model time step as the synchronization time. Both the increase of rainfall recurrence period and the synchronization time bring in better performance when synchronization time is less than 120 s. However, the error from drainage process increases with the increase of the synchronization time, especially the discharge peak is reduced when the synchronization time is greater than 120 s. Last but not least, there is almost no change in surface runoff process and drainage volume under different synchronization times. This work could provide reference for model application in the future.

Keywords: Synchronization time; GPU; 1D-2D coupled model; time improvement, rainfall runoff simulation

1 Introduction

Urban floods and inundations have become major disasters that severely affect development of urban and lives of citizens, and they show a tendency to occur frequently with the acceleration of urbanization, the change of land use and the increase of extreme weather in recent years(Hou et al., 2020; Li et al., 2020). The hydrodynamic models have been an important tool to simulate the flood and inundation process as the improvement of numerical methods and the perfection of basic data(Xia et al., 2019). Especially, the development of the 1D(one-dimensional)-2D(two-dimensional) coupled model provides great convenience for the simulation of urban flood and inundation process, since it can describe the underground drainage system and urban water system(Chen et al., 2018; Fan et al., 2017; Zhang et al., 2020).

The 1D model is used to simulate the drainage system process of the urban and river flood process, such as Infoworks CS(Wallingford, 2006) and SWMM(Rossman, 2009), they have been widely used. And due to the complex terrain, it is necessary to use the 2D model to simulate the surface runoff process(Hou et al., 2013). As for the coupled of the drainage system model and the surface model, the vertical linkage is the primary exchange mode(Chen et al., 2018; Hsu et al., 2000), that is, the exchange of water flow at manholes and rainwater inlets. And it is mainly the lateral exchange and the forward exchange for the exchange between the river channel and the surface.

Regarding the synchronization time, most coupled model used the time step of the 2D model as the synchronization time. Because the step times in 2D model is generally smaller than those of 1D model, although the they are both limited by numerical stability(Echeverribar et al., 2020). In recent years, several coupled 1D-2D models with the similar coupled patterns have been developed and applied(Chen et al., 2018; Salvan et al., 2016; Zhang et al., 2020). A coupled 1D and 2D model is first developed by(Hsu et al., 2000) for urban rainfall and inundation simulation. Chen et al.(2018) developed a hydrodynamic model that couples the SWMM and a shallow water model, the smallest step times in 1D and 2D are used as the synchronization time. Echeverribar et al.(2020) presented a hybrid CPU/GPU coupled 1D-2D model to analysis runtimes and efficiency of different model and hybrid configurations, and he pointed out that data transfer between the different model and configurations will have an impact on computing efficiency. Especially when the data amount is large or the model is carried out on different platforms and in various type of codes.

Then, it is necessary to improve computational efficiency by reducing the number of data exchanges with minimal loss of accuracy. The synchronization time is the main factor that affects the number of data exchanges. So, we can change the synchronization time to reduce the number of data exchanges. In this study, a new coupled model consisting of SWMM and a GPU accelerated shallow water model (GSWM) that can custom the synchronization time was developed. The fixed synchronization times with 5 s, 10 s, 30 s, 60 s, 120 s, 180 s and 300 s are simulated systematic to evaluate its effect for the rainfall-inundation process. Based on this overall scheme, the remainder of this work is organized as follows: Section 2 describes the methodology of hybrid CPU/GPU 1D-2D coupled model. Section 4 provides a case studies: rainfall-runoff process in an idealized urban catchment. Finally, Section4 draws conclusions.

2 Methodology of Hybrid CPU/GPU 1D-2D Coupled Model

2.1 The 1D sewer flow model

The SWMM(Rossman, 2009) developed by the USEPA(United States Environmental Protection Agency) is one of the most widely dynamic rainfall-runoff simulation models used to simulate of runoff from primarily urban areas. SWMM consists of two major components including the runoff component and the routing component. In this study, only the routing component were integrated into the coupled model as the 1D sewer flow module. Moreover, dynamic wave analysis solves the complete form of the St. Venant flow equations and therefore produces the most theoretically accurate results. It can account for channel storage, backwater effects, entrance/exit losses, culvert flow, flow reversal, and pressurized flow. Thus, the dynamic wave theory was selected to analysis the routing process in the sewer system. In the transport compartment, SWMM solves the St. Venant equation using the implicit finite difference method and successive approximation. The St. Venant equation consist of mass conservation and momentum conservation equations and can be expressed as

$$\frac{\partial A}{\partial t} + \frac{\partial Q}{\partial x} = 0 \tag{1}$$

$$\frac{\partial Q}{\partial t} + \frac{\partial (Q^2/A)}{\partial x} + gA\frac{\partial H}{\partial x} + gAS_{f1D} = 0 \tag{2}$$

where A is the flow cross-sectional area, t denotes time, Q is the flow rate, g is the acceleration of gravity, H is the hydraulic head of water in the conduit ($Z+Y$), Z is the conduit invert elevation, Y is the conduit water depth, S_{f1D} is the friction slope (head loss per unit length).

2.2 The governing equations for 2D GPU accelerated shallow water model

The numerical inundation model was developed by solving the 2D SWEs numerically within the framework of a well-balanced cell-center Godunov-type finite volume method (Liang and Marche, 2009). The 2D pre-balanced SWEs were chosen as the governing equations, as shown Eq. (3):

$$\frac{\partial \boldsymbol{q}}{\partial t} + \frac{\partial \boldsymbol{f}}{\partial x} + \frac{\partial \boldsymbol{g}}{\partial y} = \boldsymbol{S}$$

$$\boldsymbol{q} = \begin{bmatrix} \eta \\ q_x \\ q_y \end{bmatrix}, \boldsymbol{f} = \begin{bmatrix} uh \\ u^2h + g(\eta^2 - 2\eta z_b)/2 \\ uvh \end{bmatrix}, \tag{3}$$

$$\boldsymbol{g} = \begin{bmatrix} vh \\ vuh \\ v^2h + g(\eta^2 - 2\eta z_b)/2 \end{bmatrix}, \boldsymbol{S} = \begin{bmatrix} R \\ -\frac{gh\partial z_b}{\partial x} - C_f u\sqrt{u^2+v^2} \\ -\frac{gh\partial z_b}{\partial y} - C_f u\sqrt{u^2+v^2} \end{bmatrix}$$

where t is time, x and y are the cartesian coordinates, $\boldsymbol{q}$ is a vector of conserved flow variables containing η, q_x and q_y, which are the free surface water level and the unit-width discharges in the x and y-directions, respectively. Here, $q_x = uh$ and $q_y = vh$, h and v are the water depth and the depth-averaged velocities in the x and y-directions, respectively. z_b is the bed elevation and $z_b = \eta - h$; $\boldsymbol{f}$ and $\boldsymbol{g}$ are the flux vectors in the x and y-directions, respectively. $\boldsymbol{S}$ is the source vector, R is the source or sink of mass caused by rainfall, infiltration and the water exchange with sewer, C_f is the bed roughness coefficient, determined herein as $gn^2/h^{1/3}$, where n is the Manning roughness coefficient.

The numerical model for solving the SWEs within the framework of a Godunov-type cell-centered finite volume scheme is based on structured grids. The fluxes of mass and momentum are computed by the Harten Lax van Leer Contact (HLLC) approximate Riemann solver (i.e., the Harten, Lax and van Leer approximate Riemann solver with the contact wave restored). The slope source terms and the values at the midpoints of the cell edges are required. The two-stage explicit Runge-Kutta approach is applied to update the flow variables to a new time period. These values are evaluated by a novel 2D edge-based Monotonic Upwind Scheme for Conservation Laws (MUSCL) scheme (Hou et al., 2015). The code was written in C++ and Computed Unified Device Architecture (CUDA), which can run on GPUs to substantially accelerate the computation.

2.3 Coupled SWMM/GSWM

The rainfall-runoff-drainage process in the sewer networks and surfaces are simulated by coupling the GSWM and SWMM models. The two models were connected through appropriate linkages to exchange water level and flow, after the two models were executed individually to a suitable time. The rainfall is simulated at the 2D surface model, and this mode is more in line with the actual process. Generally, only the vertical exchange occurs between the sewer pipe system and the 2D surface. In this study, the inlets are assumed to be the channel where the vertical linkage occurs. When the water flows through inlets, the water would enter the drainage-pipe system at the inlets. In contrast, when the water depth in inlets exceeds the surface water elevation, the overflow would occur that the flow will move from the 1D sewer system to the 2D surface system (Chen et al., 2018). Thus, only the vertical linkage is considered for the sewer pipe system simulation.

2.3.1 *Vertical linkage*

When considering the surface water flow to the inlet, the model uses the weir equation or the orifice equation [Eq. (4)] to calculate the inflow from the surface ground.

$$Q_{in} = \begin{cases} c_w C_i h_{2D} \sqrt{2gh_{2D}} & \text{if } Z_{1D} \leqslant Z_{b2D} \leqslant Z_{2D} \\ c_o A_i \sqrt{2gh(Z_{2D} - Z_{1D})} & \text{if } Z_{b2D} \leqslant Z_{1D} \leqslant Z_{2D} \end{cases} \tag{4}$$

where, the Q_{in} is the discharge from 2D surface to the sewer, c_w represents the weir coefficient, c_o represents the orifice coefficient, c_i denotes the perimeter of the inlet, g is the acceleration of gravity, h_{2D} represents the water level at the surface and $h_{2D} = Z_{2D} - Z_{b2D}$, Z_{b2D} is the bed elevation of the surface, Z_{2D} is the surface water elevation, Z_{1D} is the inlet water elevation, A_i denotes the inlet area.

The orifice equation [Eq. (5)] is used to calculate the overflow from the sewer pipe to the surface when the water depth in inlets exceeds the surface water elevation.

$$Q_{over} = c_o A_i \sqrt{2g(Z_{1D} - Z_{2D})} \tag{5}$$

where, Q_{over} represents the overflow from the sewer system to the surface.

In this study, the source term method is used to couple the 1D model and 2D model. And the R is the source or sink of mass caused by rainfall, infiltration and the water exchange with sewer. Hence, the rainfall source term can be expressed as the Eq. (6):

$$R = i - f - c(Q_{in} - Q_{over}) \tag{6}$$

where, the i is the rainfall; f represents the infiltration; the $c(Q_{in} - Q_{over})$ denotes the water exchange with sewer, and the c is the correction factor due to the different of unit and grid size.

2.3.2 *Linking method*

SWMM is a free and open source model, and it provides Dynamic Link Library (DLL) that can be called by the external models (Rossman, 2009). This means that the two-dimensional model can be connected with SWMM through the DLL. This is also one of the important reasons why SWMM is used worldwide. Although the original DLL contains several interface functions for external calls, some additional interface functions need be added to establish linkages between the SWMM and the GSWM. And with the improved source codes, the SWMM and the GSWM can be well coupled.

2.4 Synchronization time

The time step of the SWMM model is generally fixed, and needs to be set. The time step of the 2D shallow water model is updated using the two-stage explicit Runge-Kutta scheme (Hou et al., 2015; Hou et al., 2013). Thus, the Courant-Friedrichs-Lewy (CFL) condition must be satisfied to ensure solution stability. In this study, the CFL condition [Eq. (7)] were used to estimate the time steps on quadrilateral grids:

$$\Delta t_{2D} = \text{CFL} \min\left(\frac{d_i}{\sqrt{u_i^2 + v_i^2} + \sqrt{gh_i}}\right) \quad (7)$$

where d_i is the minimum distance from cell centroid to the edges. The CFL number may take any value between 0 and 1, and CFL=0.5 is adopted in this work for all of the simulations.

There are two main methods of synchronization time for the 1D-2D coupled model (Chen et al., 2018). One is to synchronize according to the minimum time step of 1D model and 2D model, and the other is to use a defined synchronization time. For the second method, the time step of 1D model can be used as the synchronization time, due to it is generally larger. Or the fixed value can be used for the synchronization time. The schematic diagram of different synchronization time methods is shown in Fig. 1.

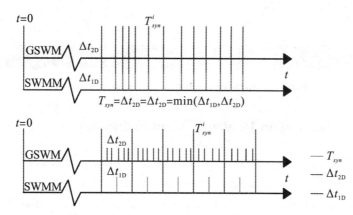

Figure 1. Schematic diagram of different synchronization time methods

The former is generally widely used in the coupled model, since it conforms to the actual physical process. However, the computational efficiency of this method is severely reduced, especially when the data amount is large or the model is carried out on different platforms and in various type of codes. And the latter may improve computational efficiency without affecting the calculation effect. In this study, the coupled model with customized synchronization time were proposed to simulate the rainfall-surface-drainage process. Both the 1D model and the 2D model use their own time steps when using customized time. The customized time can be any value, but it cannot be too large due to the limitation of the practical situation. And the customized time is generally considered larger than the time step of the 1D model and the 2D model. The sequence diagram of the simulation process is sketched in Fig. 2, where the operation sequence is outlined, distinguishing between the CPU and GPU codes. In this work, the 5 s, 10 s, 30 s, 60 s, 120 s, 180 s and 300 s are used as the synchronization time to compare with the 2D step time.

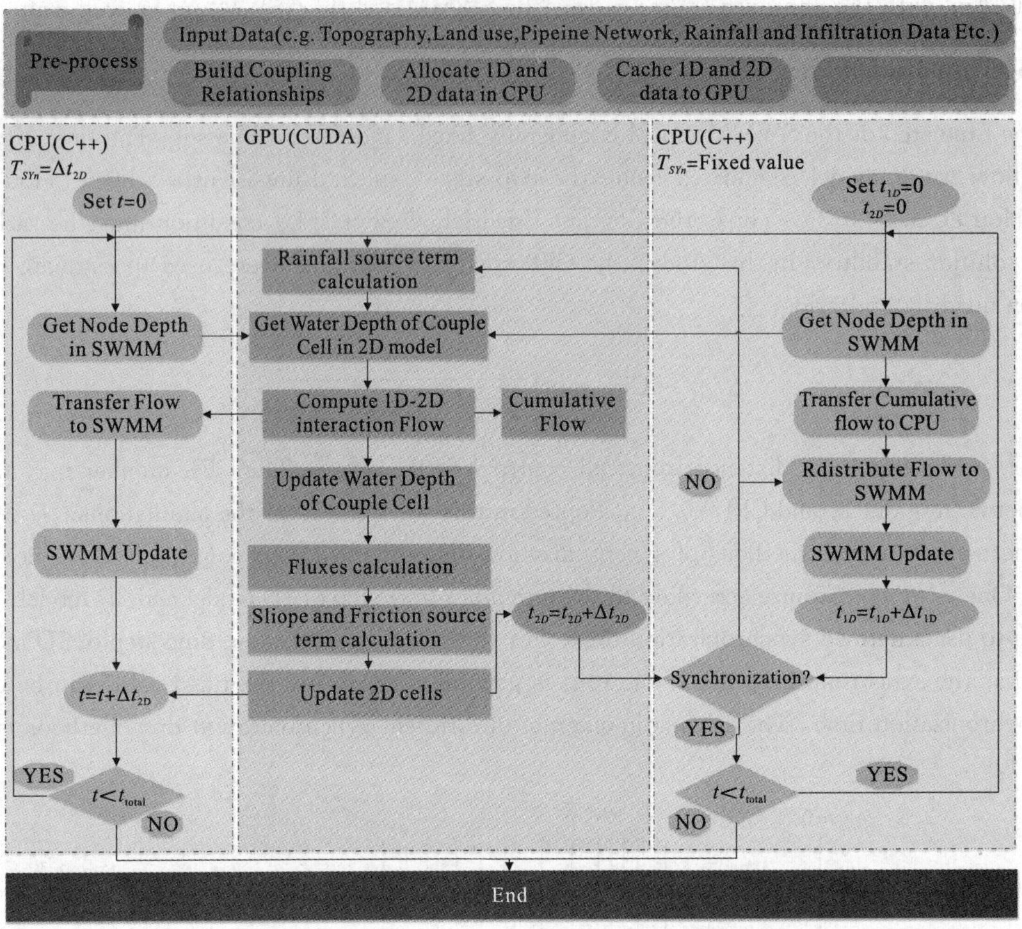

Figure 2. Sequence diagram of the simulation process

3 Rainfall-Runoff Process in an Idealized Urban Catchment

In this section, the rainfall-runoff-drainage process in an idealized urban catchment [Fig. 3(a)] is simulated under the rainfall with different years return period. The idealized urban catchment based on an actual catchment is 284 m×240 m, and the road wide is 24 m and the confluence area wide is 130 m on both sides, as shown in Fig. 3(b). The horizontal and vertical slopes of the catchment are 0.005 and 0.003, respectively. And there is also a slope of 0.02 from the middle to the sides of the road, which can be clearly seen in the Fig. 3. The drainage system of the catchment consisting of 20 inlets, 1 outfall and 20 pipelines is also depicted in Fig. 3(a). The inlet is 0.4 m×0.7 m, and the length and diameter of the pipelines are 35 m and 0.8 m, respectively(Li et al., 2020).

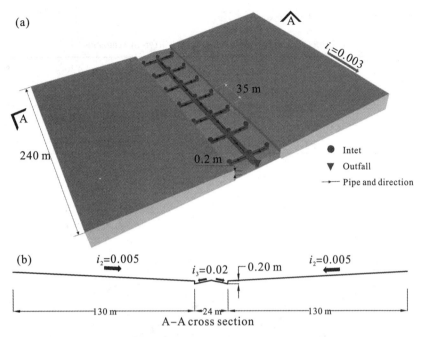

Figure 3. Idealized urban catchment and drainage system distribution

The design storm[Eq. (8)] of Gu Yuan City(Wang et al., 2020) is employed as rainfall input data for the case and the duration time is 2 hours. And the rainfall process of the 2, 5 and 30 years return periods is obtained based on the Chicago rainfall pattern.

$$q_r = \frac{5.2211 + 5.9357 \times \lg p}{(t_r + 7.9754)^{0.7688}} \tag{8}$$

where, the q_r denotes the rainfall intensity [L/(s·hm²)]; the p is the return periods, year; the t_r denotes rainfall duration, min.

The whole catchment is considered impermeable, and the infiltration is 0 mm/h. Regarding the boundary condition, all boundaries except for road exits are closed boundaries. The Manning of the road and the sides are 0.014 and 0.03, respectively. And the pipeline manning is 0.013. The surface grid size is 2 m.

Hence, the rainfall-runoff-drainage process is simulated with the different synchronization time under the rainfall with 2, 5 and 30 years return periods. And the duration time of simulation process is 4 h. The calculate time under different scenarios is presented in Table 1. Under different rainfall return periods, the calculation time is relatively long when the synchronization time is Δt_{2D}. And the time improvement between the Δt_{2D} and other fixed time is shown in the Fig. 4. The time improvement compared with the Δt_{2D} is 8.70% to 27.37%, and it increases as the return period increases under the same synchronization time. The improvement effect also increases with the increase of the synchronization time when the synchronization time is less than 120 s. However, the time improvement is slightly reduced when the synchronization time exceeds 120 s. The reason is that the frequency of data exchange between the SWMM and the GSWM is very limited as the synchronization time increases.

For this case, the number of synchronizations is 120 when the synchronization time is 120 s, and the number is 48 when the synchronization time is 300 s. It has only been reduced by 72 times, which is trivial for the computer. The slightly reduced of the time improvement may be affected by other factors when the synchronization time exceeds 120 s.

Table 1. The calculate time under different scenarios

T_{syn}	Δt_{2D}	5 s	10 s	30 s	60 s	120 s	180 s	300 s
Calculate time(s)	69.00	63.00	61.81	60.80	58.76	57.42	58.53	58.24
	82.74	70.30	68.88	67.34	67.04	63.73	65.94	66.43
	114.98	96.20	94.58	90.57	89.59	83.51	84.68	85.18

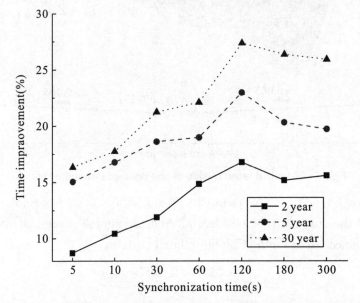

Figure 4. Time improvement of the fixed synchronization time under rainfall with different return periods

The surface runoff process under different scenarios is presented in Fig. 5. There is no difference in the runoff process under the different synchronization time, it indicates that the surface runoff is not sensitive to the synchronization time under different rainfall.

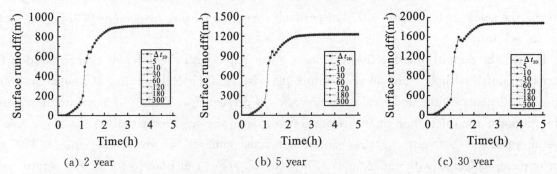

Figure 5. The effect of synchronization time on surface runoff under different rainfall return period

There is almost no difference in the discharge process at different synchronization time, except for the peak runoff(Fig. 6). This means that the synchronization time will not have a serious impact on the discharge process. However, it can be seen from the details that the discharge peak is significantly reduced when the synchronization time exceeds 180 s. The discharge peak is flattened when the synchronization time is too long, as the drainage process is averaged within a synchronization time. The difference between Δt_{2D} and fixed synchronization time(Fig. 7) also confirms once again that the error would increase significantly as the synchronization time exceeds 180s. It is not significant that the effect of different synchronization time under 2, 5 and 30 years retune periods. The discharge volume under different scenarios also be calculated, it only has an error of about 0.1‰, indicating that the synchronization time has a very small effect on the discharge volume.

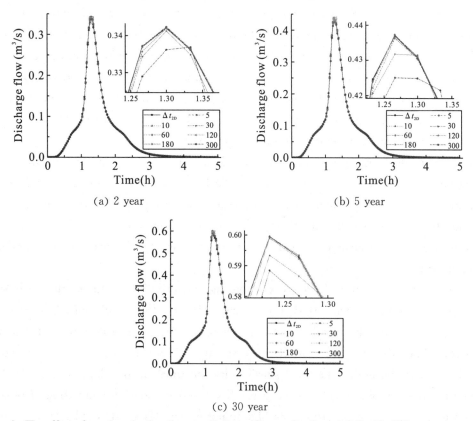

Figure 6. The effect of synchronization time on discharge flow under the rainfall with different return periods

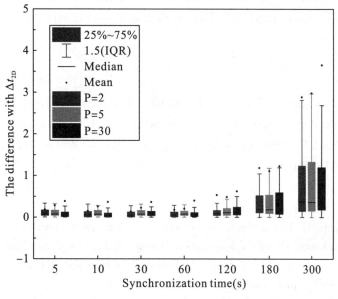

Figure 7. The difference between the Δt_{2D} and different synchronization time

4 Conclusions

In this work, a new coupled model that coupled the SWMM and GSWM is proposed to investigate the influence of different synchronization time. The coupled model can conveniently simulate flood and inundation process with fully hydrodynamic method. The rainfall-runoff-drainage-inundation process in an idealized urban catchment with drainage system is simulated under the rainfall with recurrence period in 2, 5 and 30 years. Then the inundation volume process, sewer discharge process, outfall volume and calculation time are compared thought visual and relative error analysis that applies to the test case. Based on the results of the urban case, the following conclusions can be drawn.

(a) The time improvement of fixed synchronization time can be improved by 8.70%~27.37% in different scenarios compared with the method using the Δt_{2D}. Moreover, the improvement effort increases as the rainfall return period increases, and increases with the extension of the synchronization time when the synchronization time less than 120 s.

(b) The error of discharge process between the Δt_{2D} and fixed synchronization time increases as the synchronization time increases, especially the discharge peak is significantly reduced after the synchronization time is greater than 180s. However, the difference between the Δt_{2D} and fixed synchronization time is very small when the synchronization time is less than 120 s.

(c) There is almost no change in surface runoff process and drainage volume under different synchronization time.

In summary, computational efficiency can be significantly improved as the extension of the synchronization time with a small loss of accuracy. Hence, the synchronization time cannot be extended indefinitely, otherwise it will affect the drainage process, especially when the sewer system is complicated. And the surface runoff process and drainage volume are not affected by synchronization time. Therefore, the fixed synchronization time can be adopted in the 1D-2D coupled model to improve the computational efficiency for flood and inundation simulating. Considering the simulation accuracy, the synchronization time should not exceed 120s. In the idealized urban test case, the circumstances are relatively simple. If the cases are complicated, the synchronization time may be reduced, which requires further research. This work could provide reference for model application in the future.

Acknowledgements

This work is partly supported by the National Key Research and Development Program of China (2016YFC0402704), National Natural Science Foundation of China (Grant No. 51609199) and Visiting Researcher Fund Program of State Key Laboratory of Water Resources and Hydropower Engineering Science(2016HLG01).

References

[1] Chen W, Huang G, Zhang H. Urban inundation response to rainstorm patterns with a coupled hydrodynamic model: A case study in Haidian Island, China[J]. Journal of Hydrology, 2018, 564: 1022−1035.

[2] Echeverribar I, Morales-Hernández M, Brufau P. Analysis of the performance of a hybrid CPU/GPU 1D-2D coupled model for real flood cases[J]. Journal of Hydroinformatics, 2020, 22.

[3] Fan Y, Ao T, Yu H. A coupled 1D-2D hydrodynamic model for urban flood inundation [J]. Advances in Meteorology, 201: 2819308.

[4] Hou J, Liang Q, Zhang H. An efficient unstructured MUSCL scheme for solving the 2D shallow water equations[J]. Environmental Modelling & Software, 2015, 66: 131−152.

[5] Hou J, Simons F, Mahgoub M. A robust well-balanced model on unstructured grids for shallow water flows with

wetting and drying over complex topography[J]. Computer Methods in Applied Mechanics and Engineering, 2013, 257: 126−149.

[6] Hou J. Effects of the temporal resolution of storm data on numerical simulations of urban flood inundation[J]. Journal of Hydrology, 2020, 589: 125100.

[7] Hsu M H, Chen S H, Chang T J. Inundation simulation for urban drainage basin with storm sewer system[J]. Journal of Hydrology, 2000, 234(1): 21−37.

[8] Li D. An efficient method for approximately simulating drainage capability for urban flood[J]. Frontiers in Earth Science, 2020, 8.

[9] Liang Q, Marche F. Numerical resolution of well-balanced shallow water equations with complex source terms[J]. Advances in Water Resources, 2009, 32(6): 873−884.

[10] Rossman, L. A. Storm water management model user's manual[M]. US EPA: 2009.

[11] Salvan L, Abily M, Gourbesville P J. Drainage System and Detailed Urban Topography: Towards Operational 1D-2D Modelling for Stormwater Management[J]. Procedia Engineering, 2016, 154: 890−897.

[11] Wallingford. Wallingford Software[M]. Oxfordshire: NfoWorks, 2006.

[12] Wang C, Li C, Shen R. Precipitation variation characteristics analysis of Xiji county in recent 67 years[J]. China Rural Water and Hydropower, 2020, 9(1007−2284): 152−156, 162.

[13] Xia X, Liang Q, Ming X. A full-scale fluvial flood modelling framework based on a High-performance Integrated Hydrodynamic Modelling System(HiPIMS)[J]. Advances in Water Resources, 2019, 132: 103392.

[14] Zhang W. Assessment of flood inundation by coupled 1D/2D hydrodynamic modeling: a case study in mountainous watersheds along the coast of southeast China[J]. Water, 2020, 12: 822.

Longitudinal Distributions of Phytoplankton of the Pengxi River in Low Water Level

Jiaren Chen[1], Zhaowei Liu[1], Yongcan Chen[1,2], Xiao Chen[1], Jiabei Liu[1]

[1] State Key Laboratory of Hydroscience and Engineering, Tsinghua University, Beijing 100084, China

[2] Southwest University of Science and Technology, Mianyang 621010, China

Abstract

After the impoundment of the Three Gorges Reservoir(TGR), the velocity of the tributaries were reduced, and the problems of eutrophication became more serious. Algal blooms have been found in the Pengxi(Xiaojiang) River(PXR) frequently. The Modao River(MDR) is close to its geographical location, water environment and watershed characteristics, while there are few reports of blooms. To study the longitudinal distributions and main factors of phytoplankton in the PXR and the MDR tributaries in flood season, algae and water quality samples were collected along the PXR and the MDR in September 2020. Results showed that the concentrations of nutrients in two tributaries were similar. Chlorophyll a(Chla) and cells density of algal in the PXR were higher than MDR. The dominant species in the backwater area of the PXR were Cyanophyta while in the MDR were Bacillariophyta. The composition of the algal of the Hanfeng Lake(HFL) overflow is similar to the water samples in the middle and lower reaches of the PXR. It was first found in the PXR that Cylindrospermopsis became the dominant species in the middle and lower reaches. *Cylindrospermopsis raciborskii* showed obvious low light adaptability and competitiveness in the vertical direction. Compared with the MDR, the overflow of HFL may be the important factor for the algal blooms of the PXR in flood season.

Keywords: Low water level; Pengxi River; Functional Groups; *Cylindrospermopsis raciborskii*

1 Introduction

The Three Georges Reservoir located at the middle reach of Yangtze River, is the largest man-made reservoir in China with a total capacity of 39.3 billion m³, surface area 1084 km² (Huang et al., 2006). TGR has functions such as flood control, hydro-power generation. However, the TGR has brought tributary bays environmental problems like harmful algal blooms (HABs). As the flow velocity slows down, water age extend and water depth increases, algal blooms have been observed in several tributaries every year(Qiu et al., 2008).

Algal blooms can be influenced by several factors like nutrients(Elser and Hassett, 1994; Litchman et al., 2004; Daines et al., 2014; Domingues et al., 2011), environmental factors(light, water temperature, radiation)(Huisman, 1999; Robarts, 1987; Edwards et al., 2016; Noyma et al., 2020), hydrodynamics factors(hydraulic retention time, mixing, stratification)(Peretyatko et al., 2007; Mahadevan et al., 2012; Townsend et al., 1992) and biological factors(species interaction between phytoplankton)(Huisman, 2004).

After the construction of the Hanfeng Dams, the HFL became the source of the PXR, which turned into special eutrophic lake-to-tributary systems. HFL has been observed eutrophic as well as HABs(Hu et al., 2019). At the same time, algal blooms also often occurred in Gaoyang Lake, which is in the lower reaches of the PXR(He et al., 2016). The estuaries of the PXR and MDR were about 30km, however, the backwater area of the MDR was rarely reported HABs.

Cylindrospermopsis was first reported in Java Island, Indonesiain 1912(Komárková, 1998), and was once classified as tropical to subtropical species(Briand et al., 2004). *C. raciborskii* could produce cylindrospermopsin(CYN) and caused 148 people poisoned in Australia(Byth, 1980). As the climate changes like greenhouse effect and its strong adaptability, such as resistance to weak dituebance and the ability of vertical regulation like some Cyanophyta(Shafik, 2003; Xue, 2020), adaptability of wide range of temperature and light, which can keep alive in 12℃~35℃ and 30~2500 $\mu mol/(m^2 \cdot s)$(Dokulil, 2016; Kovács, 2016; Dokulil, 1996; Bouvy, 1999;). *C. raciborskii* would be in rapid growth when the water temperature is over 25℃(Recknagel, 2014). And in light-, nitrogen deficient conditions, *C. raciborskii* demonstrated a competitive advantage with other species (Pierangelini, 2015). Recent years, *C. raciborskii* which can produce CYN was reported in several provinces and reservoirs in China as an invasive species(Wu et al., 2011; Lei et al., 2014).

Therefore, in this research, we set sampling sites and analyzed the longitudinal variations of phytoplankton compositions. To(1) analyze the relationships of the HFL-to-PXR system;(2) explore the main factors of the new dominant species *C. raciborskii.* in the PXR.

2 Materials and Methods

2.1 Study area

The Pengxi River, the largest tributary on the north side of the TGR, drains a watershed of 5172.5 km². Its annual runoff could be 2.7 billion m³. It located between latitude 30°56′~31°42′ N, 107°56′~108°54′ E in the middle reach of the TGR. It has a length over 182.4 km, while it's about 68 km long from estuary after the Hanfeng Dam was built. The backwater reach extend to 57.5 km when the water level of the TGR is 145 m(Li, 2009), which extended to Yanglu County. There were three branches in the area, the Puli River, the Shuangshui River and the Dongxi River. The Gaoyang Lake, which is the largest lake connected to the Yangtze River directly of about 5 km², lies in the lower reaches of the PXR.

2.2 Sampling and analysis

Sampling was done on September 11 and 12, 2020 in PXR and MDR. We set 15 sampling sites along the PXR from the HF Dams to estuary and 7 sampling sites for the MDR. The distance from sampling sites to each corresponding estuary were as Table 1. And the location of each site is showed in Fig. 1. The water of PX00 was taken from the spillway of HF Dams, which was the surface water of HFL near the dams. While the water depth of MD00 was less than 1 m when sampling. Samples were taken from the sub-surface(0.5 m) from near the talweg of the river. Algal samples were immediately preserved in Lugol's solution, and each sample needed 1 liter water at least. More than 48 h of standing without any disturbance, the supernate of samples was siphoned and we remained 50 mL mixed liquids for further quantitative identification. Chla and nutrients samples were preserved in low temperature without light and then sent to refrigerator, the latter also needed to keep their pH less than 2. Water temperature and conductivity were measured by YSI CTD.

Phytoplankton species were identified and counted with the help of Nikon E100 biological microscope(10×40) and the software Shineso S300 according to Hu and Wei(2006). Chla and nutrients samples were tested in laboratory. Main genera were classified into corresponding functional groups Reynolds et al.(2006), Borics et al.(2007) and Padisák et al.(2009). We classified dominant genera into corresponding groups to analyze their features.

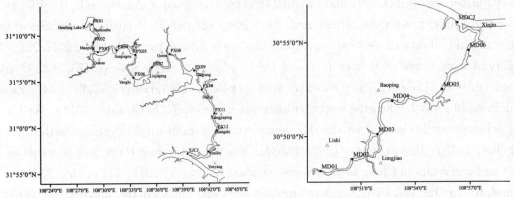

Figure 1. The location of sampling sites(a) is the PXR and(b) is the MDR

Table 1. The distances of each sampling site to the corresponding estuary

Sampling Sites	Distance (km)	Sampling Sites	Distance (km)	Sampling Sites	Distance (km)	Sampling Sites	Distance (km)
PXR		PX05	48	PX11	14	MD02	24
PXHF	67	PX06	42.1	PX12	10.5	MD03	20.3
PX01	64.5	PX07	37	PX13	5.22	MD04	15.7
PX02	60.1	PX08	29.7	PXCJ	0	MD05	9.55
PX03	57.6	PX09	23.7	MDR	—	MD06	4.8
PX04	52.7	PX10	19.6	MD01	27.9	MDCJ	0

3 Results

3.1 Environmental parameters

Table 2 illustrated the environmental parameters about the two rivers, each cell showed "the minimum, the average, the maximum". All the units of these parameters were "mg/L" except the Chla, which is "μg/L". The water temperature was fit for several phyla algal(Tilman et al., 1986; Evans, 1997; Wang Z H et al., 2005). About the nutrients in the PXR, the dissolved total phosphorus(DTP) and nitrogen(DTN) was over the threshold of algal blooms(DTP≥0.02 mg/L and DTN≥0.2 mg/L)(Cui, 2017).

Table 2. The environmental parameters

	DTN	DTP	WT	NH_4^+—N	NO_3^-—N	Chla
PXR	0.69, 1.06, 2.13	0.048, 0.059, 0.077	27.8, 28.5, 29.2	0.21, 0.38, 0.76	0.07, 0.33, 1.15	10, 31, 56
MDR	0.97, 1.19, 1.64	0.024, 0.046, 0.035	27.0, 28.1, 29.3	0.23, 0.33, 0.49	0.37, 0.56, 0.79	2, 15, 31

3.2 Phytoplankton spatial distribution

Fig. 2(a) illustrated the algal cells density for each site. For the PXR, samples in the upper reaches of have cell density over 5×10^7 cells/L while over 10^8 in the lower reaches in the PXR. We found seven phyla, Cyanophyta, Chlorophyta, Bacillariophyta, Dinophyta, Cryptophyta, Charophyta and Euglenophyta, and 65 genera in total. Among all the samples, filamentous Multicellular cyanobacteria accounted for alager portion of Cyanophyta, which made Cyanophyta contribute the most to the total abundances of cells density, followed by Bacillariophyta and Chlorophyta. Cyanophyta have an absolute advantage in the lower reaches of the PXR in both cells density and biomass. However, there were few algal of the source and the estuary of the MDR, and Bacillariophyta dominant in the middle reaches according to the biomass. The peak of the density was PX09 up to 3.77×10^8 cells/L. The biomass of PX09, PX10 and MD05 were all over 2×10^4 μg/L as Bacillariophyta contributed for the two sites.

The abundances of phytoplankton varied in these sites. PXHF and PX06 were similar at the phylum level in both density and biomass. The sample of PXHF was collected from the Hanfeng Lake surface water which overflow through the dam spillway. The dominant genera, *Cylindrospermopsis* spp. and *Ulnaria* (*Synedra*) spp., were also similar for the two samples. The total density declined by only 9.2% and the biomass reduction is 9.4%. We found that from PXHF(67 km) to PX01 (64.5 km), the proportion of Bacillariophyta increased sharply[Fig. 2(b)]. Then, the percentage of Bacillariophyta decreased while the percentage of Cyanophyta increased along the PXR. Cyanophyta dominated from PX06 to PX13, and the proportion of Bacillariophyta became to increase from PX10, which means the lower reaches of the PXR may affected by the Yangtze River. While the Bacillariophyta dominated in the middle and lower reaches(MD03-MD05) of the MDR.

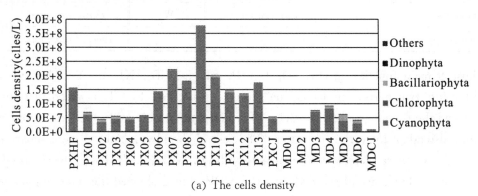

(a) The cells density

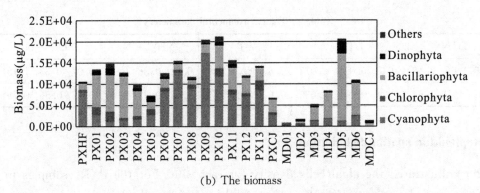

(b) The biomass

Figure 2. Phytoplankton abundance of the PXR and the MDR

3.3 Longitudinal distributions of functional groups

Table 3 lists the functional groups of algal genera which appeared in several sites or took a great proportion in some site.

Table 3. The genera and the corresponding functional groups of the PXR and the MDR

Genera	Functional Groups	Genera	Functional Groups	Genera	Functional Groups
Cyanophyta		**Bacillariophyta**		**Chlorophyta**	
Arthrospira spp.	S2	*Aulacoseira* spp.	B	*Chlamydomonas* spp.	X2
Cylindrospermopsis spp.	S_N	*Cyclotella* spp.	A	*Chlorella* spp.	X1
Leptolyngbya spp.	T_C	*Fragilaria* spp.	P	*Closterium* spp.	P
Limnothrix spp.	S1	*Melosira* spp.	P	*Coelastrum* spp.	J
Merismopedia spp.	L_O	*Navicula* spp.	P	*Cosmarium* spp.	N
Microcystis spp.	M	*Nitzschia* spp.	D	*Pediastrum* spp.	J
Oscillatoria spp.	MP	*Skeletonema* spp.	D	*Scenedesmus* spp.	J
Planktothrix spp.	S1	*Ulnaria (Synedra)* spp.	D	**Dinophyta**	
Pseudanabaena spp.	S1	**Euglenophyta**		*Peridinium* spp.	L_O
Cryptophyta		*Euglena* spp.	W1	*Ceratium* spp.	L_O
Cryptomonas spp.	Y			*Gymnodinium* spp.	L_O

We chose 9 main function groups, D, L_O, P, S1, S_N, T_C, X1 and Y. Then we divided all phytoplankton into two parts: the lacustrine group (L_O, P, S1, S_N, T_C) which just fit for lenitic environment and the fluvial group (D, J, X1, Y) which can not only multiplied in eutrophic lakes and reservoirs but also adapt for turbulent water. The rest of the phytoplankton were separated into the Others. Considering there were few algal in sites MD00, MD01 and MDCJ, we ignored them and analyze the relative abundance of the PXR and the MDR. Fig. 3 showed the relative abundance of the three groups.

There was obvious transition zone from PX01 to PX07, the lacustrine group continuously increased. In contrast, the percentage of the fluvial group decreased obviously according to Fig. 3(b). The tendency came to the peak in both density and biomass in PX09 (Gaoyang Lake). Then the percentage of the fluvial group began to rise along the river. The proportion of the fluvial group dominated from MD03 to MD06 along the MDR.

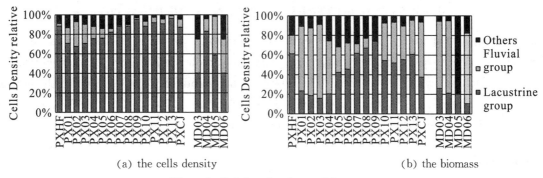

(a) the cells density (b) the biomass

Figure 3. Relative abundance of the groups

4 Discussion

4.1 Special phytoplankton structures in the HFL-to-PXR system

As the sampling time was September 11 and 12. We chose 14 days(Reynolds, 2006) as the time scale to analyze the hydrodynamic condition(Fig. 4). The water level was 155m when sampling. According to Li's(2006) one dimension hydrodynamic model, the average velocity of the cross-section from the estuary to PX09(25 km, Gaoyang Lake) was about 0.1 m/s. And the velocity increased up to 0.6 m/s from 25 km to 45 km. Thus, the HFL-to-PXR system was lake-river-lake. The HFL was the main phytoplankton source for phytoplankton. The relative abundance of the lacustrine group was declined between PXHF(67 km) and PX03(57 km), which corresponded to the end of the backwater area, as the transition area (Fig. 3). Then the abundance began to increase until PX09, as the lacustrine group was suitable in weak turbulent water(Mitrovic, 2011). The lower reaches(PX09-PXCJ) may influenced by the Yangtze River, which needed further verification.

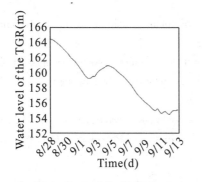 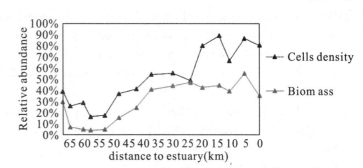

Figure 4. Water level of the TGR for two weeks   Figure 5. Relative abundance of C. raciborskii along the PXR

4.2 The Factors of C. raciborskii bloom in the lower reaches of the PXR

In the study, nutrients and water temperature were all suitable for C. raciborskii and some other Cyanophyta(Table 2), which were not the limitation. Along the river, relative abundance of C. raciborskii were similar with the lacustrine group in the upper and middle reaches(to PX09). C. raciborskii was weak in resistance to flushing(Reynolds, 2006).

While in the vertical direction. The average depth of photic zone of the PXR in flood season was

(3.27±0.18) m, or calculated by the empirical formula(Fang et al., 2010),

$$SD = 0.2555D_{eu} - 0.0725 (R^2 = 0.8256, N = 115, P \leqslant 0.005) \qquad (1)$$

where SD is the transparency, D_{eu} is the depth of photic zone. In PX10, $SD = 0.6$ m, thus $D_{eu} = 2.63$ m. So the depth was no more than 3.45 m. The two samples in PX10 were collected from 0.5 m and 3.5 m, which below the photic zone. The relative density abundance both occupied the percentage of about 80%. The relative biomass abundance increased from 42.5% to 63.7%. The cells density increased by 46%. *C. raciborskii* adapted for light deficient environment in the PXR.

5 Conclusion

With the study, we found that water from the HFL was the main source of phytoplankton for the PXR. The functional groups can help analyze the hydrodynamic condition. As for the hydrodynamic condition near the estuary, it needs further research to know how the Yangtze River influenced the lower reaches of the PXR.

Compared the two rivers, nutrients and water temperature were not the limitation for algal growth in summer. Turbulence and light were the main factor of *C. raciborskii* blooms in the middle and lower reaches in flood season. The former restrained the growth, while light deficient environment help *C. raciborskii* occupy the competitive advantage.

Overall, the risk of *C. raciborskii* blooms in the PXR, it may replace *Microcystis*, which was the most common HABs species, in flood seasons.

Acknowledgements

This work was supported by National Key Research and Development Program of China(No. 2016YFC0502204).

References

[1] Borics G, Várbíró G, Grigorszky I, et al. A new evaluation technique of potamo-plankton for the assessment of the ecological status of rivers[J]. Large Rivers, 2007, 17(3-4): 466-486.

[2] Bouvy M, Molica R, De Oliveira S, et al. Dynamics of a toxic cyanobacterial bloom(Cylindrospermopsis raciborskii) in a shallow reservoir in the semi-arid region of northeast Brazil[J]. Aquatic Microbial Ecology, 1999, 20(3): 285-297.

[3] Briand J F, Leboulanger C, Humbert J F, et al. Cylindrospermopsis raciborskii(cyanobacteria) invasion at mid-latitudes: selection, wide physiological tolerance, or global warming? [J]. Journal of Phycology, 2004, 40(2), 231-238.

[4] Byth S. Palm Island mystery disease[J]. Medical Journal of Australia, 1980, 2(1): 40-42.

[5] Cui Y J. The Sensitive Ecological Dynamic Processes and their Simulations of Algal Growth of Xiangxi Bay in the Three Gorges Reservoir[D]. Wuhan: Wuhan University, 2017.

[6] Daines S J, Clark J R, Lenton T M. Multiple environmental controls on phytoplankton growth strategies determine adaptive responses of the N:P ratio[J]. Ecology letters, 2014, 17(4): 414-425.

[7] Dokulil M T, Mayer J. Population dynamics and photosynthetic rates of a Cylindrospermopsis-Limnothrix association in a highly eutrophic urban lake, Alte Donau, Vienna, Austria[J]. Algological Studies/Archiv für Hydrobiologie, 1996, 83: 179-195.

[8] Dokulil M T. Vegetative survival of Cylindrospermopsis raciborskii(Cyanobacteria) at low temperature and low light[J]. Hydrobiologia, 2016, 764(1): 241-247.

[9] Domingues R B, Anselmo T P, Barbosa A B, et al. Nutrient limitation of phytoplankton growth in the freshwater tidal zone of a turbid, Mediterranean estuary[J]. Estuarine, Coastal and Shelf Science, 2011, 91(2): 282-297.

[10] Edwards K F, Thomas M K, Klausmeier C A, et al. Phytoplankton growth and the interaction of light and temperature: A synthesis at the species and community level[J]. Limnology and Oceanography, 2016, 61(4): 1232-1244.

[11] Elser J J, Hassett R P. A stoichiometric analysis of the zooplankton-phytoplankton interaction in marine and freshwater ecosystems[J]. Nature, 1994, 370(6486): 211-213.

[12] Evans J H. Spatial and seasonal distribution of phytoplankton in an African Rift Valley lake(L. Albert, Uganda, Zaire)[J]. Hydrobiologia, 1997, 354(1): 1-16.

[13] Fang F, Zhou H, Li Z, et al. Spatiotemporal variations of euphotic depth and its causing factors in Xiaojiang River backwater area of Three Gorges[J]. Advances in Water Science, 2010, 21: 113-119.

[14] He B, Li Z, Feng J, et al. Study on the phosphorus-algal ecological modelling during high water level period in Lake Gaoyang of Pengxi River, Three Gorges Reservoir[J]. Journal of Lake Sciences, 2016, 28(6):1244-1255.

[15] Hu H J, Wei Y X. The Freshwater Algae of China-Systematics, Taxonomy and Ecology[M]. Beijing: Sci. Press, 2006.

[16] Hu X, Yin J, Tang H, et al. Phytoplankton community structure and its redundancy correlation with environmental factors in Hanfeng Lake during the early stage of the operation of Wuyang Regulating Dam[J]. Freshwater Fisheries, 2019, 49(6): 48-55. (in Chinese)

[17] Huang Z L, Li Y L, Chen Y C, et al. Water quality prediction and water environmental carrying capacity calculation for Three Gorges Reservoir[M]. Beijing: China Water Power Press, 2006. (in Chinese)

[18] Huisman J, Jonker R R, Zonneveld C, et al. Competition for light between phytoplankton species: experimental tests of mechanistic theory[J]. Ecology, 1999, 80(1): 211-222.

[19] Huisman J, Sharples J, Stroom J M, et al. Changes in turbulent mixing shift competition for light between phytoplankton species[J]. Ecology, 2004, 85(11): 2960-2970.

[20] Komárková J. The tropical planktonic genus Cylindrospermopsis(Cyanophytes, cyanobacteria). In Anais do IV Congresso Latino-Americano de Ficologia(MTP Azevedo, DP Santos, LSC Pinto, M. Menezes, MT Fujii, NS Yokoya, PAC Senna & SMPB Guimarães, eds.)[J]. Sociedade Ficológica da America Latina e Caribe, São Paulo, 1998, 1: 327-340.

[21] Kovács A W, Présing M, Vörös L. Thermal-dependent growth characteristics for Cylindrospermopsis raciborskii (Cyanoprokaryota) at different light availabilities: methodological considerations[J]. Aquatic Ecology, 2016, 50 (4): 623-638.

[22] Lei L, Peng L, Huang X, et al. Occurrence and dominance of Cylindrospermopsis raciborskii and dissolved cylindrospermopsin in urban reservoirs used for drinking water supply, South China[J]. Environmental monitoring and assessment, 2014, 186(5): 3079-3090.

[23] Li Z. On the Habitats Transition & Succession Traits of Phytoplankton Assemblages in the Backwater Area of Xiaojiang River during Preliminary Operation Stage of the Three Gorges Reservoir[D]. Chongqing: Chongqing University, 2009.

[24] Litchman E, Klausmeier C A, Bossard P. Phytoplankton nutrient competition under dynamic light regimes[J]. Limnology and Oceanography, 2004, 49(2): 1457-1462.

[25] Mahadevan A, D'asaro E, Lee C, et al. Eddy-driven stratification initiates North Atlantic spring phytoplankton blooms[J]. Science, 2012, 337(6090): 54-58.

[26] Mitrovic S M, Hardwick L, Dorani F. Use of flow management to mitigate cyanobacterial blooms in the Lower Darling River, Australia[J]. Journal of Plankton Research, 2011, 33(2): 229-241.

[27] Noyma N P, Mesquita M C, Roland F, et al. Increasing Temperature Counteracts the Negative Effect of UV Radiation on Growth and Photosynthetic Efficiency of Microcystis aeruginosa and Raphidiopsis raciborskii[EB/OL]. [2021-02-04]. https://doi.org/10.1111/php.13377.

[28] Padisák J, Crossetti L O, Naselli F L. Use and misuse in the application of the phytoplankton functional classification: a critical review with updates[J]. Hydrobiologia, 2009, 621(1): 1−19.

[29] Peretyatko A, Teissier S, Symoens J J, et al. Phytoplankton biomass and environmental factors over a gradient of clear to turbid peri-urban ponds [J]. Aquatic conservation marine and freshwater ecosystems, 2007, 17 (6): 584−601.

[30] Pierangelini M, Stojkovic S, Orr P T, et al. Photo-acclimation to low light—Changes from growth to antenna size in the cyanobacterium Cylindrospermopsis raciborskii[J]. Harmful algae, 2015, 46: 11−17.

[31] Qiu G, Tu M, Ye D, et al. General investigation of eutrophication for tributaries in TGP reservoir area[J]. Yangtze River, 2008, 039(013): 1−4. (in Chinese)

[32] Recknagel F, Orr P T, Cao H. Inductive reasoning and forecasting of population dynamics of Cylindrospermopsis raciborskii in three sub-tropical reservoirs by evolutionary computation[J]. Harmful Algae, 2014, 31: 26−34.

[33] Reynolds C S. The ecology of phytoplankton[M]. Cambridge: Cambridge University Press, 2006.

[34] Robarts R D, Zohary T. Temperature effects on photosynthetic capacity, respiration, and growth rates of bloom - forming cyanobacteria[J]. New Zealand Journal of Marine and Freshwater Research, 1987, 21(3): 391−399.

[35] Shafik H M. Morphological characteristics of Cylindrospermopsis raciborskii(Woloszynska) Seenayya et Subba Raju in laboratory cultures[J]. Acta Biologica Hungarica, 2003, 54(1): 121−136.

[36] Tilman D, Kiesling R, Sterner R, et al. Green, bluegreen and diatom algae: taxonomic differences in competitive ability for phosphorus, silicon and nitrogen[J]. Archiv für Hydrobiologie, 1986, 106(4): 473−485.

[37] Townsend D W, Keller M D, Sieracki M E. Spring phytoplankton blooms in the absence of vertical water column stratification[J]. Nature, 1992, 360(6399): 59−62.

[38] Wang Z H, Chen J F, Xu N, et al. Dynamics on cell densities of diatom, dinoflagellate and relationship with environmental factors in Aotou area, Daya Bay, South China Sea[J]. Oceanologia Et Limnologia Sinica, 2005, 2: 186−192.

[39] Wu Z, Shi J, Xiao P, et al. Phylogenetic analysis of two cyanobacterial genera Cylindrospermopsis and Raphidiopsis based on multi-gene sequences[J]. Harmful Algae, 2011, 10(5): 419−425.

[40] Xue X G, Fang G H, Zou C J, et al. Diel vertical distribution patterns of Raphidiopsis raciborskii in Dashahe Reservoir[J]. Chinese Journal of Ecology, 2014, 39(7): 2348.

[41] Zhou W, Yuan X, Long A, et al. Different hydrodynamic processes regulated on water quality(nutrients, dissolved oxygen, and phytoplankton biomass) in three contrasting waters of Hong Kong[J]. Environmental monitoring and assessment, 2014, 186(3): 1705−1718.

Long-term Water Security of Metropolitan Regions: Assessment of Water Supply Systems

Walter Manoel Mendes Filho, Wilson Cabral de Sousa Junior, Paulo Ivo Braga de Queiroz

Instituto Tecnológico de Aeronáutica, São José dos Campos/SP, Brazil

Abstract

This paper presents the development of a generic reservoir model for systemic analysis of operational policies for water transfer between reservoirs in water crisis scenarios in metropolitan regions. The systemic understanding of the complex interactions triggered by these transfers in the water supply system of these regions, especially in the long-term horizon, is essential to define more efficient operational strategies for these systems. The System Dynamics approach is widely used in water resources management and global change analysis. The model developed uses this approach to analyze the consequences of operational strategies in water crisis scenarios. For application, the Cantareira system (basin of the rivers Piracicaba, Capivari, and Jundiaí) was considered, which supplies approximately 9 million inhabitants in the metropolitan region of São Paulo, Brazil. For the period between the years 2009 and 2016, which comprises the worst water crisis experienced in the region (2013 to 2015), a scenario of water transfer from the Jaguari reservoir (Paraíba do Sul river basin) was simulated, and a scenario applying a reduction factor to the system's target demand. Considering the volume stored 95% of the period in the system as reference, the Cantareira reserve increased by 16% with the transferred volume. This percentage reaches almost 40% by reducing the target demand by 10% since 2013. These results suggest that System Dynamics simulation can improve the assessment of operational strategies for water supply systems in metropolitan regions and under the effects of external drivers such as changing hydroclimatic and demand trends.

Keywords: Water supply; Operating strategies; Water security; Hydroclimatic trends; Integrated modelling

1 Introduction

Over the 21st century, climate change will most likely impact hydrological drought trends around the world, triggering temporary reductions in water availability, leading to water shortages, and consequently, societal responses to water shortages can result in a series of cascading effects (Di Baldassare et al., 2018; Wanders and Wada, 2015).

In water supply systems, traditional approaches to system planning and water resources management, do not consider the feedback relationships between the water and socio-economic dimensions in the water supply-demand cycle. According to Garcia et al. (2016), these approaches typically cannot provide insight into how different patterns of natural variability or human-induced changes can propagate through this coupled system, since there is no consensus on universally accepted laws of human behavior like the laws that exist for physical systems.

In South America, the southeastern region of Brazil suffered a severe water crisis in the years 2013 to 2015(Marengo et al., 2020; Nobre et al., 2016). As a strategy to face this crisis, after this period, the inter-basin connection of the Jaguari reservoir(operated by CESP—Energy Company of the State of São Paulo), located in the Paraíba do Sul river basin(PS), began to operate and the Atibainha reservoir(operated by SABESP—Water and Sanitation Company of the State of São Paulo) in the basin of the Piracicaba, Capivari, and Jundiaí rivers(PCJ), a component of the main water supply system of the São Paulo Metropolitan Region — RMSP, the Cantareira system.

Adopted for case study analysis, this inter-basin water transfer-supply was planned for the recovery of the stored volume of Cantareira reservoirs, the reduction of systemic risk in water supply in RMSP, and in the municipalities located downstream of the PCJ basin, in case of the severe drought (years 2013 to 2015) extended for prolonged periods, according to SABESP(2015).

The objective of the present study is to incorporate external drivers to water balance integrated with hydroclimatic and socio-economic trends, into a generic reservoir model. For this purpose, some main feedback relationships that can influence the supply-demand cycle in metropolitan regions are structured. In a regional context, these improvements are applied to the Cantareira system, considering the connection with the Jaguari reservoir(PS), dynamically updating the vulnerability associated with periods of higher reservoir dependence, adopting the System Dynamics(SD) approach (Forrester, 1961; Sterman, 2000).

2 Study Area

To establish the main mechanisms to be incorporated in the water balance structure, the study is delimited at the regional level(area of influence of the Jaguari-Atibainhainter-basin connection), focusing on the water balance, hydroclimatic and socio-economic trends, associated with the vulnerability of the water supply system.

In the hydrological dimension, the report SABESP(2015) highlights the Paraíba do Sul basin (PS), the basin of the Piracicaba, Capivari, and Jundiaí rivers(PCJ), and the Alto Tietê basin(AT), which receives water produced by the Cantareira System. In the socio-economic dimension, the system involves the metropolitan regions of São Paulo, Campinas, and Rio de Janeiro, besides other regions belonging to the PS and PCJ basins.

Operated by SABESP, a series of channels and reservoirs form the Cantareira System—among them the Atibainha(Fig.1). These structures capture and redirect water from some rivers in the PCJ to the AT basin, establishing the principal water supply system in the RMSP, and can serve approximately 9 million inhabitants(ANA, 2016; Nobre et al., 2016).

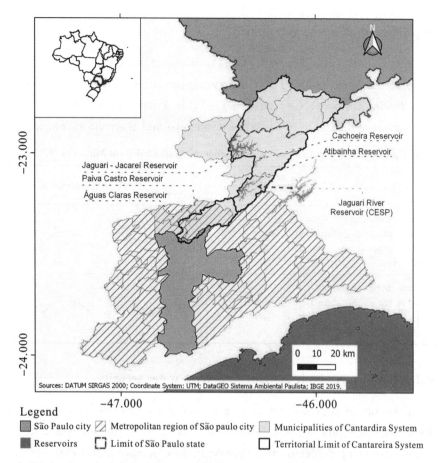

Figure 1. Regional context of the study area(Highlighted in the red arrow is the connection between the Jaguari and Atibainha reservoirs, in the Cantareira system)

The Jaguari reservoir, on the other hand, belongs to the PS basin(Fig. 1) and is managed by CESP, which includes power generation. According to SABESP(2015), the operational volume of Jaguari is 793 hm^3 (equivalent to 81% of the total useful volume of Cantareira) and its long-term average flow rate(MLT) is 28 m^3/s(63% of the average natural inflow of Cantareira), besides having a high detention time(10.8 months).

Destination of the largest portion of the flow captured in Cantareira(ANA, 2017), RMSP is located in the state of São Paulo in southeastern Brazil, approximately 600 km southwest of Rio de Janeiro and 80 km inland from the Atlantic Ocean(Fig. 1). Almost 60% of the population lives in the city of São Paulo, resulting in one of the highest population densities in the country, about 7220 inhabitants per square kilometer. The city accounts for 12% of Brazil's national GDP(\$1,648,870) and twice as much as Brazil's GDP per capita(Marengo et al., 2020).

3 Methodology

In this section, a generic reservoir model is developed for the systemic analysis of operational policies. The modeling strategy proposed by Jiang and Simonovic(2020) will allow assessing water supply systems. Involves a causal diagram of the long-term water security of metropolitan regions to map and analyze the main interactions between the components of the reservoir, and the development of the simulation model using a stock and flow diagram.

3.1 Causal Loop Diagram of a reservoir system

To structure the Causal Loop Diagram (CLD), combinations of positive and negative causal relationships can form feedback loops. There are two types of fundamental feedback loops: balance (B — negative) and reinforcement (R — positive). Table 1 includes a summary with the graphical notation of reinforcement and balance of causal relationships and their interpretation.

Table 1. Notation and polarity of causal relationships (adapted from Mirchi et al. 2012)

Connection	Causal relationship	Mathematical definition	Examples		
A ⌒+→ B	Any change in the state of A cause the state B to change in the same direction.	$\frac{\partial B}{\partial A} > 0$	Temperature	⌒+→	Evaporation
A ⌒−→ B	Any change in the state of A cause the state B to change in the opposite direction.	$\frac{\partial B}{\partial A} < 0$	Infiltration	⌒−→	Runoff

3.2 Operational policy of a reservoir system: long-term water security

The continuous supply of abundant water in a region in a water shortage can generate a message to users about its water development potential (Mirchi et al. 2012). Considering a long-term analysis, while water resources are being depleted, the increase in development and demand for water may cause a more severe shortage (reinforcement loop — R, dominates the balance loop — B).

These characteristics of the system obey the archetype of the type "Fixes that Backfire" (Fig. 2). According to Gohari et al. (2013), the persistent water shortage is mainly due to the presence of an unresolved reinforcement (R) loop, which creates a vicious supply-development-demand cycle.

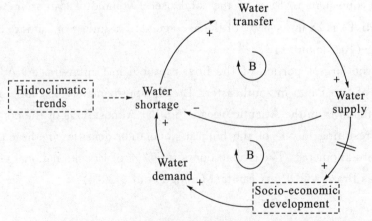

Figure 2. Under external drivers (hidroclimatic and socio-economic trends), the feedback loop with short-term water shortage reduction (B), and long-term water security (R). Double bars indicate presence of time delay

3.3 Stock and Flow Diagram of a reservoir system

Stocks (X) represent the state variable, indicated by boxes (Jiang and Simonovic, 2020; Sterman, 2000). These Stocks integrate flows (rates of change or derivatives), represented by pipes (dX) and (dZ). Auxiliary variables(Y) are used to break the flow equations into manageable segments with a clear meaning(Fig. 3).

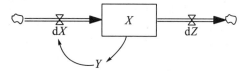

Figure 3. Notation adopted for the Stock and Flow Diagram(SFD)

The relationship between stocks and flows can be defined by Eqs.(1)(2)(3), as follows:

$$X(t) = \int_{t0}^{t} [dX - dZ] dt + X(to) \tag{1}$$

$$d(X) = f(Y(t)) \tag{2}$$

$$Y(t) = g(X(t)) \tag{3}$$

3.3.1 *Reservoir components*

The development of a generic reservoir model considers Jaguari-Jacarei reservoirs, Cachoeira, and Atibainha, as the single reservoir. The Cantareira operation considers the target demand(31 m³/s) for the RMSP, the minimum required flow downstream of the system, in the PCJ basin for the period (2009 to 2016). To incorporate external drivers, the reservoir model considers the simplified water balance equation used in Cantareira ANA(2016), according to Eq.(4).

$$V(t) = V(t-1) + I(t) - RS(t) - RP(t) \tag{4}$$

where $V(t-1)$ = volume stored at the end of the month$(t-1)$; $V(t)$ = volume stored at the end of month(t); $I(t)$ = turnout during month(t); $RS(t)$ = water volume to supply the RMSP in the month(t); $DP(t)$ = volume that follows downstream of Cantareira(PCJ basin) in a month(t), according to(ANA, 2017).

3.3.2 *Configuration of the simulation model and scenarios*

Assessing the operating policies at the reservoir, the conceptual model in Fig. 4 represents the characteristics adopted for development of a generic reservoir model.

Considering the water crisis scenario, two scenarios were considered, thus named: Water transfer-supply, and Reduction factor. In the first case, the flow contribution from the Jaguari reservoir(PS) to Cantareira was considered. And in the second, a reduction factor in the target demand is admitted. In both scenarios, the operational strategies are applied in the period(2013 to 2015). The reference scenario considered exclusively the inflow of the PCJ basin and the meeting of the target demand.

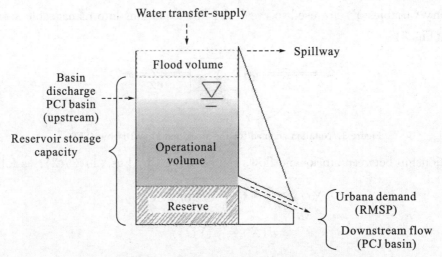

Figure 4. Conceptual model: water balance

"Basin discharge" and "Water transfer-supply" considers observed data [GET XLS DATA( )]. The downstream control conditions parameters are: "Minimum required flow downstream" ($=10$ m^3/s); spillway conditions; target demand ($=31$ m^3/s); and "Average flow" considered is ($=20$ m^3/s).

The reservoir behavior for the proposed operation scenarios was analyzed from simulation using the Software Vensim PLE Plus 6.4, the numerical approximation method adopted was the Runge-Kutta of fourth-order (details of the model are presented in Table 2).

4 Results

Based on these configurations, the proposed reservoir model is presented in Fig. 5. The variables "Water transfer-supply" and "Reduction factor" are highlighted. They were incorporated to analyze short and long-term water crisis scenarios. They can be activated as result of both climate variability and changes in demand patterns. In this way, the proposed model can assess water supply systems. It involves a long-term water security of metropolitan regions, the latter with periods of greater dependence due to long periods of flow deficit.

The loops represent the main interactions considered to analyze the reservoir. Loop B1 represents the control conditions for the "Release" (RMSP), as a function of "Total inflow", "Reservoir" storage, and "Downstream flow". Loop B2 represents the conditions allowed for spillway scenario. And the B3 loop establishes the relations to control the "Downstream flow", as a function of the "Minimum required flow", "Average flow" (when the storage does not present critical conditions), and "Flood volume" ($=3,396$ hm^3). In this case, it was adopted as a spill scenario all situations where the volume exceeds the "Reservoir storage capacity" ($=149,245$ hm^3).

The "Basin discharge" variable can also be influenced by climate variability, triggering changes in the storage pattern of the reservoirs.

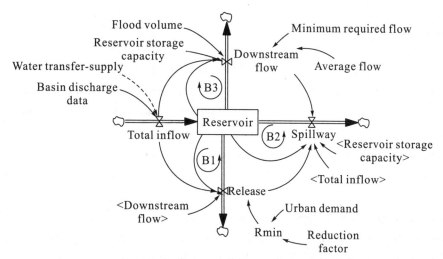

Figure 5. Reservoir model. Incorporated mecanisms for sistemic analysis(hydroclimatic and socioeconomic trends)

For the "water transfer-supply" scenario the transfer flow to the system was activated. With this operational policy, adopting as reference the volume stored 95% of the time, the reserve would be increased by 16%.

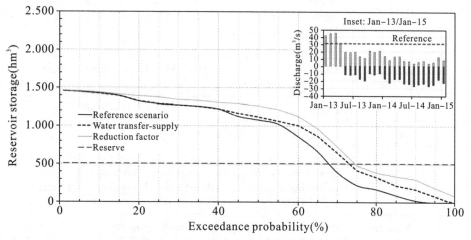

Figure 6. Duration curve for simulated scenarios

Note: Both scenarios were started in January 2013, 4 months before the system presented deficits in the supply-demand balance(considering the target demand, 31 m³/s).

Although this result represents a significant improvement in water crisis scenarios, as previously discussed(Mirchi et al. 2012; Gohari et al. 2013; Di Baldassare et al. 2018), it is postulated that there are long-term dynamics that should be considered in the supply-demand cycle and the reservoir effect. In this sense, the "Reduction factor" scenario, considering a 10% reduction in the target demand(31 m³/s), allowed an increase of almost 40% of the technical reserve by reducing the target demand by(=10%).

It is noteworthy that the discussion and identification of water resource allocation strategies were not the objects of this article. Therefore, it was considered for both scenarios a flow parceling, where an average of 15% of the "Total inflow" follows downstream(PCJ), and the rest is directed to the RMSP. (Configurations are presented in Table 2).

5 Conclusions

In the recent scenario of prolonged droughts experienced in several regions around the world, there is an important gap of systematization in the interface between integrated management of water resources and vulnerability to external drivers(hydroclimatic and socio-economic trends).

The development of integrated models for reservoir management, adopting the SD simulation can provide a systemic analysis in metropolitan regions and subsidize the future steps. This structuring allows robust numerical simulations, based on water balance components for complex and dynamic systems.

The results suggests that the systemic approach allows incorporating particularities of the new spatiality promoted by the inter-basin connection of reservoirs in different regions, besides improving the ability to deal with uncertainties associated with external drivers.

Table 2. Main model variables and their corresponding values and units

Variable	Type	Calculation	Unit
Downstream flow	Auxiliary	IF THEN ELSE((Reservoir+Total inflow)>=(Reservoir storage capacity-Flood volume), MAX((0.15 * Total inflow), Average flow), MAX((0.15 * Total inflow), Minimum required flow))	m^3/Month
Release	Auxiliary	IF THEN ELSE(Rmin <= Downstream flow, 10, MIN(Reservoir + Total inflow-Downstream flow, Rmin-Downstream flow))	m^3/Month
Reservoir	Level	INTEG(Total inflow-Downstream flow-Release-Spillway)	m^3
Rmin	Auxiliary	(Urban demand) * Reduction factor	m^3/Month
Spillway	Auxiliary	IF THEN ELSE((Reservoir + Total inflow-Downstream flow-Release)>Reservoir storage capacity, (Reservoir+Total inflow-Downstream flow-Release-Reservoir storage capacity), 0)	m^3/Month
Total inflow	Auxiliary	Basin discharge data+"Water transfer-supply"	m^3/Month
Urban demand	Auxiliary	RANDOM NORMAL(minimum=6.5536e+007, maximum=9.5536e+007, mean=8.5536e+007, sd=1.536e+006)	m^3/Month

Acknowledgements

The research is developed with support from the Coordenação de Aperfeiçoamento de Pessoal de Nível Superior-Brazil(CAPES)-Funding Code 001.

References

[1] ANA Agência Nacional de Águas. Resolução Conjunta ANA/DAEE N° 925 de 29 de maio de 2017[J]. Dispõe sobre as condições de operação para o Sistema Cantareira-SC, 2017.

[2] ANA Agência Nacional de Águas. Dados de referência acerca da Outorga do Sistema Cantareira [J]. Brasília, DF, 2016.

[3] Di Baldassarre G, Wanders N, AghaKouchak A, et al. Water shortages worsened by reservoir effects[J]. Nature Sustainability, 2018, 1(11):617−622.

[4] Forrester J W. Industry dynamics[M]. Massachusetts: Cambridge, 1961.

[5] Garcia M, Portney K, Islam S. A question driven socio-hydrological modeling process[J]. Hydrology and Earth System Sciences Discussions, 2016,12(8):8289−8335.

[6] Gohari A, Eslamian S, Mirchi A, et al. Water transfer as a solution to water shortage: a fix that can backfire[J]. Journal of Hydrology, 2013,491:23−39.

[7] Jiang H, Simonovic S P, Yu Z. A system dynamics simulation approach for environmentally friendly operation of a reservoir system[J]. Journal of Hydrology, 2020,587:124971.

[8] Marengo J A, Alves L M, Ambrizzi T, et al. Trends in extreme rainfall and hydrogeometeorological disasters in the Metropolitan Area of São Paulo: a review[J]. Annals of the New York Academy of Sciences, 2020, 1472(1): 5−20.

[9] Nobre C A, Marengo J A, Seluchi M E, et al. Some characteristics and impacts of the drought and water crisis in Southeastern Brazil during 2014 and 2015[J]. Journal of Water Resource and Protection, 2016, 8(2): 252−262.

[10] Mirchi A, Madani K, Watkins D, et al. Synthesis of system dynamics tools for holistic conceptualization of water resources problems[J]. Water resources management, 2012,26(9):2421−2442.

[11] SABESP. Estudo de Impacto Ambiental e Relatório de Impacto Ambiental-EIA/RIMA para a Interligação entre as Represas Jaguari e Atibainha. São Paulo: Sabesp. v. 1. [EB/OL]. [2021−06−07]. http://site.sabesp.com.br/site/interna/Default.aspx?secaoId=548.

[12] Sterman J. Business dynamics[M]. Irwin:McGraw-Hill, 2000.

[13] Wanders N, Wada Y. Human and climate impacts on the 21st century hydrological drought[J]. Journal of Hydrology, 2015, 526:208−220.

Modelling Sediment Transport Capacity of Loessial Slopes Based on Effective Stream Power

Chenye Gao[1,3], Jingwen Wang[2], Xile Liu[3], Shasha Han[4]

[1] Hangzhou Regional Center for Small Hydro Power, National Research Institute for Rural Electrification, Hangzhou 310012, China.

[2] State Key Laboratory of Water Resources and Hydropower Engineering Science, Wuhan University, Hubei 430072, China

[3] State Key Laboratory of Soil Erosion and Dryland Farming on the Loess Plateau, Northwest A&F University, Yangling 712100, China

[4] Yellow River Institute of Hydraulic Research, Zhengzhou 450004, China

Abstract

The sediment transport capacity is indeed not negligible because it provides a theoretical basis for accurate prediction of soil erosion. To obtain a prediction model for a variety of soils and evaluate its applicability, sandy loess and loess soil (d_{50} = 0.095 mm and $d_{50'}$ = 0.04 mm) were chosen in the indoor artificial simulated sediment transport experiments. The experimental slopes ranged from 7% to 38.4% and the unit discharges were adjusted from 0.00014 to 0.00526 m²/s. Moreover, this study combined the experimental data with cohesive soil and cohesionless sand from four scholars to analyze the response relationship between sediment transport capacity and each flow intensity parameter through dimensionless processing. Results showed that the effective stream power could be seen as an optimum indicator (R^2 = 0.9692). After considering the effective stream power and volume sediment concentration, this study derived a formula for calculating the sediment transport capacity. It was better than the ANSWERS (Areal Nonpoint Source Watershed Environment Response Simulation) model, improved WEPP (Water Erosion Prediction Project) model, Zhang's formula and Ali's model due to its superior applicability to cohesive soil and cohesionless sand. These findings lay a basis for establishing prediction models of soil erosion.

Keywords: Overland flow; Sediment transport capacity; Flow intensity parameters; Effective stream power; Volume sediment concentration

1 Introduction

Soil erosion, as a major environmental problem, has attracted public concern from all over the world (Ali et al., 2012). Although recently, this phenomenon in the Loess Plateau has been mitigated significantly as continuous soil and water conservation practices are implemented, analysing the water erosion process and sediment transport capacity helps to provide a reference for establishing soil erosion prediction models. The sediment transport capacity of overland flow mainly means the maximum flux of sediment transport under specific slopes and discharges (Zhang et al., 2009). It plays a critical role in defining areas of net erosion where the actual sediment concentration is below the transport capacity, and it also determines net deposition where the transport capacity is exceeded (Yu et al., 2015).

However, as for predicting sediment transport capacity, there were no consistent conclusions regarding the selection of flow intensity parameters, and exsiting models were aimed at a certain kind of soils like cohesive soil and sand, so the prediction model cannot be applied to each other. The experiment here is similar to that of Zhang et al. (2011), Wu et al. (2017), Ail et al. (2012) and Aziz and Scott(1989), therefore, this study synthesized the data of above four scholars. The objectives of this study were:

(1) to evaluate the relationship between dimensionless sediment transport capacity and dimensionless flow intensity parameters.

(2) to establish the sediment transport capacity for rill flow considering the volume sediment concentration and an optimum flow intensity parameter.

(3) to evaluate the applicability by comparing the prediction model in this study with four existing models.

2 Materials and Methods

2.1 Experimental set-up

Test soil materials were Shenmu sand loess from Shenmu county (110°30'E, 38°49'48"N) and Huangmian soil collected from Ansai County(109°19'23"E, 36°51'30"N). The particle size distribution of Shenmu sand loess(d_{50}=0.095mm) and Huangmian soil(d_{50}=0.04 mm).

The experimental set-up consisted of a adjustable test flume(4.5×0.3×0.1 m), a constant water tank, a sediment hopper and sediment collection devices(as shown in Fig. 1). The flow discharge was controlled by seven drain valves at the outlet of the constant water tank. A sediment hopper which was 0.5m distant from the upper side of this flume was used to provide the water flow with sediments. The rate of adding sediments was controlled by guide plates and rollers. Three kinds of slopes(10.5%, 15.6% and 20.8%) and seven unit discharges(0.14, 0.28, 0.42, 0.56, 0.69, 0.83 and 1.11×10^{-3} m^2/s) were investigated here. As for Huangmian soil materials, five types of slopes (7.0%, 10.5%, 14.1%, 17.6% and 21.3%) and six unit discharges(0.19, 0.39, 0.58, 0.75, 0.95 and 1.12×10^{-3} m^2/s) were set up in this test.

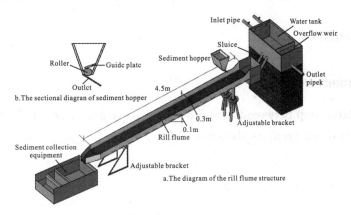

Figure 1. Schematic of the experimental set-up

2.2 Experiment procedures

Each time the velocity and depth were measured, and the deposition could be observed on the bed surface, at which time the water flow was assumed to have reached the transport capacity. When measuring the sediment transport capacity, sampling buckets with different numbers were used at the outlet of the flume. The sampling time determined by flow discharges and the size of containers could be recorded by a stopwatch. After around 30s, the sediment samples were sent to be deposited and dried, and the M value in Equation (1) was the average the mass of these sand samples. The sediment transportcapacity could be calculated by Formula (1):

$$T_c = \frac{M}{Tb} \tag{1}$$

where M is the mass of dried sand samples, kg. And the parameter b is the width of this test flume while T is the sampling time, s.

2.3 Theoretical analysis

This study combined experimental data with Zhang et al. (2011), Wu et al. (2016), Ali et al. (2012) and Aziz and Scott (1989), 348 groups in total. As different scholars chose diverse median particle sizes, it is necessary to make parameters dimensionless, as shown in Table 1.

Table 1. Non-dimensional formulas of flow intensity parameters

Flow intensity parameters	Formulas of flow intensity parameters	Dimensionless formulas of flow intensity parameters
Mean velocity	$V = \dfrac{Q}{bh}$	$V^* = V/\sqrt{(\gamma_s/\gamma - 1)gd_{50}}$
Flow stress force	$\tau = \rho g h S$	$\theta = \gamma h J / [(\gamma_s - \gamma)d_{50}]$
Unit stream power	$P = VS$	$P^* = VS/\omega$
Stream power	$W = \tau V$	$W^* = \theta V^*$
Effective stream power	$W_{eff} = W^{1.5}/h^{0.67}$	$W_{eff}^* = W^{*1.5}/(h/d_{50})^{0.67}$
Sediment transport capacity	$T_c = M/(Tb)$	$\Phi = T_c/[\gamma_s \sqrt{(\gamma_s/\gamma - 1)gd_{50}^3}]$

Notes: γ_s is particle density of sediment particles, γ_s was 2650(kg/m³) in this study. γ is water particle density. ω is the settling velocity of sediment particles; J is the energy slope. d_{50} is the median particle size of sediment particles.

3 Results and Discussion

3.1 Selection of parameters

The response relationship betweendimonsionless intensity parameters and dimonsionless sediment transport capacity Φ was analyzed, as shown in Table 2.

Table 2. **Response relationships between dimonsionless intensity parameters and sediment transport capacity**

Dimonsionless intensity parameters	Fitting formulas	Fitting coefficient
Mean velocity V^*	$\Phi = 0.0214 V^{*3.009}$	$R^2 = 0.664$ $N_{SE} = 0.664$
Flow shear stress θ	$\Phi = 0.0369 \theta^{2.0307}$	$R^2 = 0.9498$ $N_{SE} = 0.806$
Stream power W	$\Phi = 0.0209 W^{*1.3067}$	$R^2 = 0.9458$ $N_{SE} = 0.721$
Unit stream power P^*	$\Phi = 1.8927 P^{*1.1961}$	$R^2 = 0.8847$ $N_{SE} = 0.495$
Effective stream power W^*_{eff}	$\Phi = 0.0036 W^{*1.0927}_{eff}$	$R^2 = 0.969$ $N_{SE} = 0.797$

The dimonsionless effective stream power W_{eff}^* mainly refers to the residual net output power of water flow without losses due to the shear stress. The value of the effective flow power can also reflect sediment transport capacity because it represents the actual output power of water flow. The larger the effective flow power is, the greater the output energy of the water flow is. And the flow with large energy can transport more sediments, this is why the effective flow power is most highly correlated with the sediment transport capacity.

3.2 Empirical formula for sediment transport capacity

The relationship between the dimonsionless sediment transport capacity and dimonsionless effective stream power can be expressed by Equation (2):

$$\Phi = K W^{*b}_{eff} \tag{2}$$

Where K is the sediment transport coefficient and b is the power exponent.

Sediment transport coefficient K is a comprehensive coefficient representing the overall state of water flows and soils (Zhao et al., 2020). The relationship between S_v and K is analyzed as

$$K = 0.0576 S_v^{0.1551} \qquad R^2 = 0.393 \tag{3}$$

Then, the dimensionless effective flow power W^*_{eff} and the volume sediment concentration S_v are mutually independent, and the dimensionless sediment transport capacity Φ can be seen as a dependent variable. It would be

$$\Phi = 0.1742 S_v^{0.322} W^{*0.949}_{eff} \qquad R^2 = 0.989 \tag{4}$$

In Equation (4), both volume sediment concentration S_v and dimensionless sediment transport capacity are unknown parameters, so it needs to be further deduced by the implicit function method.

As the sediment transport capacity can also be expressed by

$$T_c = S_v \gamma_s q \tag{5}$$

By introducing Formula (5) into the formula in Table 1, it can be obtained that:

$$\Phi = \frac{S_v \gamma_s q}{\gamma_s \sqrt{\left(\frac{\gamma_s}{\gamma} - 1\right) g d_{50}^3}} \tag{6}$$

When considering the Equation (6) is equal to Equation (4), it can be concluded that:

$$\frac{S_v \gamma_s q}{\gamma_s \sqrt{\left(\frac{\gamma_s}{\gamma} - 1\right) g d_{50}^3}} = 0.1742 S_v^{0.322} W^{*0.949}_{eff} \tag{7}$$

The volume soil comcentration can be calculated as

$$S_v^{0.678} = \frac{0.1742 W_{eff}^{*0.949} \sqrt{\left(\frac{\gamma_s}{\gamma} - 1\right) g d_{50}^3}}{q} \tag{8}$$

If the parameter related to soil types(d_{50}, γ_s) and runoff conditions(W_{eff}^*, q) can be obtained, the volume soil comcentration S_v can be determined. Then a formula for dimonsionless sediment transport capacity is derived by integrating Equation (6) and Equation (9):

$$T_c = S_v \gamma_s q = \frac{0.076 W_{eff}^{*1.4} \gamma_s g^{0.7374} d_{50}^{2.2125} (\gamma_s - \gamma)^{0.7374}}{q^{0.475} \gamma^{0.7374}} \qquad R^2 = 0.989 \tag{9}$$

To evaluate the applicability of Equation (9), the Fig. 2 shows the relationship between the measured value and calculated value of sediment transport capacity.

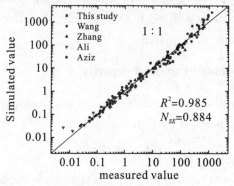

Figure 2. Comparison between simulated and measured value of Equation (10)

It can be seen in Fig. 2 that the calculated and measured value are around the 1 : 1 line, although this study included a series of data from this experiment(Wu et al. 2017; Zhang et al. 2011; Ali et al. 2012; Aziz and Scott 1989). It means that this formula is applicable to calculate the sediment transport capacity of conhesive and cohesionless soil particles. In addition, compared with four existing models in Fig. 3, the formula in this study has a wider applicability.

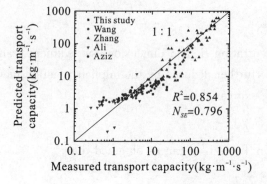 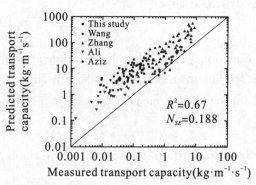

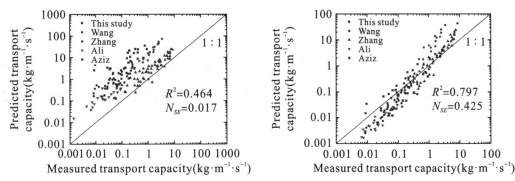

Figure 3. Comparison between simulated and measured value of ANSWERS model(Beasley et al., 1980), improved WEPP model(Zhang et al., 2008), Zhang's model(Zhang et al., 2011) and Ali's model(Ali et al., 2012)

4 Conclusion

This study chose sandy loess and loess soil with different median particle sizes to obtain an appropriate formula for the sediment transport capacity. Following are the main conclusions:

(1) Compared with some existing parameters, the compound factor stream power parameters especially the effective stream power was the best predictor.

(2) It was concluded that the volume sediment concentration is a necessary factor when calculating the sediment transport capacity.

(3) This study chose the dimensional effective stream power and the volume sediment concentration as two influence factors, and the predictional model obtained here had a wide applicability.

Acknowledgements

This research was supported financially by the Central Public-interest Scientific Institution Basal Research Fund [Grant No. Y120002] and the Yellow River Institute of Hydraulic Research, YRCC science and technology development fund(YRIHR202104).

References

[1] Ali M, Sterk G, Seeger M, et al. Effect of hydraulic parameters on sediment transport capacity in overland flow over erodible beds[J]. Hydrology and Earth System Sciences, 2012, 16(4): 591−601.

[2] Aziz N M, Scott D E. Experiments on sediment transport in shallow flows in high gradient channels[J]. International Association of Scientific Hydrology Bulletin, 1989, 34(4): 465−478.

[3] Beasley D B, Huggins L F, Monke E J. Answers: a model for watershed planning[J]. Transactions of the Asabe, 1980, 23(4), 938−0944.

[4] Bing W, Wang Z, Zhang Q. Modelling sheet erosion on steep slopes in the loess region of China[J]. Journal of Hydrology, 2017, 553.

[5] Yu B, Zhang G H, Fu X. Transport Capacity of Overland Flow with High Sediment Concentration[J]. Journal of hydrologic engineering, 2015, 20(6): C4014001.1-C4014001.10.

[6] Zhang G H, Wang L L, Tang K M, et al. Effects of sediment size on transport capacity of overland flow on steep slopes[J]. Hydrological Sciences Journal/Journal des Sciences Hydrologiques, 2011, 56(7): 1289−1299.

[7] Zhang G H, Liu B Y, Zhang X C. Applicability of WEPP sediment transport equation to steep slopes [J]. TRANSACTIONS OF THE ASABE, 2008, 51(5): 1675−1681.

[8] Zhao L, Zhang K, Wu S, et al. Comparative study on different sediment transport capacity based on dimensionless flow intensity index[J]. Journal of Soils and Sediments, 2020, 20(1): 1−17.

Non-invasive Measurements of Benthic Oxygen Exchanges of a Drinking Reservoir by Eddy Correlation Method

Yuanning Zhang[1], Bowen Sun[1], Chang Liu[2], Qingzhi Zong[1], Xiaobo Liu[2], Xueping Gao[1]

[1] State Key Laboratory of Hydraulic Engineering Simulation and Safety,
Tianjin University, Tianjin 300072, China

[2] State Key Laboratory of Simulation and Regulation of Water Cycle in River Basin,
China Institute of Water Resources and Hydropower Research, Beijing 100036, China

Abstract

Many drinking reservoirs suffer from water quality deteriorations due to a benthic oxygen deficit in its thermally stratified periods, where the sediment is the primary sink of oxygen. A non-invasive eddy correlation measurement of benthic oxygen exchanges was conducted in the Daheiting Reservoir, an important drinking reservoir in North China. In-situ measured data over 5000 h in 9 months revealed a stratification cycle variation of benthic environmental factors, including water temperature, dissolved oxygen concentrations, and hydrodynamic conditions. Furthermore, the responses of benthic oxygen fluxes to different environmental factors were discussed, and the leading factors that control fluxes were distinguished under different conditions, providing a reference to improve water quality models and plan water quality operation measures.

Keywords: Drinking reservoir; Benthic oxygen fluxes; Eddy correlation method; Thermal stratification

1 Introduction

Reservoirs have great significance in supplying water for domestic and agricultural purposes, but this benefit is being threatened by water quality deteriorations on account of the benthic oxygen deficit (Dutton et al., 2018). While the dam raised the water level, it also weakened local hydrodynamics and promoted seasonal stratifications. During the thermally stratified stage, the hypolimnetic dissolved oxygen(DO) is quickly depleted since the vertical oxygen supplement is hindered and cannot catch up with the consumption. Sediment is a primary sink for oxygen in the bottom boundary layer (BBL), and its oxygen intake provides valuable information on the diagenesis of particulate organic matter, biological respiration and oxidation of reduced compounds diffusing from the sediment-water interface(SWI)(Bryant et al., 2011). For example, the benthic oxygen flux is closely related to the release of deposited nutrients and pollutants like phosphorus, nitrogen, methylmercury, hydrogen sulphide, iron, and manganese(Beutel et al., 2014), whose releases in large quantity are likely to result in environmental hazards. Therefore, quantifying the benthic oxygen exchange is crucial for understanding the aquatic oxygen budget and improving the supplied water quality.

The benthic oxygen flux is sensitive to BBL environmental factors like near-sediment hydrodynamics and DO concentrations(Bierlein et al., 2017). However, most methods of measuring benthic oxygen fluxes either isolate or change BBL factors, such as micro-profiling, benthic chambers, and core incubations. Moreover, those approaches typically account for an enclosed and small sediment area in a short period and could bias flux estimation for fluffy and permeable sediments (Reidenbach et al., 2010). In contrast, the aquatic eddy correlation method, an adaption of a long-used method for quantifying atmospheric scalar fluxes, is a non-invasive and in-situ alternative to benthic oxygen exchange measurements. It deploys measurements directly above the SWI and determines fluxes by calculating covariance of vertical velocities and DO concentrations. While this method is more expensive and technically more challenging to apply than traditional methods, it has unique advantages. For example, this method covers a relatively large footprint ($10 \sim 100$ m^2) and integrates spatial heterogeneities found in most ecosystems. Furthermore, it obtains results with a high resolution and is applicable to study sites where traditional methods are less feasible (Berg et al., 2003). At present, the eddy correlation method has been successfully conducted in rivers, lakes, shallow bays, and deep ocean; but the reservoir, whose benthic oxygen fluxes are essential to drinking safety, is still poorly measured and discussed.

In addition to the in-situ measurements, theoretical models of benthic mass exchanges have also been developed, supplying a basic to explore the relationship between SWI oxygen fluxes and BBL environmental factors. The diffusive boundary layer (DBL) was proposed as a several-mm-thick area above the SWI, which regulates benthic oxygen exchange rates via molecular diffusions (Jørgensen and Revsbech, 1985). Then, the response of DBL thickness to hydrodynamics was experimentally determined, and other indexes like the Batchelor scale were adopted. Moreover, based on Fick's first law, mathematical models for estimating benthic oxygen fluxes were put forward, such as the thin-film, the film renewal, the shear velocity, and the Lorke and Peeters model (Bierlein et al., 2017). Those models offer valid comparisons for our research.

This study aims to reveal the variation of benthic oxygen fluxes within a stratification cycle and explore its response to BBL environmental factors. We conducted 9-months-covered eddy correlation measurements in the Daheiting Reservoir, an important drinking reservoir in north China. The typical annual variations of benthic environmental factors and oxygen fluxes were analyzed, and the leading factors that control fluxes were distinguished under different stages, providing a reference to improve water quality.

2 Study Site and Methods

2.1 Study site

The Daheiting Reservoir is a river-type reservoir in the northern Hebei Province, China, as Fig. 1 shows. It lies in the mainstream of Luanhe River and is a vital drinking water source of the Luanhe Water Diversion Project. This reservoir reversely regulates the discharged water from the Panjiakou Reservoir, which is located just 20 km away. With a total storage capacity of 337 million m^3, the Daheiting Reservoir provides domestic and industrial water for Tianjin and Tangshan, and supports agricultural irrigation along the lower reaches of the Luanhe basin.

However, the water quality deterioration events like the algal blooms often occurred in the Daheiting reservoir. From the 1990s to 2017, extensive cage fish farming was conducted in this reservoir, and large amounts of organic pesticides entered the water body with the runoff. As a result, the sediment is rich in nutrients, organics, and heavy metals, threatening the drinking water safety of this reservoir.

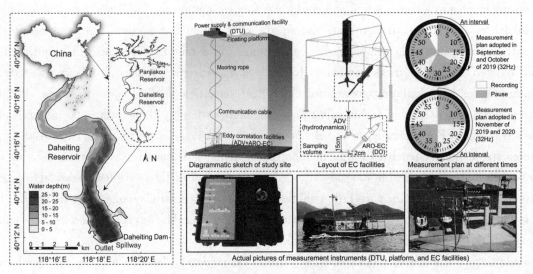

Figure 1. Study site and measurement facilities

2.2 Measurement method

The eddy correlation method was adopted to measure benthic oxygen fluxes. This method employs its measurement directly above the SWI in an open water body; thus, it does not disturb sediments and has vast footprints. Since the measurement point is in the logarithmic layer, oxygen diffuses steadily and its source/sink from the measurement point to the SWI is negligible. Moreover, assuming there is a flat and long underlying surface upstream of the measurement point, the advection and horizontal oxygen gradient can also be neglected. As a result, the benthic oxygen flux(F) was deduced as the covariance of vertical flow velocity(w) and DO concentrations(C), which could be written as $F=\overline{w'C'}$ (Berg et al., 2003).

A standard eddy correlation system consists of an Acoustic Doppler Velocimeter(ADV) and a fast-responding oxygen sensor; in this study, the ADV Vector 300m(Nortek, Norway) and the ARO-EC sensor(JFE Advantech, Japan) were selected, respectively. Furthermore, some auxiliary facilities like the platform, the power supplying module, and the data transmitting module(DTU) were also adopted to conduct a long-term measurement (Fig. 1). The eddy correlation fluxes calculation workflow is shown in Fig. 2, which was realized in batches by MATLAB programs combined with SOHFEA V2.0(http://sohfea.dfmcginnis.com).

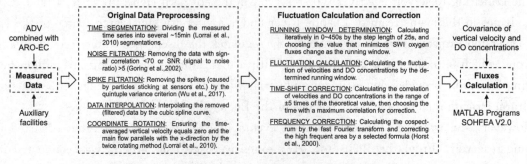

Figure 2. Workflow for processing eddy measurement data

2.3 Measurement plans

The eddy correlation measurements were conducted at a distance of fewer than 800 m from the dam and the outlet of the Dahaiting Reservoir(40°12′7.07″N, 118°18′43.49″E, Fig. 1). This station has a flat underlying surface and a robust spatial representation that meet the requirement of the eddy correlation method. Several measurements were carried out from 2019 to 2020, all of which were periodically performed by a 15 min high-frequency(32 Hz) data recording followed by a pause. A 5 min pause was adopted in September and October of 2019, while a 15 min pause was employed in the rest time(Fig. 1). A data recording and a pause constitute an eddy correlation interval.

The ADV and the ARO-EC were placed above the SWI by a tripod, as Fig. 1 shows. The ADV measures a cylindrical volume with a few centimetres in diameter and height located 15 cm below its signal transmitter, but the ARO-EC acquires DO concentrations near its tips. To avoid signal interference, the ARO-EC tip was fixed at an equal height but 2 cm away from the ADV's sampling volume. Considering the relative still and settleable benthic environment, the distance from sampling volume to the tripod was adjusted to 65 cm.

3 Results and Discussions

3.1 Variations of benthic environmental factors

A total of more than 10,000 eddy correlation intervals covering 9 months were measured in this study, as Fig. 3 shows. Each interval displayed here is a mean of a 15 min and 32 Hz recorded data string, so they are considered fully representative. The three velocities were further running averaged within a 24 h long window, which is presented by the red line in Fig. 3.

The water depth(benthic pressure) varied greatly within a year due to the cascade reservoir operations. During the flood season of early June to middle August in 2020, the water depth remained relatively stable at about 14 m with strong regulations. In contrast, it is usually much larger in the non-flood season that could reach 24 m, but the water depth at the same time of different years was not comparable.

The benthic temperature indicates the thermal stratification stage of this reservoir, which essentially controls benthic oxygen and dynamic conditions. It showed a distinct characteristic of hypolimnion from late June to early October in 2020, when the temperature stayed around 10.5℃ with minor fluctuations. However, the benthic temperature during the stratification stage in 2020 was firmly higher than that of the same stage in 2019, which should be attributed to the low water level in the flood season of 2020. After early November of 2020, the benthic temperature decreased linearly after a sharp rise and fall since the stratification was broken, and this phenomenon appeared a week later in 2019 than it did in 2020.

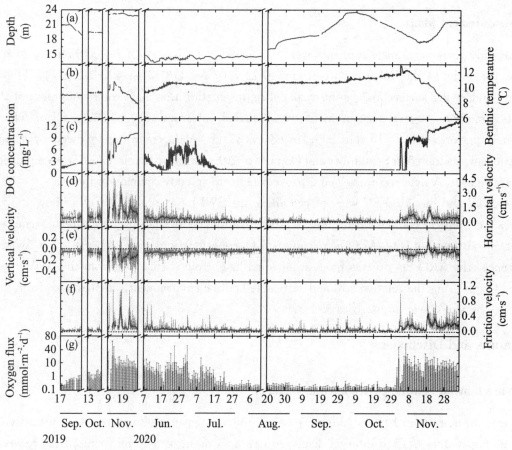

Figure 3. Eddy correlation measurement results in the Daheiting Reservoir

The DO concentrations were closely related to the thermal stratification of this reservoir since it prevents the vertical oxygen supplement from the air reaeration. In early June of 2020, the stratification was basically formed but not quite strong, so the DO was once consumed to below 1 mg · L^{-1} but later rebounded to about 5 mg · L^{-1} with significant fluctuations. After the stratification has grown to a well-developed state in July, the DO was depleted, and a continual anoxic area occurred in the BBL. Finally, as the warm water impacted the benthic environment unstably in early November when the stratification was ended, the DO repeatedly surged and then gradually increased. Comparing with the same period in 2019, DO concentrations in 2020 were relatively lower during the stratification stage but showed more dramatic changes in the mixing stage.

The hydrodynamic conditions, including the horizontal, vertical, and friction velocities, were inhibited by thermal stratifications as well, so it showed similar annual variations with DO concentrations. The thermal stratification hindered external oxygen supplies mainly by weakening hydrodynamic conditions, especially the vertical velocity; as a result, they were remarkably consistent when saltatory events happened. In addition to the thermal structure, the benthic hydrodynamics were likely to be regulated by cascade reservoir operations; the benthic environment of the study site was relatively static with flow velocities usually less than 2 cm · s^{-1}. Moreover, the horizontal and vertical velocities in 2020 were generally lower than those in the same period of 2019, but the friction velocities of those two years were similar.

Under the dual influences of DO concentrations and hydrodynamics, the benthic oxygen fluxes presented prominent annual variations that generally bigger in the mixing period while smaller in the stratification period. Since the downward exchange was set as the positive direction of oxygen fluxes, it can be concluded that the sediment always acted as an oxygen sink. The magnitude of fluxes in different periods may vary hundreds of times, which is determined by oxygen levels in a long-term scale but controlled by hydrodynamic bursts in a short time scale. Furthermore, the error bar in Fig. 3(d) refers to the standard deviation of flux variations within one day. Because the phytoplankton thrived and the water body was rich in particulate matters, the BBL received little light, and there was no apparent diurnal variation of benthic oxygen fluxes.

3.2 Responses of benthic oxygen fluxes to BBL factors

This study focuses on the response of oxygen fluxes to DO concentrations and hydrodynamics. Although temperatures also affect the DO solubility and the molecular diffusion rate, the measured benthic temperature always varied in a small range beyond the optimal living temperature of most microbial communities. As a result, the direct impact of temperature on benthic oxygen exchanges in the study site was negligible.

Controlling the hydrodynamic conditions by filtrations, the response of fluxes to DO concentrations was explored, as Fig. 4(a) shows. It was founded that the bulk oxygen concentration C_B significantly regulated the oxygen flux F_{DO} in a power pattern under an anoxic environment, indicating the benthic DO concentration was the bottleneck of flux increase at that time. However, considering the oxic condition, the average flux remained almost unchanged with DO increase, and the linearly fitting goodness of the two was relatively weak. Under such a condition, the linearly fitted intercept made no sense to discuss flux responses and could be replaced by an indefinite variable n. Therefore, it can be inferred that the benthic oxygen flux responded differently under different benthic oxygen environments. A further exploration based on partial measured data found that the DO threshold between those two kinds of respond patterns was about 2.7 mg · L^{-1}.

In a similar way, controlling the DO concentrations by filtrations, the response of fluxes to hydrodynamic conditions was explored. However, trials found that it is hard to establish a direct relationship between fluxes and velocities. Then, the DBL thickness was introduced to quantify the impact of hydrodynamics on benthic oxygen fluxes. Since the Daheiting Reservoir has a relatively static benthic environment, the DBL thickness δ_{DBL} was estimated by $\delta_{DBL}=0.96L_B+0.04$, where L_B is the Batchelor scale, and its calculation formula can be referred to Zhang et al. (2021). Moreover, considering the dimensional balance, the relation between DBL thickness and horizontal velocity U can be written as $\delta_{DBL}=aDU^{-1}+b$, where D is the molecular diffusion rate while a and b are the fitting coefficient. As Fig. 4(b) shows, the less stratified the water body, the smaller the DBL thickness and the coefficient b, but the larger the coefficient a, suggesting that the DBL thickness was increasingly sensitive to the horizontal velocity as the thermal stratification weakened.

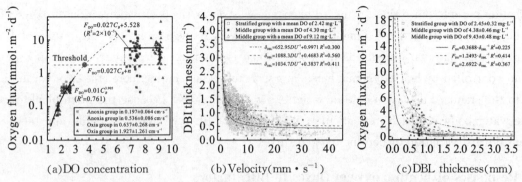

Figure 4. Responses of benthic oxygen fluxes to different BBL environmental factors

Furthermore, the response of oxygen fluxes to DBL thicknesses was hypothesized as $F_{DO}=c/\delta_{DBL}$ adapted from the thin-film model (Li et al., 2021). In this formula, c is the fitting coefficient that theoretically equals $D(C_B-C_{SWI})$ and is estimated to be around 10^0 mmol·mm^{-1}·d^{-1}. Three groups under stratified, middle, and mixed thermal conditions were fitted as Fig. 4(c) shows, whose fitting coefficients all met the anticipation. It can also be concluded that the less stratified the water body or the larger DO concentrations, the larger the coefficient c and the benthic oxygen fluxes under the same DBL thickness or hydrodynamics. However, this increase was nonlinear, which is consistent with the discussion of flux's response to DO concentrations before. Considering the fitting coefficient c has a dimension of $M·L^{-1}·T^{-1}$, it indicates the change rate of vertical linear density and describes a space-time distribution of BBL solutes. Therefore, this coefficient c is determined by the benthic oxygen circumstance and is an independent environment parameter that quantifies the response characteristics of benthic oxygen fluxes to DBL thicknesses.

To sum up, the DO concentrations coupling with hydrodynamic conditions regulates the benthic oxygen fluxes. The fluxes are depressed by oxygen shortages under anoxic conditions, while it is mainly controlled by hydrodynamics when the benthic oxygen is redundant. Likewise, the hydrodynamics affects the fluxes by changing the DBL thickness, but this effect is also restrained by the benthic oxygen circumstance.

3.3 Environmental effects of benthic oxygen fluxes

The benthic oxygen flux is closely related to the redox condition and the biochemical reaction process at the SWI, and even determines the transformation direction of pollutions in the sediments. As a result, it plays an important role in progressive and sudden environmental hazards of reservoirs, such as eutrophication, heavy metals, and persistent organic pollution. Banks et al. (2012) studied the oxygen depletion process with sediments highly polluted by heavy metals, who found that the dissolved metals precipitated back into the sediment by generating metal sulphides under an anoxic condition. However, if the chemical or hydrodynamic conditions changed, the precipitated pollutants may be reactivated and then released to the overlying water, forming a troublesome chemical timebomb that threatens ecosystem health. With the influence of benthic oxygen fluxes, the sediment could be both source and sink of pollutants, so it regulates pollutants distribution in different mediums and buffers sudden ecological hazards of the reservoir. For the Daheiting Reservoir, the sediment acted as the sink of pollutants in its cage fishing stage, which was beneficial to the water quality conservation at that time. In contrast, the sediment changed its role to the source after the cage fishing was banned in 2017 and become a chronic problem of water protection.

Moreover, different pollutants show different sink-source responses to benthic oxygen fluxes at the SWI. The migration and release of nutrients in sediments depend on the redox of ferrum. Fe^{2+} always binds with nitrogen and phosphorus after being oxidized, then precipitates into sediments and reduces nutrient fluxes(Zhao et al., 2018). That is to say, relatively high benthic oxygen fluxes could inhibit nutrient diffusions, where phosphorus is more sensitive to dissolved oxygen environment while nitrogen is more affected by the mineralization of organic matters. However, it is low oxygen fluxes that hinder the diffusion of heavy metals like cuprum and chromium (Yan et al., 2018). The occurrence form of heavy metals is also affected by the oxygen environment. For example, an oxic condition is favorable for the release of inorganic mercury, but a dark anoxic condition is favorable for methyl mercury(Zhu et al., 2018). Organic pollutant fluxes are also depressed by low oxygen fluxes since organic matter is slowly degraded with electron acceptors of NO^{3-} and SO_4^{2-} under the actions of anaerobic microorganisms. Nevertheless, the electron acceptor of this process is oxygen in an oxic condition, when the oxygen demanded microorganisms are more active and promote the decomposition of organic matters(Cantwell et al., 2002). The diffusion of different pollutants is closely related, such as there is a coupling relationship between iron and phosphorus, and the mineralization of organic matter is an important driving force for the diffusion of all materials. In addition to the benthic oxygen condition, the fluxes of various pollutants are also affected by other factors like pH and light intensity to different degrees.

It is noteworthy that the DBL is an objective physical layer above the sediment, so it regulates diffusions of all soluble substances at the SWI. In other words, the pollution flux has a similar response to the DBL thickness with the oxygen flux. As summarized in Fig. 5, the long-term in-situ measurement of benthic oxygen fluxes could not only provide valuable information on benthic biochemical processes and reservoir oxygen budgets, but also of great significance for predicting endogenous release and ensuring water quality.

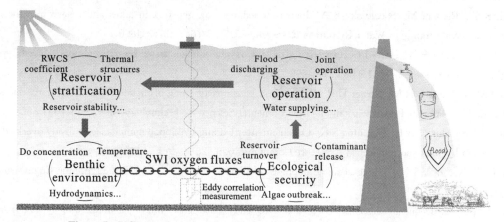

Figure 5. Influence factors and environmental effects of benthic oxygen fluxes

4 Conclusions

Non-invasive eddy correlation measurements covering 5000 h in 9 months were carried out in the Daheiting Reservoir, an important drinking water source in north China. The results indicated that the aquatic eddy correlation method is a robust technique to conduct long-term field observations of benthic oxygen fluxes, providing a reference to estimate endogenous release and protect drinking water quality.

The benthic environmental factors of stratified reservoirs varied greatly within a year. With the regulation of human operations, the water depth double increased and frequently changed. In the thermally stratified period, the benthic temperature, DO concentration, and hydrodynamics stabilized at their lower level. They all increased as the reservoir mixed, but the temperature soon declined after reached its peak.

The annual variation of benthic oxygen fluxes was determined by DO concentrations and hydrodynamics in a coupled way. The benthic oxygen exchanges were inhibited in an anoxic condition, but were controlled by hydrodynamic conditions if the underlying oxygen was abundant. The hydrodynamics influenced the oxygen flux by changing the diffusion boundary layer thickness, which is limited by DO concentrations as well.

Acknowledgements

This study was supported by the National Natural Science Foundation of China(51609166), the National Key R&D Program of China(2018YFC0407902). We thank Dr. Daniel F. McGinnis for sharing the SOHFEA Software Package V2.0 on the Internet(http://sohfea.dfmcginnis.com).

References

[1] Banks J L, Ross D J, Keough M J. Measuring hypoxia induced metal release from highly contaminated estuarine sediments during a 40day laboratory incubation experiment[J]. Science of the Total Environment, 2012, 420: 229−237.

[2] Berg P, Røy H, Janssen F. Oxygen uptake by aquatic sediments measured with a novel non-invasive eddy correlation technique[J]. Marine Ecology Progress Series, 2003, 261: 75−83.

[3] Beutel M, Dent S, Reed B. Effects of hypolimnetic oxygen addition on mercury bioaccumulation in Twin Lakes, Washington, USA[J]. Science of the Total Environment, 2014, 496: 688−700.

[4] Bierlein K A, Rezvani M, Socolofsky S A. Increased sediment oxygen flux in lakes and reservoirs: the impact of hypolimnetic oxygenation[J]. Water Resources Research, 2017, 53(6): 4876−4890.

[5] Bryant L D, Gantzer P A, Little J C. Increased sediment oxygen uptake caused by oxygenation-induced hypolimnetic mixing[J]. Water Research, 2011, 45(12): 3692−3703.

[6] Cantwell M G, Burgess R M, Kester D R. Release and phase partitioning of metals from anoxic estuarine sediments during periods of simulated resuspension[J]. Environmental Science & Technology, 2002, 36(24): 5328−5334.

[7] Dutton C L, Subalusky A L, Hamilton S K. Organic matter loading by hippopotami causes subsidy overload resulting in downstream hypoxia and fish kills[J]. Nature Communication, 2018, 9(1): 1951.

[8] Goring D G, Nikora V I. Despiking acoustic doppler velocimeter data[J]. Journal of Hydraulic Engineering, 2002, 128(1): 117−126.

[9] Horst T W. On frequency response corrections for eddy covariance flux measurements[J]. Boundary-Layer Meteorology, 2000, 94(3): 517−520.

[10] Jørgensen B, Revsbech N. Diffusive boundary layers and the oxygen uptake of sediments and detritus[J]. Limnology & Oceanography, 1985, 30(1): 111−122.

[11] Li N, Huang T L, Chang Z Y. Effects of benthic hydraulics on sediment oxygen demand in a canyon-shaped deep drinking water reservoir: experimental and modeling study[J]. Journal of Environmental Sciences, 2021, 102: 226−234.

[12] Lorrai C, McGinnis D F, Berg P. Application of oxygen eddy correlation in aquatic systems[J]. Journal of

atmospheric and oceanic technology, 2010, 27(9): 1533—1546.

[13] Reidenbach M A, Limm M, Hondzo M. Effects of bed roughness on boundary layer mixing and mass flux across the sediment-water interface[J]. Water Resource Research. 2010, 46(7): 1—15.

[14] Wu J F, Jin W D, Tang P. Survey on monitoring techniques for data abnormalities[J]. Computer Science, 2017, 44(11A): 24—28.

[15] Yan C Z, Zeng L Q, Che F F. High-resolution characterization of arsenic mobility and its correlation to labile iron and manganese in sediments of a shallow eutrophic lake in China[J]. Journal of Soils and Sediments, 2018, 18: 2093—2106.

[16] Zhang Y N, Sun B W, Ju W H. Eddy correlation measurements of benthic oxygen fluxes in a stratified and operated reservoir[J]. Journal of Hydrology, 2021:595.

[17] Zhao H C, Zhang L, Wang S R. Features and influencing factors of nitrogen and phosphorus diffusive fluxes at the sediment-water interface of Erhai Lake[J]. Environmental Science & Pollution Research, 2018, 25: 1933—1942.

[18] Zhu J S, Gao R X, Wang Y M. The diffusion flux of mercury at water/sediment interface in water-level-fluctuating zone of the three gorges reservoir[J]. Earth and Environment, 2018, 46(2): 164—172.

Real-time Water Depth Prediction of Multiple Manholes Based on Spatio-temporal Correlation

Changxun Zhan, Ting Zhang, Shuqi Li

Department of Water Resources and Harbor Engineering, College of Civil Engineering,
Fuzhou University, Fuzhou 350116, China

Abstract

The manhole water depth prediction model combined with artificial neural network(ANN) is one of the research hotspots in waterlogging disaster management. Traditional ANN research mostly focuses on single-output models, and the interaction between different outputs is separated due to the lack of consideration of the correlation between outputs. This paper uses two ANN models to predict the depth of 358 manholes. First, Analyzing the correlation between rainfall and runoff is used to select the sliding window width of the input data. Then, back-propagation neural network(BPNN) and long short-term memory(LSTM) are adopted to build the multi-output water depth prediction model with the help of the consideration containing both the time lag of rainfall/runoff and the interaction among multiple manholes in spatial dimension. The proposed model was verified by using measured data of four typhoon storms. The results show that the ANN models based on spatio-temporal correlation can effectively capture the behavior of the multi-input to multi-output nonlinear dual drainage mechanism, and the accuracies of proposed models can reach more than 90%.

Keywords: Water depth prediction of multiple manholes; Multi-output data-driven model; Spatio-temporal correlation; Back-propagation neural network; Long short-term memory

1 Introduction

As a widely used black box model, artificial neural network is a data processing system composed of many non-linear interconnected artificial neurons. It has been regarded as an effective tool for learning non-linear input and output systems. A neural network trained using a back-propagation algorithm is called a Back-Propagation Neural Network(BPNN). BPNN is a static feedforward neural network. The earliest use of BPNN to predict rainfall and runoff can be traced back to the early 1990s, such as the papers published by Daniell(Daniell T M, 1991) in 1991 and Halff(Halff A H, 1993) in 1993. Since then, many studies have used artificial neural networks to model runoff processes. In 2006, Tayfur(Tayfur G, 2008) used BPNN to make a better prediction of the peak runoff flow data. In 2008, Chua(Chua L, 2008) compared the difference between the prediction results of using the measured flow and the calculated flow of the motion wave as the input of BPNN, and also discussed the prediction results under different lag time. In 2013, Paleologos(Paleologos E K, 2013) used BPNN to simulate the daily runoff in a karst environment. In 2014, Aziz(Aziz K, 2014) used BPNN to estimate the maximum flow data at the river station.

As the research progressed, it was discovered that BPNN lost any information about the input sequence when it was used for time series analysis, and Long Short-Term Memory(LSTM) could fill this shortcoming. In recent years, many scholars have explored the potential of LSTM in rainfall-runoff simulation. In 2018, Zhang(Zhang J, 2018) used LSTM to predict the groundwater level in agricultural areas. The author compared the simulation results based on the LSTM method with the BPNN simulation results, and found that the performance of the former is better than the latter. In 2018, Zhang(Zhang D, 2018) compared the performance of different neural network architectures in simulating and predicting the water level of the combined sewer structure based on online rainfall and water level data. They confirmed that compared with the traditional architecture without explicit unit storage, LSTM is more suitable for multi-step early prediction.

We can find that the application of neural network for urban flooding prediction has attracted widespread attention in academic circles at home and abroad. Researchers at home and abroad have proposed a variety of methods for urban flooding prediction and achieved good results. However, in practical applications There are still some problems:

(1) The selection of input variables in urban flooding prediction is still a difficult problem.

(2) Research on urban flooding prediction mostly focuses on single-output models.

2 Materials and Methods

2.1 Study area and data

The study area Shengang District is located in the northwestern part of Taichung City, Taiwan. Its geographical location is shown in Fig. 1. The total catchment area of the study area is 6.85 square kilometers, and the urban planning area is 4.68 square kilometers.

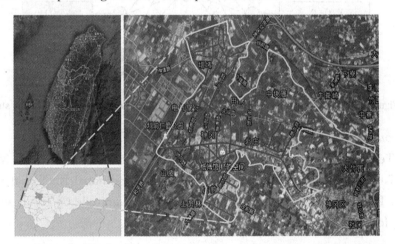

Figure 1. The location of the study area and the main road network

The manhole water depth data set is composed of two parts: the original training data set from 20 design rainstorms and the original test data set from 4 measured rainstorms. The time resolution is 10 min.

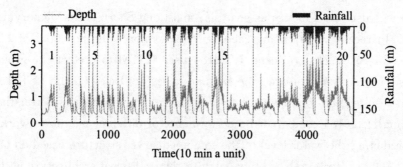

Figure 2. Manhole A3 training data set

2.2 Methods

2.2.1 *Water depth regression prediction model based on BPNN*

The input-output topology of the BPNN water depth regression prediction model is the same as that of the classification prediction, as shown in Fig. 3 The difference lies in: First, the water depth uses continuous data; second, the loss function uses Mean Square Error(MSE).

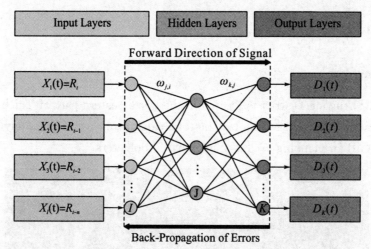

Figure 3. The BPNN water depth regression prediction model at the current time step

2.2.2 *Water depth regression prediction model based on LSTM*

Long Short-Term Memory Network(Hochreiter S, 1997)(Long Short-Term Memory Network, LSTM) is a special RNN used to overcome the weakness of traditional RNN that cannot learn long-term dependence. LSTM constructs three gates of order of magnitude on the basis of RNN's cyclic unit: input gate, forget gate, and output gate.

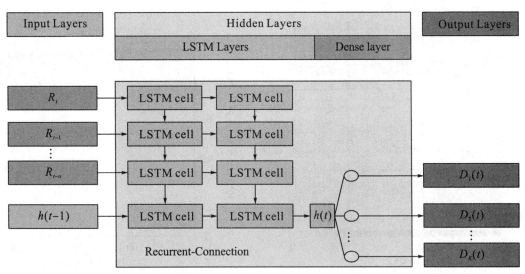

Figure 4. LSTM water depth regression prediction model

3 Model Structure

3.1 Scenarios

3.1.1 *Analysis of spatial correlation of aperture flow*

The degree of similarity of the manholes in the study area will affect the output topology of the model. In order to analyze the spatial correlation between the manholes, the PCC analysis method is used to analyze the spatial correlation of the training data set composed of 20 rains, and get different The PCC cross-correlation confusion matrix between manholes is shown in Fig. 5(a). It can be seen that the PCC of most manholes in the study area is 0.6~1.0, and only a few manholes have a PCC of less than 0.2. In addition, the proportions of the strong and weak intervals of the manhole correlation in the study area can be seen more intuitively in Fig. 5(b). the proportions of areas with PCC greater than 0.8, 0.6, and 0.4 are 0.742, 0.894, and 0.939, respectively. This shows that more than 74% of the manholes in the study area have strong cross-correlation(PCC=0.8~1.0), nearly 90% of the manholes have strong cross-correlation(PCC=0.6~1.0), and more than 93% of the manholes It has strong cross-correlation(PCC=0.4~1.0). Therefore, the manholes in the research area have strong cross-correlation, and the multi-output topology can be constructed with all manholes as a whole.

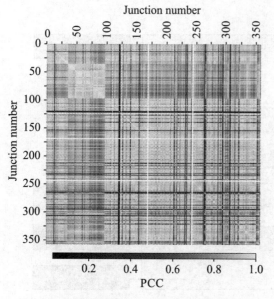
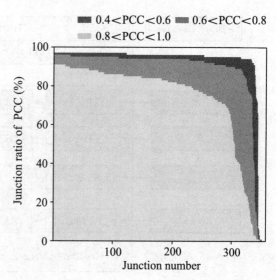

(a) Manhole cross-correlation confusion matrix　　(b) Percentage of manholes in different PCC intervals

Figure 5. **Spatial correlation of manholes in the study area**

3.1.2 *Rainfall-runoff time correlation analysis*

Fig. 6(a) shows the PCC of 358 manholes under different rainfall-water depth time differences. Each broken line represents a manhole. It can be seen from the figure that the PCC of all holes is less than 0.4 at 300 min, indicating that the correlation between rainfall and water depth is weakly correlated after a certain period of time. Therefore, to determine the effective rainfall input value is to determine the point in time when the rainfall-water depth correlation changes from strong to weak. We believe that the highly correlated part has a greater impact on water depth, so Fig. 6(b) shows the proportion of manholes with PCC greater than 0.4 and PCC greater than 0.6 under different time differences.

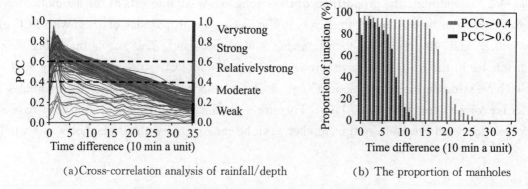

(a)Cross-correlation analysis of rainfall/depth　　(b) The proportion of manholes

Figure 6. **Rainfall-water depth time correlation analysis**

4 Results and Discussion

4.1 Overall prediction result

Fig. 7(a) and Fig. 7(b) respectively show the NSE and RMSE prediction results of BPNN and LSTM 10 min to 60 min ahead of time. It can be seen that as the number of advance steps increases, the overall NSE and RMSE prediction results of the two types of models show a downward and upward trend respectively. It can be seen from Fig. 2~10 that whether PCC is greater than 0.4 or PCC is greater than 0.6, the proportion of manholes shows a decreasing trend. Therefore, the multi-step prediction results of BPNN and LSTM in advance are directly proportional to the correlation between rainfall and runoff.

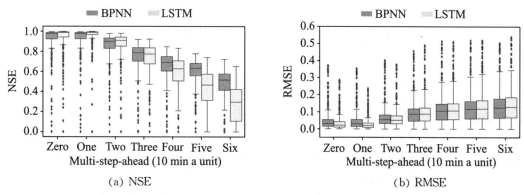

Figure 7. Comparison of the overall prediction results of BPNN and LSTM

4.2 Local prediction result

Fig. 8 shows the RMSE and NSE prediction results of 5 manholes prone to flooding by 10 min to 60 min. As the advance step increases, the prediction results of the five manholes are getting worse, the RMSE gradually rises, and the NSE gradually declines. Fig. 9 shows the prediction results of manhole RP_42-1 on BPNN and LSTM respectively. The same conclusion can be drawn from this, the more steps ahead, the greater the deviation of the predicted value from the target value.

	RP_182	RP_C5-1B	RP_42-1	RP_9	RP_14
BPNN	—	—	—	—	—
LSTM	···	···	···	···	···

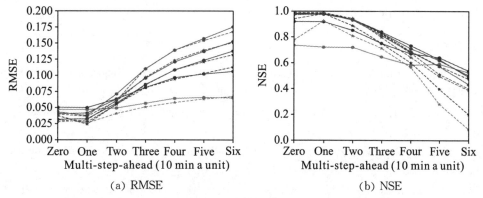

Figure 8. Comparison of BPNN and LSTM's multi-step advance prediction

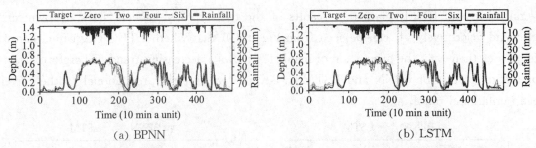

(a) BPNN (b) LSTM

Figure 9. Prediction results of manhole RP _ 42-1 multi-step in advance

5 Conclusions

In this study, two neural network models (BPNN, LSTM) were used to perform multi-output water depth regression prediction of 358 manholes in the area, and the effective learning ability and accurate prediction ability of LSTM and BPNN were tested through four measured rainstorms. The results It is summarized as follows.

(1) LSTM and BPNN both have smaller RMSE (0.02 vs. 0.03) and larger NSE (0.985 vs. 0.975). This shows that the LSTM and BPNN models proposed in this study fully extract the nonlinear spatial relationship between 358 manholes and the time lag between rainfall and runoff, which can effectively adjust the connection weights between neurons and provide reliable and accurate double drainage. System manhole water depth forecast.

(2) Compared with BPNN, LSTM has a more obvious advantage in the degree of fit of the prediction results of sudden water level and large peak water level. LSTM or its recursive connection can store the output of the hidden layer at the previous time step as the input of the next time step, which makes it more powerful in time series prediction than BPNN.

(3) LSTM has more advantages than BPNN when the advance step is less than three hours (that is, the advance time is less than 30 min). However, when the advance step is greater than three, the NSE prediction result of LSTM drops significantly, indicating that the cyclic connection of LSTM at this time has no advantage for BPNN, and the rain in the first 30 min has a decisive influence on the depth of the manhole in the study area.

References

[1] Daniell T M. Neural networks, applications in hydrology and water resources engineering [J]. Proceedings of the International Hydrology and Water Resource Symposium, 1991, 3: 797-802.

[2] Halff A H, Halff H M, Azmoodeh M. Predicting runoff from rainfall using neural networks [J]. Engineering Hydrology, ASCE, 1993: 760-765.

[3] Tayfur G, Singh V P. ANN and fuzzy logic models for simulating event-based rainfall-runoff [J]. Journal of Hydraulic Engineering, 2008, 134(9).

[4] Chua L, Wong T, Sriramula L K. Comparison between kinematic wave and artificial neural network models in event-based runoff simulation for an overland plane [J]. Journal of Hydrology, 2008, 357(3-4): 337-348.

[5] Paleologos E K, Skitzi I, Katsifarakis K. Neural network simulation of spring flow in karst environments [J]. Stochastic Environmental Research & Risk Assessment, 2013, 27(8): 1829-1837.

[6] Aziz K, Rahman A, Fang G. Application of artificial neural networks in regional flood frequency analysis: A case study for Australia [J]. Stochastic Environmental Research and Risk Assessment, 2014, 28(3): 541-554.

[7] Zhang J, Yan Z, Zhanga X. Developing a Long Short-Term Memory (LSTM) based model for predicting water table

depth in agricultural areas[J]. Journal of Hydrology, 2018: 561.

[8] Zhang D, Lindholm G, Ratnaweera H. Use long short-term memory to enhance Internet of Things for combined sewer overflow monitoring[J]. Journal of Hydrology, 2018, 556: 409−418.

[9] Hochreiter S, Schmidhuber J. Long short-term memory[J]. Neural Computation, 1997, 9(8): 1735−1780.

Reoccured Paleo-glacier Dammed Lakes and Outburst Floods in Langcang River

Xiwen Tang[1], Niannian Fan[1], Chengshan Wang[2], Xingnian Liu[1]

[1] State Key Laboratory of Hydraulics and Mountain River Engineering,
College of Water Resource & Hydropower, Sichuan University, Chengdu 610065, China

[2] School of Earth Sciences and Resources, China University of Geosciences(Beijing),
Beijing 100191, China

Abstract

Ice-dammed lakes form as glaciers block the rivers in the valley, which might be much more in glacial periods than present. We found a paleo-glacier dammed lake at the intersection of Mingyong glacier Valley and Lancang River, near Meli Snow Mountain in the middle reaches of Lancang River. Field investigation, Optically Stimulated Luminescence dating and numerical stimulation of floods are comprehensively used in our study, and we found at least three period of corresponding lacustrine and dam break flood sediments which deposited in 38 ka, 16 ka and 7 ka, respectively. The deposits demonstrate that river blockage and outburst dam break flood in Langcang River valley caused by Mingyong glacier has appeared repeatedly in Late Quaternary. The OSL age of the aeolian deposit which cover the supposed dam and lacustrine sediments near the dam demonstrate that the last retreat of Mingyong glacial from Lancang River valley can be constrained in 7ka, and the results given by Zhang et al. (2018) is in agreement with ours within a reasonable error range. The glacier dammed lake could be extended nearly 100 km upstream with volume of 7×10^9 m^3. The peak flow of the largest dam break flood since last deglaciation was 3.8×10^5 m^3/s, which was more than 80 times of recorded maximum flood. The flood submerging range and depth are simulated by using numerical simulation model based on HEC-RAS, revealing this glacier dam failure process and flood developing procedure downstream. This dam break flood was still larger than maximum recorded floods propagating 1500 km downstream of the dam. We illustrated that reoccurred glacier damming and outburst floods affected the mountainous geomorphic evolution thoroughly, from raising the erosion base level upstream, eroding river channel downstream dramatically, and prompting equilibrium rebound in surrounding areas. Under the background of climate change, future risk assessment of glacier dam failure in this region is evaluated.

Keywords: Lancang River; Mingyong Snow Moutain; Glacier dam; Outburst flood

1 Introduction

Dammed lakes are formed when rockfall, landslide, moraine, lava flow or glacier block the river and cut the water flow. The barrier lakes of well stabilization restrain the downcutting of riverbed upstream and reduce total erosion(Korup et al., 2006; Ouimet et al., 2007), while it could release a larger volume of flow in a dam break, and outburst flow could be much larger than a regular storm flood, resulting in a severe erosion downstream(Cook K L et al., 2018). In addition, dams play an important role in river ecosystem(Zhou et al., 2019) and evolution of historical civilization(Wu et al., 2016). In the world, landslide dammed lakes are very common, while it is rare to discover glacial barrier lakes in a fluvial system which block the mainstream and cause intense effect on the evolution of topography and geomorphology at a continental scale, only with very rare exception e.g., Yarlung Zangbo River(Korup et al., 2008).

Here we report an ice-dammed lake in the middle reaches of Lancang River, forming a backwater area that can extend to the upstream for nearly 100 km. Moreover, the glacier dam can be repeatedly built and block the river for many times; on the other hand, its smaller density and the property of being easily broken ensure the larger peak discharge and shorter time of failure process than the landslide dam. Therefore, it is of great significance to research the paleo-glacier dammed lake for emergency treatment of contemporary sudden dammed lake and risk assessment of reservoir.

2 Study Area

Our work mainly focuses on Three Rivers Region[Fig. 1(a)], which is located in the southeast margin of Qinghai Tibet Plateau, where Jinsha River, Lancang River and Salween River, run in parallel from north to south for more than 170 km. Geologically, the TRR is the main body of the Hengduan Mountains, which is characterized by intense crustal deformation, strong uplift, complex geological structure, dense deep faults, multi-stage structural deformation and superimposition of metamorphism, considered to be one of the areas of the highest frequency of rockfalls and landslides. The Mingyong glacier[Fig. 1(c)], which originates from Meili Snow Mountain, is still 6 km away from the main stream of Lancang River at present. However, the moraine and sediments along the river we found and tested demonstrate that glacier could block the river and form a huge dammed lake. We have investigated more than 500 km of the Lancang River valley[Fig. 1(b)], covering nearly the whole middle reaches and the residual dam we found is near the outlet of Mingyong glacier.

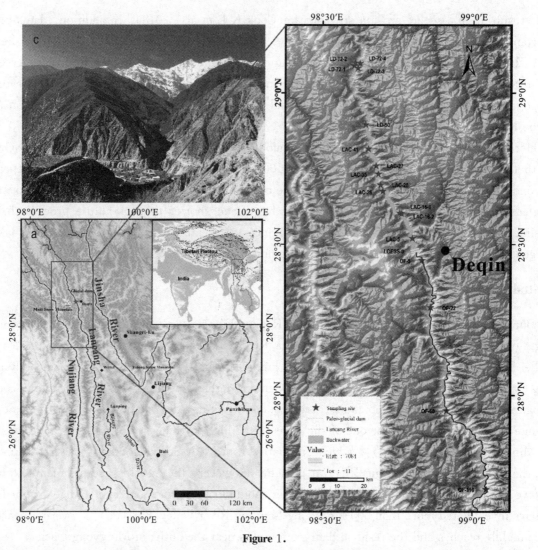

Figure 1.

Note: (a) The Three River Region in the SE Tibetan Plateau, and study area is marked with a red rectangle.
(b) All the sampling site location along the Lancang River.
(c) Viewing the Meili Snow Mountain from the top of the residual dam.

3 Results

A total of 31 samples were taken during our fieldwork, including 19 lacustrine sediments [Fig. 2(a)~(c)] in the upstream of the dam and 12 catastrophic flood deposits [Fig. 2(d)~(f)] due to the dam break in the downstream. The southernmost sample locality downstream reaches Yingpan, and Quzika is at the northern tip of our research area [Fig. 1(b)]. All optical stimulated luminescence dating results are listed below in Table 1.

Figure 2. Photographs of major depositional facies exposed in terrace

Note: (a) LAC-28, (b) LAC-26, (c) LAC-16-1 are the lacustrine sediments, and (d) OF-5, (e) OF-22, (f) OF-110 are the dam break flood deposits.

Table 1. Optically stimulated luminescence dating results of the Lancang River

Sample name	Latitude	Longtitude	Altitude (km)	Depth (m)	Grain size (μm)	U (μg/g)	Th (μg/g)	K (%)	Dose (Gy)	Dose rate (mGy/a)	Age (ka)
OF-110	27.7	99	1.777	2	125~180	3.27±0.5	17.9±0.3	3.02±0.03	74.38±7.87	4.64±0.2	16±1.8
OF-60	27.9	98.9	1.976	15	125~200	2.34±0.5	12.8±0.3	2.06±0.03	>716±	4.38±0.26	>163.4±9.64
OF-22	28.3	98.8	2.082	4.6	150~250	2.56±0.5	8.39±0.3	1.31±0.03	92.15±7.9	2.48±0.13	37.2±3.8
OF-5	28.4	98.8	2.038	12.2	90~180	2.31±0.5	11.1±0.3	2.14±0.03	60.71±6.69	3.47±0.18	17.5±2.1
LOESS-0	28.5	98.8	2.246	0.65	64~90	3.54±0.5	13.5±0.3	2.3±0.03	28.91±0.6	4.37±0.21	6.6±0.3
LAC-5	28.5	98.8	2.235	30	64~125	2.26±0.5	12.2±0.3	1.79±0.03	1274.77±252.72	3.58±0.21	356.4±73.8
LAC-16-1	28.6	98.7	2.123	4	90~125	2.32±0.5	11.4±0.3	1.62±0.03	26.84±1.09	3.1±0.16	8.7±0.6
LAC-16-2	28.6	98.7	2.103	6	125~180	2.51±0.5	12.5±0.3	1.64±0.03	43.38±3.53	3.18±0.16	13.6±1.3
LAC-26	28.6	98.7	2.151	3	64~90	2.2±0.5	10.4±0.3	1.55±0.03	55.94±1.91	3.03±0.16	18.4±1.2
LAC-28	28.7	98.7	2.122	16	64~90	2.35±0.5	11.9±0.3	1.45±0.03	35.58±1.52	2.89±0.15	12.3±0.8
LAC-35	28.7	98.7	2.211	8.2	64~90	1.72±0.5	8.33±0.3	1.71±0.03	48.92±3.18	2.77±0.15	17.7±1.5
LAC-37	28.8	98.6	2.176	315	90~150	2.95±0.5	12.6±0.3	1.68±0.03	72.21±2.08	3.28±0.17	22.1±1.3
LAC-41	28.8	98.6	2.166	3	64~90	4.36±0.5	14.9±0.3	1.42±0.03	11.96±0.31	3.76±0.18	3.2±0.2
LD-50	28.9	98.6	2.206	5	150~250	3.79±0.5	18.3±0.3	2.68±0.03	145.5±7.11	4.69±0.21	31.1±2.1
LD-72-1	29.1	98.6	2.392	3.2	64~200	2.81±0.5	14.9±0.3	2.98±0.03	118.3±14.18	4.83±0.24	24.5±3.2
LD-72-2	29.1	98.6	2.336	23	64~250	2.69±0.5	14.2±0.3	2.79±0.03	80.22±5.54	4.3±0.22	18.7±1.6
LD-72-3	29.1	98.6	2.321	2	64~150	3.07±0.5	15.8±0.3	2.77±0.03	184.45±69.71	5.48±0.32	33.7±12.9
LD-72-4	29	98.6	2.313	11.5	125~200	4.23±0.5	18.4±0.3	2.69±0.03	87.87±6.61	4.75±0.21	18.5±1.6

Note: OF— outburst flood, LAC — lacustrine sediments, LD — lacustrine delta deposits, the numbers after the sample name represent the distance between the sample site and the supposed glacier dam. The location of all sampling points are shown in Fig. 1(b).

The water level-storage curve (Fig. 3) is drawn by the analysis which come from the Area and Volume Statistics technique in ArcGIS based on the 12.5 m solution DEM from NASA. According to the elevation of moraine, lacustrine and deltaic deposit obtained from our field investigation, combined with the topographic data, we can conclude that the full paleo-dammed lake may inundate the elevation of LD-72-1~4 and the maximum dam height can be constrained with 350 m. In addition, the volume of the glacier dammed lake may reach 7 billion cubic meters according to dam height and the elevation of top layer of lacustrine sediment.

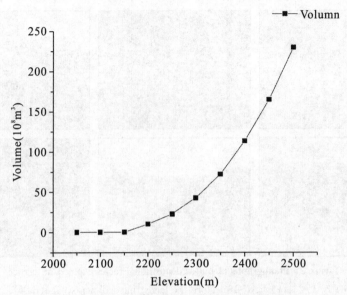

Figure 3. Water level-storage curve of the glacier dam lake

4 Outburst Flood Reconstruction and Numerical Simulation

The U. S. Army Corps of Engineers Hydrologic Engineering Center's River Analysis System (HEC-RAS) software was used to perform a series of one-dimensional unsteady flow simulations in reaches from dammed lake to Vientiane, covering nearly 2200 km downstream. Two-dimensional simulation will be used for the river reaches with obvious bend, channel widening or narrowing, which also have sediment retention, during our next work.

Due to the high erosion of the river valley, prehistoric flood channel geometry may differ a lot from present which not only increase the difficulty but also have poor accuracy in simulation, we choose relatively younger floods which appeared at 16 ka for reconstruction. This flood was the largest flood since last deglaciation, moreover, we found deposits in different reaches from the same flood, which could reconstruct the whole propagation process of this flood more robustly.

With the elevation indicated from dam break flood sediment OF-110, the peak discharge(Q_p) of this flood has been calculated to be 3.8×10^5 m³/s using Manning equation. Using long profile evolution method of breaching channel suggested by Capart(2013), we can obtain the breaking dam flow hydrograph and then set it as the upstream boundary condition. The peak flow attenuation results are shown in Table 2.

The intersection of flow attenuation hydrograph and investigated peak flow hydrograph show that this outburst flood flow will affect Lancang River valley as far as 1500 km downstream(Fig. 4).

Table 2. Peak flow attenuation of the glacier dam break

Profile code	Distance from dam(km)	Peak discharge(m³/s)	Time to peak discharge(h)
1	0.0	379000	2.5
2	66	330000	4
3	148	266000	6.5
4	281	190000	12
5	572	130000	23.5
6	695	98000	31
7	903	76300	42
8	1010	67600	50
9	1110	38900	66
10	1210	37400	74
11	1310	32900	84
12	1410	23700	108
13	1460	17000	120
14	1500	10500	—

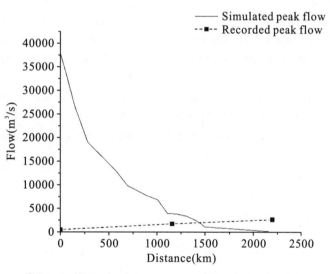

Figure 4. Water level-storage curve of the glacier dam lake

5 Conclusions

(1) The OSL age of the aeolian deposit which cover the supposed dam and dam break deposits demonstrate that last retreat of Mingyong glacial from Lancang River can be constrained in 7ka, The results given by Zhang et al. (2018) is in agreement with the OSL age of sample LAC-16−1 that we took from Gushui profile within a reasonable error range, which is a reaffirmation of our hypothesis for the time of glacier moving back from river valley.

(2) We have found at least three period of the Last Glacial Maximum (LGM) dam break flood sediment which deposited in 38 ka, 16 ka and 7 ka, respectively. That may be the evidence that the

glacier's clogging up the river and retreat can take place repeatedly in Lancang River valley.

(3) The maximum dam are calculated to be 350 m height using Manning equation, and the full glacier dammed lake are estimated to be 7×10^9 m³. We simulated the process of dam break at 16 ka based on HEC-RAS and analyzed the flood routing attenuation results, demonstrating that the outburst flood caused by dam break will still have strong effect in Lancang River valley 1500 km downstream the supposed dam.

(4) The occurrence of glacier blockage will raise the erosion datum and weaken bedrock cutting, but the dam break outburst flood could give a boost to downstream river channel erosion(Korup et al., 2008). How the upstream impoundment and aggradation shaping the river channel and accelerate the erosion rate downstream will be studied quantitatively in our next work.

References

[1] Capart, Hervé. Analytical solutions for gradual dam breaching and downstream river flooding[J]. Water Resources Research, 2013, 49: 1968−1987.

[2] Cook K L, Andermann C. Glacial lake outburst floods as drivers of fluvial erosion in the Himalaya[J]. Science, 2018, 362(6410): 53−57.

[3] Kong P, Na C, Fink D. Moraine dam related to late Quaternary glaciation in the Yulong Mountains, southwest China, and impacts on the Jinsha River[J]. Quaternary Science Reviews, 2009, 28: 3224−3235.

[4] Korup O, Montgomery D R. Tibetan plateau river incision inhibited by glacial stabilization of the Tsangpo gorge[J]. Nature, 2008, 455: 786−789.

[5] Korup, Oliver. Rock-slope failure and the river long profile[J]. Geology, 2006, 34(1): 45−48.

[6] Ouimet W B, Whipple K X, Royden L H. The influence of large landslides on river incision in a transient landscape: Eastern margin of the Tibetan Plateau(Sichuan, China)[J]. Geological Society of America Bulletin, 2007, 119: 1462−1476.

[6] Yu J, Zhang J. Spatiotemporal variation of late Quaternary river incision rates in southeast Tibet, constrained by dating fluvial terraces[J]. Lithosphere, 2018, 10: 662−675.

[7] Zhou L Q, Liu W M, Lai Z P. Geomorphologic response of river damming[J]. Quaternary Science, 2019, 39(2): 366−380.

Research on Influence Factors of Sedimentation of Three Gorges Reservoir

Yue Zeng, Sichen Tong, Qi Shao

School of River and Ocean Engineering, Chongqing Jiaotong University,
Chongqing, 400074, China

National Engineering Research Center for Inland Waterway Regulation,
Chongqing, 400074, China

Abstract

With the rise of the water level of the Three Gorges reservoir (TGR), the movement characteristics of water and sediment inflow have changed greatly. From the view of long run, the flood control volume would be decreased, and the sedimentation may reduce the regulation capacity of the reservoir, affect the power generation and even cause navigation obstruction. Therefore, the problem of sedimentation is one of the most important issues in the study of the TGR. The mechanism and processes of sedimentation is complex and the sedimentation under the influence of various factors needs to be explored. In recent years, under the new water and sediment condition, it is necessary to study the comprehensive influence of inflow and sediment discharge on the sedimentation of the TGR, which can help to provide a reference for slowing down the sedimentation for better operation. Based on the measured data from 2003 to 2017, this paper studies the relationship between the sedimentation of the TGR and influence factors under the condition of new water and sediment by the method of theoretical research and statistical analysis. Results show that the main factors affecting the sedimentation amount are different in each period, and also have a great influence on the sedimentation.

Keywords: TGR; Sedimentation; Sediment concentration; Water and sediment condition; Influence factors

1 Introduction

The great benefits of flood control, power generation, shipping and so on of the TGR are brought into full implementation since its impounding in 2003. With the rise of reservoir water level, the sediment carrying capacity of flow is reduced, and the sediment in the reservoir area is gradually deposited. Reservoir sedimentation is the result of sediment movement, which is determined by the water and sediment conditions, water level of reservoir and the topographic conditions. Influence factors are mainly including water and sediment inflow discharge, water level of reservoir and river boundary conditions. The influence of various factors on reservoir sedimentation is complex. Comprehensive analysis of the influence of various factors on the sedimentation of the TGR and the qualitative conclusions obtained are of great significance to reservoir operation and sediment release.

2 Water and Sediment Conditions of the TGR

The larger the inflow discharge is, the greater the sediment transport capacity is(Fig. 1). There is a good relationship between the inflow sediment concentration and the inflow discharge. And the change process of the monthly mean value is similar.

According to the statistics of the period of the maximum sediment concentration and discharge in the TGR from 2003 to 2017, the period of the maximum sediment concentration and discharge is mainly concentrated in the main flood season. The maximum sediment concentration and the maximum discharge are all occurred in flood season. However, the period of the maximum sediment concentration is relatively scattered, which reflects that the movement of sediment is slightly behind the water movement, and the movement of water and sediment is dissimilar.

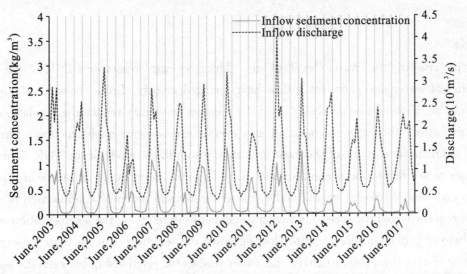

Figure 1. Variation of monthly average sediment concentration and monthly average inflow flow of the TGR(2003—2017)

3 Relationship Between Sediementation and Influence Factors

3.1 Main influence factors

Reservoir sedimentation is closely related to the change of runoff and sediment load, reservoir operation and natural river conditions. The water and sediment flow from upstream are the entry conditions of sedimentation and the water level in front of the dam is the exit condition affecting the characteristics of sedimentation. Due to the experience of reservoir operation, the main factors considered are the inflow sediment concentration and inflow discharge. Based on the measured water and sediment data after impoundment of the TGR, relationship between sedimentation and different influence factors from 2003 to 2017 was analyzed.

3.2 Sediment concentration

The variation process of yearly sedimentation and annual inflow sediment concentration over the years since the impoundment of the TGR is almost the same (Fig. 2). When the inflow sediment concentration increases, the sedimentation also increases, and vice versa. The average inflow sediment concentration in flood season is much greater than that in the averaged year, and its variation trend is similar to that of sedimentation volume in flood season. Relationship between monthly sedimentation volume and average inflow sediment concentration in flood season from 2003 to 2017 is shown in Fig. 3. The relationship between monthly sedimentation and monthly average inflow sediment concentration in flood season keeps a good variation tendency.

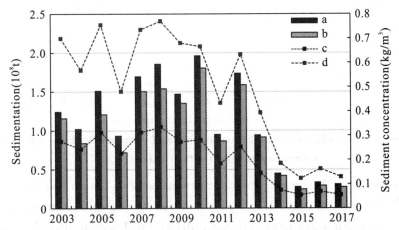

Figure 2. Variation of yearly sedimentation and average sediment concentration of the TGR over the years(2003—2017)

Note: a. Yearly sedimentation; b. Sedimentation in flood season; c. Annual average inflow sediment concentration; d. Inflow sediment concentration in flood season.

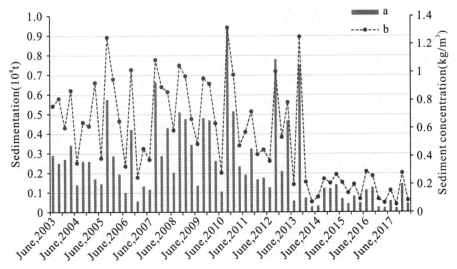

Figure 3. Monthly sedimentation volume and average inflow sediment concentration of the TGR in flood season(2003—2017)

Note: a. Monthly sedimentation in flood season; b. Monthly average inflow sediment concentration in flood season.

Since the impoundment of the reservoir, the correlation between annual and flood season monthly sedimentation and monthly average inflow sediment concentration is high, and the correlation coefficient R^2 is about 0.94 and 0.93 respectively. The annual correlation is slightly higher than that in flood season(Fig. 4). However, the water inflow and sediment concentration in 2010 were relatively large, and the flood peak and sediment peak appeared in July and August. The sedimentation was affected by many factors to reach the maximum value since the impoundment of the TGR, and the flood season point value was slightly dispersed than other points.

201

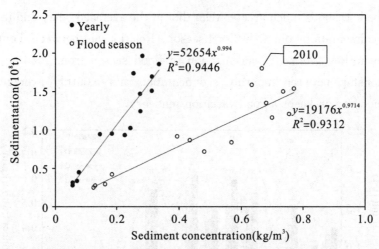

Figure 4. Relationship between sedimentation volume of yearly and flood season and inflow sediment concentration of the TGR(2003—2017)

Sedimentation of the TGR is mainly concentrated in the flood season and the change range of the sediment in the flood season is large. Relationship between the monthly sedimentation volume and the monthly average inflow sediment concentration in different flood periods after the impoundment of the TGR is shown in Fig. 5. The relationship of monthly sedimentation volume and sediment concentration is good and the R^2 are greater than 0.82. So, the inflow sediment concentration is a very important factor affecting the sedimentation. When the sediment concentration is less than 0.6 kg/m³, the sedimentation volume in flood season in different periods is relatively concentrated. During this period of the water level of reservoir in different reservoir operation periods on the sediment concentration is not obvious, and the runoff and sediment load conditions are the main influence factors. When the average inflow sediment concentration is more than 0.6 kg/m³, with the increase of sediment concentration, the correlation points become dispersed, and the sedimentation corresponding to the same sediment concentration changes greatly, which is mainly caused by the change of water level of reservoir. When the sediment concentration is small, the sedimentation volume is not sensitive to the change of water level of reservoir. When the sediment concentration is large, the sensitivity of sedimentation to the water level of reservoir increases.

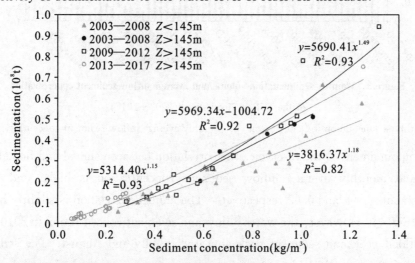

Figure 5. Relationship between monthly sedimentation volume and monthly average inflow sediment concentration of the TGR in flood season(2003—2017)

3.3 Inflow discharge

The dynamic force of the water scour and sedimentation is different under different discharge levels, and the inflow discharge is also the main influencing factor of the sedimentation. The variation trend of sedimentation and inflow is similar(Fig. 6). The average inflow discharge in flood season is much greater than that in the whole year, and the variation range of inflow in flood season is more similar with the sedimentation. The sedimentation volume shows a large reduction trend since 2013 due to the influence of water storage and sediment detention of upstream reservoirs. The change trend of monthly sedimentation volume and inflow discharge is also similar(Fig. 7).

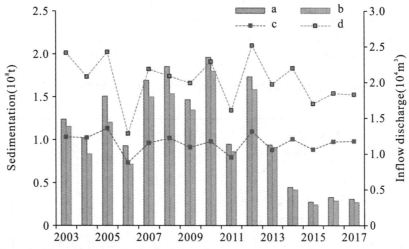

Figure 6. The sediment and average inflow flow of the TGR over the years(2003—2017)

Note: a. Yearly sedimentation; b. Sedimentation in flood season; c. Annual average inflow discharge; d. Average inflow discharge in flood season.

The statistical relationship between the annual and flood season sedimentation and inflow discharge is shown in Fig. 8. Generally speaking, the annual and flood season sedimentation is positively correlated with the inflow, but the distribution of the correlation points is scattered, and the correlation coefficients are only about 0.1 and 0.3 respectively. When the inflow flow is constant, the corresponding sedimentation volume changes greatly, which indicates that it is also affected by other factors.

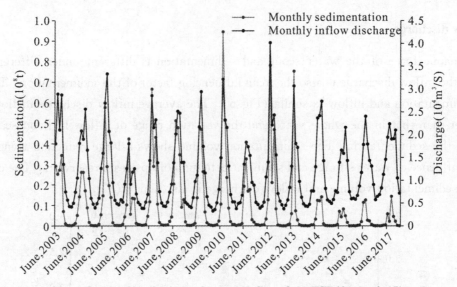

Figure 7. Monthly siltation and average inflow of the TGR(2003—2017)

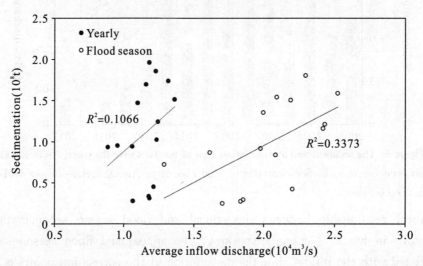

Figure 8. Relationship between annual and flood sedimentation and inflow discharge of the TGR(2003—2017)

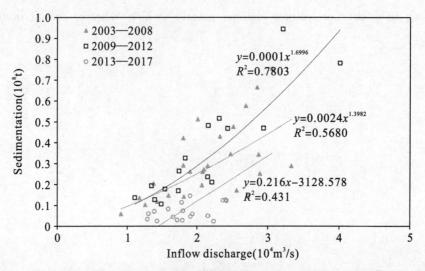

Figure 9. The relationship between monthly sedimentation and monthly average inflow discharge of the TRG in flood season(2003—2017)

The relationship between the monthly sedimentation and the average inflow discharge of the TGR in the flood season in different operation periods is shown in Fig. 9. The correlation coefficients of 2003—2008, 2009—2012 and 2013—2017 are about 0.57, 0.78 and 0.43 respectively. With the increase of the inflow discharge, the sedimentation volume in flood season from 2003 to 2012 is increasing rapidly, and the correlation points become more scattered. The inflow discharge in flood season affects the change of sedimentation volume, and other factors also affect the sedimentation volume, such as the inflow sediment concentration analyzed above.

4 Conclusions

Based on the analysis of the relationship between the sedimentation volume of the TGR and the sediment concentration and the inflow discharge, the process of sedimentation changes is complex. The main factors affecting the sedimentation in different periods are different, and the obvious factors will have a great influence on the sedimentation in this period. The relationship between the reservoir sedimentation and each single influencing factor is discussed from different aspects. The conclusions can provide reference for the operation of the TGR.

(1) From different time scales, there is a good positive correlation between sedimentation and sediment concentration, and the change of sediment directly affects the sedimentation in the reservoir.

(2) From different periods 2009 to 2012, the relationship between sedimentation of the TGR and the inflow sediment concentration and inflow discharge are the best compared with other periods.

(3) Due to the concentrated flood peak process in the flood season, the inflow discharge and sediment of the TGR are large relatively and fluctuate in a large range. During this period, the distribution of the relationship between the inflow sediment concentration, the inflow discharge and the sedimentation volume are scattered. More factors should be considered in the analysis of the influence factors of sedimentation.

Acknowledgements

This study was partially supported by the National Key Research and Development Program of China (2018YFB1600400) and the Basic and Frontier Research Programs of Chongqing, China (cstc2018jcyjAX0534) and Model Development and Application of Water Environment in the Yangtze River Basin(Grant No. 2019-LHYJ-0102).

References

[1] Yang L X. Outlet for reservoir silting of three gorges project[J]. Science&Technology Review, 2006(1): 65−66.

[2] Ren S, Liu L. Sediment deposition and countermeasures in the Three Gorges Reservoir[J]. Journal Of Sediment Research, 2019, 44(6): 40−45.

[3] Tang X Y, Tong S C, Xu G X. Simulation of groundwater flow and environmental effects resulting from pumping[J]. Advances in Water Science, 2019, 30(4).

[4] Han J Q, Sun Z H, Huang Y. Features and causes of sediment deposition and erosion in Jingjiang reach after impoundment of the Three Gorges Project[J].Journal Of Hydraulic Engineering, 2014, 45(3).

[5] Huang L. Mechanism research of water& sediment load variation impact on transversal profile in Jianli River reach [D]. Wuhan: Changjiang River Scientific Research Institute, 2008.

[6] Tian Z. Study on the reservoir sedimentation with tributary[D]. Chongqing: Chongqing Jiaotong University, 2015.

[7] Xu Q X, Yuan J, Dong B J. Sediment change and river bed erosion and deposition in the Yangtze River [J].

Technology And Economy Of Changjiang, 2019, 3(3).

[8] Gao J. Preliminary studies on the operation regulation of the sustainable use[D]. Chongqing: Chongqing Jiaotong University, 2010.

[9] Hu C H, Fang C M, Xu Q X. Application and optimization of "storing clean water and discharging muddy flow" in the Three Gorges Reservoir[J]. Journal Of Hydraulic Engineering, 2019, 50(1).

[10] Xiao J H. Study on effects of dams on river ecosystem service functions and its evaluation[D]. Nanjing: Hohai University, 2007.

Research Progress on the Formation Mechanism and Numerical Modeling of Cyanobacteria Blooms

Chenhui Xu[1], Sichen Tong[2], Guoxian Huang[3], Zhengze Lv[4], Qinghuan Zhang[5]

[1,2,4] School of River and Ocean Engineering, Chongqing Jiaotong University, Chongqing 400074, China

[1,3,4] Chinese Research Academy of Environmental Sciences, Beijing 100012, China

[3,5] State Key Laboratory of Plateau Ecology and Agriculture, Qinghai University, Xining 810016, China

Abstract

In recent years, due to the increase of nutrient discharge and climate change, the excessive growth of phytoplankton such as cyanobacteria has resulted in cyanobacteria blooms of different degrees in some lakes, reservoirs and estuaries. However, the mechanism and driving factors of cyanobacteria blooms are still not clear. Based on the investigation of the recent literatures, this paper reviewed some key advances on the mechanism, key driving factors of controlling cyanobacteria blooms and related modeling technology. The main conclusions are as follows: (1) the mechanism of cyanobacteria blooms is complex and diverse, which is mainly controlled by the bioavailable nitrogen and phosphorus mobilization, flow field and water temperature etc. The accumulation of key nutrients driven by different conditions plays an important role in the accumulation and growth of algae species; (2) whether the bloom process of cyanobacteria is mainly controlled by P or N has not been concluded and still exist disputes. However, as far as the current research results are concerned, there have been many cases with P as the limiting condition, while few cases with N controlling the eutrophication process. In the future, a large number of case studies are still needed to clarify the inner linkage between the algae bloom and ratios of Chla : TP, Chla : TN and TN : TP at different level of water temperature level in each water body, so as to gradually clarify whether P control or N control; (3) Recently, the eco-environment numerical modeling has made rapid progress, but are still needed to strengthen technology aspects including the three-dimensional flow and nutrients transport, coupling between the sediment particles and nutrient process in overlying water, multivariate dynamic coupling between the water environment variables and growth and reproduction of cyanobacteria, together with the high efficient parallel simulation during the simulating of cyanobacteria blooms.

Keywords: Cyanobacterial blooms; Forming mechanism; Numerical modeling; Recent advance reviews

1 Introduction

Cyanobacteria blooms have gradually become a general disaster in the ecological environment with the rapid development of modern society. It may causes a large number of economic losses and environmental hazards, absorbing sunlight and releasing toxins by taking advantage of its own physiological and ecological characteristics to occupy favorable terrain, which would change the ecological environment and food web of the water body.

In recent years the domestic water blooms involve Taihu Lake, Chaohu Lake, West Lake, Dianchi Lake, Xuanwu Lake, Huaihe River and Haihe River, etc(Wu, 2008). The eutrophication characteristics of different degrees have been reported in coastal estuaries(Li, 2016), due to the large amount of nutrient flowing into the sea through the river basin. Lake Victoria in Africa, the Baltic Sea in Europe, Lake Erie in North America, the Caspian Sea in West Asia, and Lake Kasumigaura in Japan have been affected by the blooms in the world(Paerl, 2011) and it is indispensable to indentify the mechanism of cyanobacteria blooms to predict or prevent the ecological disaster. However, till now, the mechanism of cyanobacterial blooms is complex, which cannot be fully explained by the existing theories and related studies, together with the related influencing factors have not been preliminarily understood. Water temperature, water volume, flow field, nutrients, sunlight, wind speed and CO_2 are all the influencing factors of cyanobacterial blooms. Eutrophication caused by excessive nitrogen, phosphorus and other nutrients flowing into the basin is also one of the reasons for the outbreak of water bloom(Wu, 2012). As early as 1970s, Schindler put forward the theory of whether one or two nutrients should be limited to control the lake eutrophication, and the outbreak of water bloom(Schindler, 2008). Historically, phosphorus had been used as the priority to limit the eutrophication of the upstream freshwater ecosystem, while nitrogen had been the main limiting condition of the coastal waters(Paerl, 2009). This kind of research and measures had indeed improved the quality of some waters, and the problem of cyanobacteria bloom had also been alleviated. However, the formation mechanism of cyanobacterial bloom and the coupling effect of several key factors on its development are still unclear.

2 The Influence of Natural Climatic Factors on Cyanobacteria

In earlier studies, many researchers focused on the impact of a single environmental factor on cyanobacteria blooms. Under sufficient nutrients, the *Anabaena* and *Microcystis* blooming in Biwa Lake in Japan, were found that they grew most vigorously at 28℃~32℃(Nalewajko, 2001). In the pure culture system, the optimal growth temperature for *Microcystis* was 25℃ (Chen, 2010). Temperature affects the solubility of various substances in water, the concentration of nutrients, and the photosynthesis of phytoplankton indirectly. It will also directly controls the enzymatic activity and nutrient absorption efficiency of phytoplankton (Li, 2013), which is the leading factor in environmental controlling cyanobacteria blooms. Yang Yan investigated the superimposed effect of temperature and phosphorus concentration, and found that with the increase of phosphorus concentration, the growth rate of *Microcystis* aeruginos also increased with the augment of temperature(Yang, 2013). The superimposed effect was the most obvious at 26℃ where the biomass peak appeared. The *Anabaena* was subject to the superimposed effect of temperature and phosphorus was more obvious at concentration of 0.3 mg/L. Results showed that the growth of *Microcystis aeruginosa* was sensitive to the increase in temperature as well as nutrient, and the superposition of the two showed a logical response pattern. Yin Zhikun did a similar study, but they explored the effects of phosphorus, light and temperature on the growth of *Anabaena* at the same time(Yin, 2015). Through the fitting model, they found that the optimal temperature and light intensity of *Anabaena* were 21.03℃±1.55℃ and 2675.12 lx±262.93 lx, respectively. In addition, under the influence of temperature, the UV irradiated *Anabaena* and *Nostoc* will recover the photosynthetic activity due to the increase of temperature(Giordanino, 2011).

The change of climate conditions in recent years is another reason for cyanobacterial blooms around the world. Due to the accumulation of greenhouse effect, the global warming caused by the increase of CO_2 emissions worldwide, the spring temperature is higher year by year, which causes cyanobacteria to reach the recovery temperature earlier. The higher the temperature, the earlier the cyanobacteria recover from the substrate. The increase of the average winter temperature also makes the dormant period of cyanobacteria delayed(Xie, 2016), and the risk of cyanobacterial blooms around the world is greatly increased. In the context of global warming, although the inflow of nutrients into the water column has been restricted to achieve the purpose of reducing nutrient input to limit the crazy growth of cyanobacteria, the frequent occurrence of extreme weather induces the release of nutrients inside the sediment of the water column to float or replenish nutrients to the water column through runoff and wet deposition, which stimulates the growth of cyanobacteria and causes the outbreak of cyanobacterial blooms(Yu, 2016). The occurrence of water blooms caused by winds, heavy rainfall, and storms is ultimately the result of nutrient deposition and cyanobacterial resuspension at the bottom of the water column. In fact, in an earlier study by Ian T. Webster, it has been found that when the wind speed at the water surface is greater than the critical wind speed, the wind will involve the cyanobacteria floating on the water surface underwater, and when the wind speed is less than the critical wind speed, the horizontal distribution of phytoplankton populations on the water surface will be changed and algal accumulation will be formed on the shore, and this critical wind speed range is approximately 2~3 m/s(Hutchinson, 1994). In a study on the effect of wind and waves on phytoplankton, Sun Xiaojing found that phytoplankton density, cyanobacteria density and biomass increased with wind speed when the wind speed was less than 4 m/s, and the indicators decreased when the wind speed was exceeded, which indicated that small wind and waves would contribute to the growth of cyanobacteria(Sun, 2007).

Nutrients such as nitrogen and phosphorus are the key driving factors of cyanobacterial blooms. Among the studies on the mechanism of cyanobacterial blooms, there are numerous studies on the relationship between nitrogen, phosphorus and total nitrogen to total phosphorus ratio(TN∶TP) and cyanobacterial growth. The increase of nutrients is often accompanied by the occurrence of eutrophication in the water column, which is now generally considered to be the cyanobacterial bloom caused by nitrogen and phosphorus(Conley, 2009; Hai, 2010). The specific values of nitrogen and phosphorus triggering cyanobacterial blooms have been widely discussed by experts and scholars. Xu Hai concluded that the threshold values for nitrogen are in the range of 0.3 mg/L to 0.8 mg/L and for phosphorus are in the range of 0.02 mg/L to 0.2 mg/L(Xu Hai, 2010). Lin believed that the threshold value for nitrogen in freshwater systems are in the range of 0.5 mg/L to 1.0 mg/L and for phosphorus are in the range of 0.02 mg/L to 0.1 mg/L(Lin, 2008).

To investigate the coupling effect of nitrogen and phosphorus with other influencing factors on the growth of cyanobacteria, Wang took Taihu Lake as the object and found that the Chla(chlorophyll a) concentration corresponding to 1000 ppm CO_2 was greater than the Chla concentration corresponding to 400 ppm CO_2 after CO_2 aeration treatment, concluding that under high carbon conditions, the concentration of Chla and utilization efficiency of the nitrogen and phosphorus were obtained, which indicate that the elevated atmospheric CO_2 level played a significant supportive role in the development of cyanobacterial blooms(Wang, 2021). North explored the co-limiting effects of nitrogen, phosphorus and Fe on phytoplankton and found that the addition of Fe and phosphorus promoted the uptake of NO_3^- (nitrate) by phytoplankton and alleviated the nitrogen limitation by phytoplankton and made them more responsive to the change of concentration of phosphorus(North, 2007). For the total nitrogen to total phosphorus ratio, Smith found that the occurrence of cyanobacterial blooms was more favorable when the TN : TP of the water column did not exceed 29 (Smith, 1983). However, Xie concluded that TN : TP was not the cause of cyanobacterial blooms, but the result after cyanobacterial blooms(Xie, 2003).

3 Discussion on Whether N or P Controls the Cyanobacterial Bloom Process

Since cyanobacterial blooms are directly influenced by N and P, the current views of experts and scholars on the key drivers controlling cyanobacterial blooms are divided into three main aspects: controlled by N, controlled by P, and controlled by both N and P. In early study, Schindle identified P as the main limiting nutrient responsible for the occurrence of cyanobacterial blooms through long-term experimental manipulation of Lake 227(Schindler, 2008). Subsequently, Conley suggested that a focus on limiting only phosphorus or nitrogen should not be considered unless there is clear evidence or strong reasoning that focusing on only one nutrient in that ecosystem is justified and will not harm downstream ecosystems (Conley, 2009). Paerl stated that "eutrophication in lakes cannot be controlled by reducing nitrogen inputs" and this conclusion is a narrow view of biogeochemical cycling and eutrophication dynamics and lacks consideration of the physical, chemical and biological constraints on the ability of microorganisms to meet N demand at the ecosystem level along the freshwater to marine continuum through N_2 fixation(Paerl, 2009). If only P is restricted in the upstream, and the constraint on N is reduced, the discharge of N to the downstream will stimulate N-controlled phytoplankton such as cyanobacteria downstream, leading to enhanced downstream bloom hazard. The idea of simultaneous N and P control of eutrophication can be confirmed in the current study, but there is a lack of complete ecological data to confirm that N limitation is necessary and it is difficult to achieve N control in the short term, while there have been many successful cases of P limitation alone(Schindler, 2009).

In fact, in upstream freshwater systems, N and P nutrients are mainly derived from high N and P ratio precipitation and high N and P ratio runoff from undisturbed soils, and N concentrations are higher than P concentrations, hence the phenomenon that cyanobacterial blooms in upstream freshwater systems are mainly controlled by P. With the water moving downstream, it is gradually enriched by high P and low N and P ratio runoff from terrestrial ecosystems, leading to N limitation in primary production and cyanobacterial bloom outbreaks(Downing, 1997).

In the process of evaluating and analyzing the temporal and spatial changes of the nutrient state of reservoirs, Mamun quantified whether reservoirs are controlled by P, N, or both by Chla : TP, Chla : TN, and TN : TP were statistically collected from 60 reservoirs with different uses and set the rule: P controls when Chla : TP is higher or TN : TP>20, N controls when Chla : TN is higher or TN : TP<1 and both control when TN : TP is between 10 and 20. It was concluded that in the agricultural reservoirs, P is more predictive of algal growth. The multipurpose reservoirs are more influenced by N and in the estuarine reservoirs algal growth is jointly influenced by TP and TN, but the relationship between TP and Chla is uncertain(Mamun, 2020). This result may not be sufficient to indicate future judgment criteria. Ann-Kristin Bergström concluded that DIN : TP showed higher variability than TN : TP in evaluating whether the ratio of TN and TP is a better indicator to distinguish nitrogen and phosphorus limitation of phytoplankton, In short-term bioassay experiments, DIN∶TP showed nitrogen and phosphorus limitation better than TN∶TP, and phytoplankton moved from net nitrogen limitation to net phosphorus limitation when DIN∶TP increased from 1.5 to 3.4. Higher DIN∶TP indicated that phytoplankton were P-limited(Bergström, 2010).

4 Overview of Water Bloom Model and Key Research Progress

4.1 EFDC model

In the study of cyanobacterial blooms and water eutrophication, Wu, Guozheng applied the EFDC model to the prediction of cyanobacterial blooms, taking Daoxiang Lake in Beijing as the research object, the initial concentration conditions and boundary conditions including flow boundary, climate, water temperature, and pollution load were input into the EFDC model(Wu, 2010). The simulation value of chlorophyll and the variation trend of chlorophyll in Daoxiang Lake were obtained, and compared with the actual value, the accuracy of the final prediction was 63.43%, which could be applied to the prediction and simulation of cyanobacteria bloom. Wang Yaning explored the corresponding spatial and temporal distribution of exogenous load reduction in Taihu Lake water quality by establishing a water quality model of Taihu Lake, and results showed that the improvement response of COD, ammonia and nitrogen in each lake area was characterized by decreasing from the reduction boundary to the periphery(Wang, 2020). While exploring the solution of eutrophication in water bodies, it is also necessary to achieve a balance in economic benefits. Luo conducted different water supply strategies for artificial ponds and established EFDC model to explore the eutrophication degree and Chla concentration of artificial ponds under different sources of water supply, and found that the water supply scheme from reclaimed water had the highest eutrophication status and higher algal diversity(Luo, 2018). The study attempted to integrate eutrophication risk, eutrophication status and economic cost to optimize the water supply strategy from various aspects. At the same time, EFDC model simulation also has some defects. Firstly, the growth rate of algae can only be set as a constant in EFDC model and cannot be changed with time, so that the simulation effect is better in the short term, but the long-term simulation may have larger errors. Secondly, the flow rate and water depth are set constant in the simulation process, and the effect of wind speed is not considered, which may have a large impact on the prediction results(Wu, 2010). The study of multivariate coupling on the co-limitation of cyanobacterial blooms has always been a long-term topic to explore the mechanism of cyanobacterial blooms.

In order to cope with the outbreak of cyanobacterial blooms, vertical hydrodynamic disturbance machines have emerged. Zou Rui aimed at the fact that the vertical hydrodynamic disturbance can inhibit cyanobacteria, quantitatively analyzed the influence of the quantity and location distribution of the machines on chlorophyll, total nitrogen and total phosphorus, comprehensively considered the inhibition effect and economic benefit, and ensured the minimum economic cost to achieve the maximum inhibition effect(Zou, 2012). Zhang Liwei simulated the feasibility of reservoir optimization scheduling on algal bloom suppression and reducing the inflow of upstream nitrogen and phosphorus nutrients to influence the growth of cyanobacteria in reservoirs by EFDC, which provides some reference basis for daily management of reservoir water environment and algae prevention and control under long-term stage(Zhang, 2016).

4.2 Delft3D model

The current water bloom models have made great progress in simulation accuracy and wholeness. Liu adopted Delft3D as the eutrophication model with Taihu Lake as a typical example, and proposed that the data assimilation of nonlinear system by ensemble Kalman filter could improve the simulation accuracy of eutrophication model, which proposed a useful method to improve the prediction accuracy of bloom model(Liu, 2017). The improvement of the accuracy of the bloom model is beneficial to the research on the prediction and control of bloom situation. Chen used Delft3D and meta-cellular automata to build a water bloom prediction model based on light and nutrient as the main conditions, and used the water quality module of the model to calculate the nutrient concentration of water bodies and predict algal biomass(Chen, 2006). In the study on the prevention and control of bloom in the Three Gorges Reservoir area, Yang Pan concluded that the changes of hydrodynamic conditions in the Three Gorges reservoir are the main inducing factors for the outbreak of water bloom in tributaries. However, reservoir ecological scheduling has a certain preventive and control effect on bloom in tributaries. Yang used a planar two-dimensional hydrodynamic numerical model to simulate tidal scheduling with different starting levels, different water level variation and different duration days (Yang, 2019). The results of the study provide a theoretical basis for reservoirs to carry out ecological scheduling to regulate runoff to change hydrodynamic conditions, and then provide a theoretical basis for preventing and controlling the occurrence of water bloom. In reservoirs, rivers, and other water ecosystems, changes in hydrodynamic conditions are one of the causes of water blooms, and the root cause is mainly in the input of nutrients and the migration and exchange of pollutants deposited in the bottom sediment to the overlying water. Li Yongqing analyzed the factors affecting water quality in Xinlicheng, established a water quality model for eutrophication evaluation and analysis, and for exogenous pollution and bottom sediment that is the internal source of reservoir pollution, proposed the use of ecological measures to establish artificial wetland ecosystems to solve the problem of nitrogen and phosphorus pollution(Li, 2017).

4.3 FVCOM model

In the study of the influencing factors of cyanobacterial blooms, the factors affecting the growth and extinction of cyanobacteria include biological mechanistic processesand kinetic factors. The kinetic factors of lake water bodies are complex and variable, so it is difficult to accurate prediction of cyanobacterial biomass. Therefore, Ren generalized the influencing factors of cyanobacterial blooms, used multiple regression methods to construct a statistical model of cyanobacterial growth and extinction, combined with the FVCOM model to analyze the correlation and significance between cyanobacterial biomass and influencing factors, achieved the prediction simulation of the 72 h spatial and temporal distribution characteristics of cyanobacteria successfully in Taihu Lake(Ren, 2019). Many studies are not based on a single model for simulation calculation or prediction. For example, Fu Lei established a hydrodynamic water quality model of Chaohu Lake based on the FVCOM model and dynamically coupled PCLAKE ecological model through the FABM framework. it assessed comprehensively from the aspects of nutrients, water bloom, and water retention time of Chaohu Lake, considered the wind speed and the runoff pollutants sharing rate of the lake. The higher the sharing rate means the greater the concentration of pollutants inflowing from the location, which provides some basis and ideas for lake pollution management and traceability(Fu, 2020). At present, although numerical simulation methods have been applied to various aspects in the exploration of eutrophication and water blooms, the mechanism of cyanobacterial water blooms is still unclear, and some parameters are difficult to be accurately determined, therefore, the numerical model of water blooms is difficult to be accurately portrayed.

5 Conclusion

(1) Nitrogen and phosphorus are the key driving factors affecting the growth of cyanobacteria. In addition to nitrogen and phosphorus, water temperature and flow field are the leading factors in eutrophication environmental factors. The increase of water temperature has an important effect on the early recovery period and the delay of hibernation period of cyanobacteria, and the accumulation of nutrients in the sediment plays an important role in the growth and bloom outbreak of cyanobacteria.

(2) Both nitrogen and phosphorus play a controlling role in the cyanobacterial bloom process, so we should not only focus on one of these factors to achieve the purpose of limiting the bloom, but also consider the eutrophication of the freshwater to marine ecosystem continuum. If we want to investigate what nutrients are dominant in a certain water ecosystem, quantitative analysis the ratio relationship of Chla : TP, Chla : TN and TN : TP is necessary to figure out which is the dominant factor.

(3) Although great progress has been made in water ecosystem model calculations, the accuracy of model simulations, the three-dimensional flow and nutrients transport, coupling between the sediment particles and nutrient process in overlying water, Multivariate dynamic coupling between the water environment variables and growth and reproduction of cyanobacteria, together with the efficient parallel simulation during the simulating of cyanobacteria blooms needs to be further enhanced.

Acknowledgements

This study was partially supported by the National Key Research and Development Program of China(2018YFB1600400) and the Basic and Frontier Research Programs of Chongqing, China (cstc2018jcyjAX0534) and Model Development and Application of Water Environment in the Yangtze River Basin(Grant No. 2019-LHYJ-0102).

References

[1] Bergström A. The use of TN : TP and DIN : TP ratios as indicators for phytoplankton nutrient limitation in oligotrophic lakes affected by N deposition[J]. Aquatic Sciences, 2010, 72(3).

[2] Qiuwen Chen, Arthur E M. Modelling algal blooms in the Dutch coastal waters by integrated numerical and fuzzy cellular automata approaches[J]. Ecological Modelling, 2006, 199(1).

[3] Chen J C, Meng S L, Hu G D. Effects of temperature on interspecific competition between two cyanobacteria[J]. Chinese Journal of Ecology, 2010, 29(3): 454−459.

[4] Daniel J C, Hans W P, Robert W. Howarth, controlling eutrophication by reducing both nitrogen and phosphorus[J]. Science, 2009, 323(5917).

[5] Downing J A. Marine nitrogen: Phosphorus stoichiometry and the global N : P cycle[J]. Biogeochemistry, 1997, 37(3).

[6] Fu L. Simulation of hydrodynamic force and water quality processes of Chaohu Lake based on FVCOM model[D]. Dalian: Dalian University of Technology, 2020.

[7] Giordanino M V F, Strauch S M, Villafañe V E. Influence of temperature and UVR on photosynthesis and

morphology of four species of cyanobacteria[J]. Journal of Photochemistry & Photobiology, B: Biology, 2011, 103(1).

[8] Ian T. Webster, Paul A. Hutchinson. Effect of wind on the distribution of phytoplankton cells in lakes revisited[J]. Limnology and Oceanography, 1994, 39(2).

[9] Li J H, Li Y P, Tang C Y. Water age distribution characteristics of Lake Star(Zhaoqing, Guangdong) influenced by the Water Diversion Project and wind field[J]. Journal of Lake Sciences, 2021, 33(2): 449−461.

[10] Li J L, Zheng B H, Liu L S. Phytoplankton community characteristics and their response to environment in the Changjiang Estuary[J]. Research of Environmental Sciences, 2013, 26(4): 403−409.

[11] Li J L, Zheng B H, Zhang S L. Eutrophication characteristics and differences of major estuaries in China[J]. China Environmental Science, 2016, 36(2): 506−516.

[12] Li Y Q. Research of water quality evolution law and protectioncountermeasures of Xinlicheng Reservoir[D]. Changchun: Jilin University, 2017.

[13] Lin Y J, He Z L, Yang Y G. Nitrogen versus phosphorus limitation of phytoplankton growth in Ten Mile Creek, Florida, USA[J]. Hydrobiologia, 2008, 605(1): 247−258.

[14] Liu Z, Li Z J, Hu L M. Ensemble Kalman filter based data assimilation in the Delft3D-BLOOM lake eutrophication model[J]. Journal of Lake Sciences, 2017, 29(5): 1070−1083.

[15] Luo X, Li X. Using the EFDC model to evaluate the risks of eutrophication in an urban constructed pond from different water supply strategies[J]. Ecological Modelling, 2018, 372.

[16] Ma J J, Wang P F. Effects of rising atmospheric CO_2 levels on physiological response of cyanobacteria and cyanobacterial bloom development: a review[J]. Science of the Total Environment, 2021, 754.

[17] Mamun M D, Kwon S, Kim J E. Evaluation of algal chlorophyll and nutrient relations and the N : P ratios along with trophic status and light regime in 60 Korea reservoirs[J]. Science of the Total Environment, 2020, 741.

[18] Nalewajko C, Murphy T P. Effects of temperature, and availability of nitrogen and phosphorus on the abundance of anabaena and microcystis in Lake Biwa, Japan: an experimental approach[J]. Limnology, 2001, 2(1).

[19] North R L, Guildford S J, Smith R E H. Evidence for phosphorus, nitrogen, and iron co-limitation of phytoplankton communities in Lake Erie[J]. Limnology and Oceanography, 2007, 52(1).

[20] Paerl H W. Controlling eutrophication along the freshwater-marine continuum: dual nutrient(N and P) reductions are essential[J]. Estuaries & Coasts, 2009, 32(4):593−601.

[21] Paerl H W, Hall N S, Ca Landrino E S. Controlling harmful cyanobacterial blooms in a world experiencing anthropogenic and climatic-induced change[J]. Science of the Total Environment, 2011, 409(10):1739−1745.

[22] Ren K. Biomass model simulation of cyanobacteria in Taihu Lake based on growth model[D]. Nanjing: Nanjing Normal University, 2019.

[23] Schindler D W, Hecky R E. Eutrophication: more nitrogen data needed[J]. Science, 2009, 324(5928).

[24] David W, Schindler R E, Hecky D L. Eutrophication of lakes cannot be controlled by reducing nitrogen input: results of a 37-year whole-ecosystem experiment[J]. Proceedings of the National Academy of Sciences of the United States of America, 2008, 105(32).

[25] Val H, Smith. Low nitrogen to phosphorus ratios favor dominance by blue-green algae in lake phytoplankton[J]. Science, 1983, 221(4611).

[26] Sun X J, Qin B Q, Zhu G W. Effect of wind-induced wave on concentration of colloidal nutrient and phytoplankton in Lake Taihu[J]. Environmental Science, 2007,(03): 506−511.

[27] Wang Y N, Li Y P, Cheng Y. Sensitivity analysis of boundary road reduction of large shallow lake[J/OL]. Environmental Science: 1−16. [2021−06−07]. https://doi.org/10.13227/j.hjkx.202010049.

[28] Wu F, Zhan J Y, Deng X Z. Influencing factors of lake eutrophication in China-based on an empirical analysis of 22 lakes in China[J]. Ecology and Environmental Sciences, 2012, 21(1): 94−100.

[29] Wu G, Xu Z. Prediction of algal blooming using EFDC model: case study in the Daoxiang Lake[J]. Ecological Modelling, 2010, 222(6).

[30] Wu Y. Eutrophication: a management problem for lakes in China[J]. Ecological Economy, 2008, 201(9): 14−19.

[31] Xie L Q, Xie P, Li S X. The low TN∶TP ratio, a cause or a result of Microcystis blooms?[J]. Water Research, 2003, 37(9).

[32] Xie X P, Li Y C, Hang X, et al. Effects of temperature on the recovery and dormancy process of cyanobacteria in Taihu Lake[J]. Journal of Lake Sciences, 2016, 28(4): 818−824.

[33] Xu H, Paerl H. W, Qin B Q. Nitrogen and phosphorus inputs control phytoplankton growth in Eutrophic Lake Taihu, China[J]. Limnology and Oceanography, 2010, 55(1).

[34] Yang P, Lu L, Wang J B. Influence of reservoir tidal dispatch on tributary hydrodynamics: a case study of Xiaojiang River, a tributary in Three Gorges Reservoir area[J]. Yangtze River, 2019, 50(1): 191−197.

[35] Yang Y, Zhu X Z, Zhang M. Responses of different algae to the superposition of temperature and phosphorus[J]. Journal of Lake Sciences, 2015, 28(4): 843−851.

[36] Yin Z K, Li Z, Wang S. Effects of light and temperature on the growth kinetics of Anabaena flosaquae under phosphorus restriction[J]. Environmental Science, 2015, 36(3): 963−968.

[37] Yu Y, Ma R, Kong F X. Nutrient reduction magnifies the impact of extreme weather on cyanobacterial bloom formation in large shallow lake Taihu(China)[J]. Water Research, 2016, 103.

[38] Zhang L W. The effects of water level fluctuation and quality of pumping water on algae in pumped storage reservoir based on EFDC model[D]. Changsha: Hunan University, 2016.

[39] Zou R, Zhou J, Sun Y J. Numerical experiment study on the algae suppression effect of vertical hydrodynamic mixers[J]. Environmental Science, 2012, 33(5): 1540−1549.

Risk Assessment of Landslide Dams with Different Internal Structure

Chuke Meng, Zhipan Niu, Yi Long

Institute for Disaster Management and Reconstruction, Sichuan University, Chengdu 610065, China

Risk Assessment of Landslide Dams with Different Internal Structure

Abstract

Sliding slopes form landslide dams to block the river. The gradation and distribution of soil in different landslide dams are pretty different. The internal structure of landslide dams has a significant influence on its slope stability and seepage stability. Therefore, the accuracy of risk assessment of landslide dams can be improved by considering landslide dams' internal structure. Because of the different initial conditions (dam height, dam width, discharge, etc.) of landslide dams in the prototype case, it is challenging to control initial conditions to study the risk of landslide dams with different internal structures. In this paper, under the same initial conditions, fourteen groups of model dams with different internal structures are constructed based on two kinds of soil with different gradations. Model experiments obtain the breaking characteristics (the water storage time, the breaking time, peak discharge) of different structures. These indexes are analyzed using Analytic Hierarchy Process (AHP) to evaluate the risk of landslide dams with different internal structures. Finally, the quantitative evaluation of the risk of landslide dams with different internal structures could provide a scientific basis for engineering application.

Keywords: Landslide Dam; Risk assessment; Model experiment; Analytic hierarchy process

1 Introduction

Mountainous regions in the world account for about 30% of the land area. Among them, the geological structure of the watershed in southwestern China is the most complex. The cross-sections of many canyons are obviously "U" or "V"-shaped, with well-developed faults, steep valley slopes, and structural. The effects of unloading and gravity are highly detrimental to the stability of the bank slope. In addition, many river basins are in strong seismically active zones, which are more likely to induce landslides or mudslides and form barriers and dams to block the river. There have been many occurrences in these areas since ancient times. Landslide dams blocking the river and breaking disasters have caused significant damage to production and living facilities, residents' safety, and engineering facilities in the upstream and downstream river valleys. With the gradual increase in mountainous engineering projects, these areas will face more severe landslide dams blocking the river and dam break disasters(Costa and Schuster, 1988; Fan et al., 2020).

In the prototype landslide dam, its material composition has a substantial heterogeneity. With the dam material composition, the breaking characteristics of the landslide dam, such as the breaking form of the dam, the duration of breaking, the magnitude of the peak discharge, the time of occurrence, and the evolution of the dam shape, will change accordingly. Due to the non-uniformity of the material composition of landslide dams, the particle size and physical-mechanical properties are related to the depth of the location. For example, in the Tangjiashan dammed lake produced after the 2008 Wenchuan earthquake, the soil structure of the dam is mainly composed of upper and lower parts. The upper part is covered with a layer of loose gravel soil, and the bottom is accumulated with strongly weathered debris. Debris and weakly weathered clastic rocks are also one of the most dangerous dams caused by the Wenchuan earthquake and one of the most severe dam failures(Li Shou-Ding et al., 2010; Shi Zhen-Min et al., 2016). In 2014, a magnitude earthquake caused Hongshiyan landslide dam on the Niulan River. The material composition of the dam is very complex, mainly composed of upper and lower parts. The upper soil layer is mainly composed of boulders, rocks with crushed rocks, and a small amount of sand and soil. For the diameter of the solitary rock block, due to the loose accumulation of large stones and noticeable overhead, the permeability coefficient will be more significant. Three soil layers cover the lower part of the dam body. The lithology of the soil layers is different, but the permeability coefficient is generally low, and it is not prone to seepage failure(Liu Ning, 2014). In 2018, Baige Landslide Dam was formed by two landslides. The material comprising the landslide dam is mainly comprised of fine particles, but the upstream slope and the downstream soil body mainly consist of coarse-grained particles(Cai Yaojun et al., 2019). After the material inside the landslide dam is eroded by water seepage, its physical and mechanical properties will change greatly. For example, the shear strength, plasticity index, and liquidity index will change greatly, thereby the physical and mechanical properties of the entire dam body. It will change accordingly, and the stability of the dam body will be quite different from that before infiltration. The formation and breaking mechanisms of landslide dams with different internal structures need to be studied in depth, and the risk assessment of landslide dams with different internal structures needs to be accelerated.

This paper combines model experiments and risk assessment decision-making methods to assess the risk of landslide dams with different internal structures. This paper's risk assessment would play a significant guiding role in the corresponding engineering practice.

2 Methods and Data

The research method adopted in this paper is a comprehensive evaluation method combining objective data and subjective evaluation. Analytic hierarchy process(AHP) is a theory put forward by Saaty(1990), which is the most commonly used comprehensive decision-making method(Mikhailov and Tsvetinov, 2004; Hu et al., 2021). The data source is the model experiment data made by Meng (2021) on 14 groups of landslide dams with different internal structures. The specific parameters of the internal structure of the model are shown in Fig. 1, and the data obtained through the experiment is shown in Table 1.

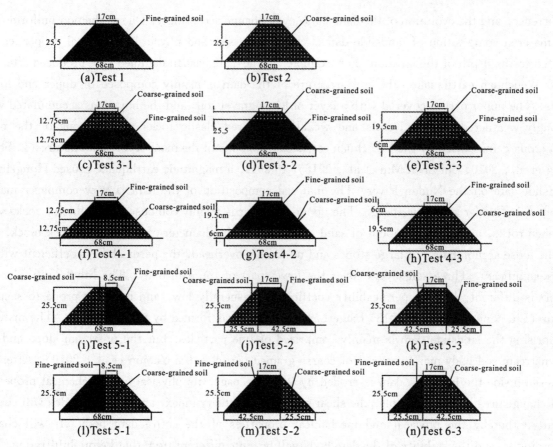

Figure 1. Schematic diagram of the internal structure of the landslide dam

Table 1. Data from model experiment

Test number	Lifespan(s)	The duration of dam break(s)	Peak flow(L/s)
1	53	50	2.38
2	35	11	6.96
3-1	37	80	2.23
3-2	50	70	1.44
3-3	49	63	3.79
4-1	45	18	15.41
4-2	35	20	10.31
4-3	35	10	8.8
5-1	59	50	2.41
5-2	50	75	3.06
5-3	60	55	2.76
6-1	60	53	3.28
6-2	50	60	1.85
6-3	45	60	2.22

3 Risk Assessment

3.1 Hierarchical structure

Liu Ning(2013) believes that the risk assessment of landslide dams is based on the stability and vulnerability assessment of landslide dams. The water storage time, the breaking time, and peak discharge of landslide dams of different structures obtained in the experiments in this paper could reflect the landslide dam's stability and vulnerability. As a result, these three parameters are used as evaluation indicators to evaluate the risk of the landslide dam. The hierarchical structure is shown in Fig. 2.

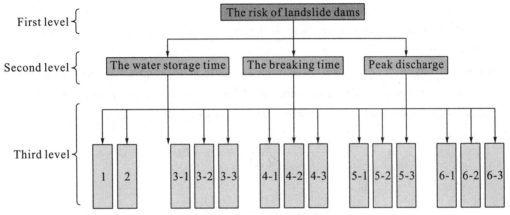

Figure 2. Hierarchical structure

3.2 Second-level weight

According to the general AHP method, the weight determination of the three parameters of the water storage time, the breaking time, and peak discharge is mainly based on subjective expert scores. However, after investigation, experts have different opinions on the weights of these three parameters. Some experts believe that the water storage time has the most significant impact on the risk of a dam. The shorter the water storage time, the faster the dam will break, and the easier it is to cause damage without countermeasures. And if the water storage time is infinite, the dam would not break, so there is no risk. Another part of the experts believes that the breaking time has the most significant impact on the risk of the dam. A longer breaking time means that floods will wash away the downstream for a long time. The remaining experts believe that the peak discharge has the most significant impact on the risk of the dam. The greater the peak discharge, the easier it is to break the design flood level of downstream water conservancy facilities. The impact on downstream water conservancy facilities Operation and safety will bring significant threats. When the flood flow is too large, it will overflow the river course and cause considerable losses to the lives and properties of downstream residents.

Due to the difficulty of unifying expert opinions, this paper proposes three schemes for determining Second-level weights. The first, second, and third schemes take the three parameters of the water storage time, the breaking time, and peak discharge as the most important influencing factors and make the weights of the other two influencing factors equal. The weight ratio of the most important factor and the minor factor is subject to the golden section method.

The weight vectors are

$$W_1 = \begin{bmatrix} 0.67 & 0.33 & 0.33 \end{bmatrix} \stackrel{\text{Normalized}}{=} \begin{bmatrix} 0.50 & 0.25 & 0.25 \end{bmatrix} \quad (1)$$

$$W_2 = \begin{bmatrix} 0.33 & 0.67 & 0.33 \end{bmatrix} \stackrel{\text{Normalized}}{=} \begin{bmatrix} 0.25 & 0.50 & 0.25 \end{bmatrix} \quad (2)$$

$$W_3 = \begin{bmatrix} 0.33 & 0.33 & 0.67 \end{bmatrix} \stackrel{\text{Normalized}}{=} \begin{bmatrix} 0.25 & 0.25 & 0.50 \end{bmatrix} \quad (3)$$

where W_1, W_2, and W_3 are the weight vectors when the water storage time, the breaking time, and peak discharge are the most important factors, respectively.

3.3 Third-level weight

Unlike the traditional AHP weight determination method, this research is based on the data obtained from the model experiment to calculate the weight of the three-tier indicators to each of the second-tier indicators.

The pairwise comparison of indicators also uses a scale of 1~9, and the score value is determined according to the comparison of the index size of different working conditions. V_{ij} determines the pairwise comparison value of different working conditions of a specific index:

$$V_{ij} = \begin{cases} \dfrac{8(Q_i - Q_j)}{Q_{\max} - Q_{\min}} + 1, & Q_i - Q_j \geqslant 0 \\ \\ \dfrac{8(Q_i - Q_j)}{Q_{\max} - Q_{\min}} - 1, & Q_i - Q_j < 0 \end{cases} \quad (4)$$

where V_{ij} is the index comparison value between the element in row i and the element in row j, Q_i is the index value obtained in the experiment represented by the element in row i, Q_j is the index value obtained in the experiment represented by the element in row j, $Q_{\max}$ Is the largest indicator value among all elements, and $Q_{\min}$ is the largest indicator value among all elements.

Based on the above formula and calculated based on test data, the pairwise comparison matrix of the fourteen sets of tests on the water storage time is shown in Table 2.

Table 2. The pairwise comparison matrix of the fourteen sets of tests on the water storage time

The water storage time	1	2	3-1	3-2	3-3	4-1	4-2	4-3	5-1	5-2	5-3	6-1	6-2	6-3
1	1.00	0.15	0.16	0.51	0.44	0.28	0.15	0.15	2.92	0.51	3.24	3.24	0.51	0.28
2	6.76	1.00	1.64	5.80	5.48	4.20	1.00	1.00	8.68	5.80	9.00	9.00	5.80	4.20
3-1	6.12	0.61	1.00	5.16	4.84	3.56	0.61	0.61	8.04	5.16	8.36	8.36	5.16	3.56
3-2	1.96	0.17	0.19	1.00	0.76	0.38	0.17	0.17	3.88	1.00	4.20	4.20	1.00	0.38
3-3	2.28	0.18	0.21	1.32	1.00	0.44	0.18	0.18	4.20	1.32	4.52	4.52	1.32	2.28
4-1	3.56	0.24	0.28	2.60	2.28	1.00	0.24	0.24	5.48	2.60	5.80	5.80	2.60	1.00
4-2	6.76	1.00	1.64	5.80	5.48	4.20	1.00	1.00	8.68	5.80	9.00	9.00	5.80	4.20

(To be continued)

(Continue)

The water storage time	1	2	3-1	3-2	3-3	4-1	4-2	4-3	5-1	5-2	5-3	6-1	6-2	6-3
4-3	6.76	1.00	1.64	5.80	5.48	4.20	1.00	1.00	8.68	5.80	9.00	9.00	5.80	4.20
5-1	0.34	0.12	0.12	0.26	0.24	0.18	0.12	0.12	1.00	0.26	1.32	1.32	0.26	0.18
5-2	1.96	0.17	0.19	1.00	0.76	0.38	0.17	0.17	3.88	1.00	4.20	4.20	1.00	0.38
5-3	0.31	0.11	0.12	0.24	0.22	0.17	0.11	0.11	0.76	0.24	1.00	1.00	0.24	0.17
6-1	0.31	0.11	0.12	0.24	0.22	0.17	0.11	0.11	0.76	0.24	1.00	1.00	0.24	0.17
6-2	1.96	0.17	0.19	1.00	0.76	0.38	0.17	0.17	3.88	1.00	4.20	4.20	1.00	0.38
6-3	3.56	0.24	0.28	2.60	0.44	1.00	0.24	0.24	5.48	2.60	5.80	5.80	2.60	1.00

By calculating the largest eigenvalue of the matrix, obtained $\lambda_{max} = 14.7308$, and consistency test parameter CR=0.0356<0.1, the weight vector of the fourteen tests to the water storage time is

$$W_{wst} = [0.03 \ 0.17 \ 0.14 \ 0.04 \ 0.05 \ 0.06 \ 0.17 \ 0.17 \ 0.01 \ 0.04 \ 0.01 \ 0.01 \ 0.04 \ 0.06] \tag{5}$$

where W_{wst} is the weight vector of the fourteen tests to the water storage time.

The weight vectors of the fourteen tests to the breaking time and peak discharge are obtained as the same:

$$W_{bt} = [0.05 \ 0.01 \ 0.19 \ 0.13 \ 0.09 \ 0.02 \ 0.02 \ 0.01 \ 0.05 \ 0.16 \ 0.06 \ 0.06 \ 0.08 \ 0.08] \tag{6}$$

$$W_{pd} = [0.03 \ 0.09 \ 0.03 \ 0.02 \ 0.05 \ 0.31 \ 0.16 \ 0.13 \ 0.03 \ 0.04 \ 0.03 \ 0.04 \ 0.02 \ 0.03] \tag{7}$$

where W_{bt} is the weight vector of the fourteen tests to the breaking time, W_{pd} is the weight vector of the fourteen tests to peak discharge.

3.4 Risk assessment quantification

The third-level elements and the second-level elements are weighted, and finally, the risk assessment values of landslide dams with different internal structures are calculated by the following formula. The risk assessment values are shown in Table 3.

A larger R_i means a higher risk of a landslide dam. The histogram of risk assessment under the three schemes is shown in Fig. 3. Under different second-level weights, there are specific differences between R values, but the overall difference is not significant. Researchers and emergency managers can select the appropriate R values for reference according to their needs.

$$R_i = W_i \times \begin{bmatrix} W_{wst} \\ W_{bt} \\ W_{pd} \end{bmatrix}, i = 1,2,3 \tag{8}$$

where R_i is the risk assessment values under different second-level weights (W_1, W_2 and W_3).

Table 3. The risk values under different second-level weights

Test number	R_1	R_2	R_3
1	0.0373	0.0319	0.0314
2	0.0728	0.0929	0.113
3-1	0.1388	0.0968	0.1247
3-2	0.0775	0.051	0.0547
3-3	0.0689	0.0574	0.0576
4-1	0.1012	0.1743	0.1134
4-2	0.0926	0.129	0.1317
4-3	0.0813	0.1103	0.1216
5-1	0.0348	0.0299	0.0261
5-2	0.0969	0.0664	0.0663
5-3	0.0419	0.0347	0.0296
6-1	0.0406	0.0364	0.0298
6-2	0.0544	0.0409	0.0436
6-3	0.0612	0.0483	0.0565

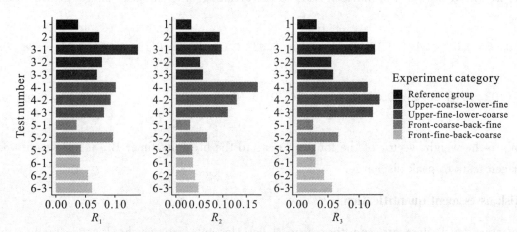

Figure 3. The histogram of risk assessment under the three schemes

4 Discussion and Conclusions

This paper evaluates the risks of landslide dams with different internal structures based on comprehensive decision-making methods. The traditional AHP method to evaluate the weight of each lower-level element is usually a method of scoring by experts. To unify expert opinions and avoid the risk assessment being too subjective, this study uses golden section methods to determine the weights of the three elements of the second layer (the water storage time, the breaking time, and peak discharge) and obtained three different weights vectors. For the third layer, the weight evaluation of each element of this layer is calculated in the formula proposed in the paper according to the specific process data of each plan in the model experiment. Finally, quantitative indicators for risk assessment of landslide dams with different internal structures can be obtained. In this way, the element weights of the third layer obtained by objective experiments are more convincing than expert scoring.

Based on model experiments and comprehensive evaluation theories, combined with subjective analysis method and objective experiment data, the risk of landslide dams with different internal structures is quantitatively evaluated, which has a significant guiding role in the emergency management of landslide dams. This cross-combination research method can also serve as a reference for researchers in related fields and promote landslide dam emergency management.

Acknowledgements

This work was financially supported by the National Key Research and Development Plan of China(Grant No. 2019YFC1510704) and the Sichuan Province Applied Basic Research Project(Grant No. 2020YJ0321).

References

[1] Costa J E, Schuster R L. The formation and failure of natural dams[J]. Geological Society of America Bulletin, 1988, 100: 1054−1068.

[2] Fan X, Dufresne A, Siva Subramanian, et al. The formation and impact of landslide dams-State of the art[J]. Earth-Science Reviews, 2020, 203: 103116.

[3] Hu J, Xu B, Chen Z, et al. Hazard and Risk Assessment for Hydraulic Fracturing Induced Seismicity Based on the Entropy-Fuzzy-AHP Method in Southern Sichuan Basin, China[J]. Journal of Natural Gas Science and Engineering, 2021, 90(B1): 103908.

[4] Meng C K, Chen K T, Niu Z P, et al. Influence of Internal Structure on Breaking Process of Short-Lived Landslide Dams[J]. Frontiers of Earth Science, 2021, 9: 604635.

[5] Mikhailov L, Tsvetinov P. Evaluation of services using a fuzzy analytic hierarchy process [J]. Applied Soft Computing, 2004, 5: 23−33.

[6] Saaty T L. Multicriteria Decision Making: The Analytic Hierarchy Process [M]. Pittsburgh: RWS Publications, 1990.

[7] Liu N. Hongshiyan landslide dam danger disposal and coordinated management[J]. Strategic Study of CAE, 2014, 16(10): 39−46.

[8] Liu N, Cheng Z L, Cui P, et al. Dammed lake and risk management[M]. Beijing: Science Press, 2013.

[9] Li S, LI X, Zhang J, et al. Study of geological origin mechanism of Tangjiashan landslide and entire stability of landslide dam[J]. Chinese Journal of Rock Mechanics and Engineering, 2010, 2908−2915.

[10] Shi Z, Zheng H, Peng M, et al. Breaching mechanism analysis of landslide dams considering different spillway schemes—a case study of Tangjiashan landslide dam[J]. Journal of Engineering Geology, 2016, 741−751.

[11] Cai Y, Luan Y, Yang Q, et al. Study on structural morphology and dam-break characteristics of Baige barrier dam on Jinsha River[J]. Yangtze River, 2019, 50(3): 15−22.

Simulation of Colloids Transport and Retention by Pore-network Model: Influence of Particle Size and Flow Velocity

Dantong Lin[1], Scott Bradford[2], Liming Hu[3], Xinghao Zhang[4], Irene Lo[5]

[1,3,4] State Key Laboratory of Hydro-Science and Engineering, Department of Hydraulic Engineering, Tsinghua University, Beijing 10084, China

[2] USDA-ARS, Sustainable Agricultural Water System Unit, Davis, California 95616, United States

[3] Department of Civil and Environmental Engineering, The Hong Kong University of Science and Technology, Hong Kong, China

Abstract

Contaminated groundwater remediation is of importance for the sustainable development of water environment. Colloid transport and retention is a common phenomenon during the remediation process, i. e. the migration of nano-particles including environmental remediation materials, clay particles, bacteria, humic acids, and so on. Previous experimental observation reports that particle size and flow velocity have a great impact on the mobility of colloids. However, the mechanism behind the experimental observation is still not well understood. The pore-network model (PNM) is an effective method to represent the pore structure and study the mass transfer process in porous media, which is recently used to simulate the transport and retention of colloids. In this study, a PNM is developed to study the transport and retention of colloids in porous media. The size-related mechanisms including effective porosity reduction and size exclusion are considered in the model. Pore-scale deposition parameters are determined under favorable conditions. Breakthrough curves and retention profiles are provided by the PNMs under different conditions. Results show that under unfavorable conditions, size-related mechanisms can cause the early breakthrough of colloids than conservative solute. An increase in colloid size leads to earlier breakthroughs. Under favorable conditions, the mobility of colloids increases first and then decreases with particle size due to the combined effects of gravity sedimentation and diffusion. Lower flow velocity leads to more retention. The result can provide scientific evidence to understand the mechanism of colloid transport and surface deposition during contamination control and in-situ remediation for groundwater.

Keywords: Groundwater remediation; Colloid transport; Surface deposition; Particle size; Pore-network model

1 Introduction

Prediction of transport parameters of colloid in porous media is of great significance in both nature and industry. For example, the transport of viruses and bacteria in the underground system will influence the quality of the groundwater during extraction. Recently, nano- or microscale particles like zero-valent iron or bubbles have been used as in-situ remediation materials for groundwater(Hu and Xia, 2018).

The transport of colloids is usually studied by column tests and the breakthrough curves of colloids are often compared with that of conservative tracers(Yu et al., 2018). Previous work has reported the early breakthrough of colloids compared to tracers in column tests. This phenomenon is generally explained by two size-related mechanisms: (a) colloids cannot get into the pores smaller than their size, which is also called effective porosity reduction(b) colloids are excluded from the low-velocity regions near the solid-water interface(SWI) due to its size(Scheibe and Wood, 2003), which is also called size exclusion. According to previous experimental observation, the early breakthrough of colloids is significantly influenced by the particle size of colloids under unfavorable conditions (almost on surface deposition).

Under favorable conditions, the attachment and detachment of colloids can happen on the surface of the collector(sand or glass beads)(Hilpert and Johnson, 2018). Previous experimental observations have reported that particle size and flow velocity have a great impact on the transport and retention of colloids. In the field application of nanomaterials in groundwater remediation, particle size and flow field are two key factors which influence the effective remediation area(He et al., 2010). Therefore, it is of significance to study the influence of particle size and flow velocity on the mobility of colloids.

The Pore-network model(PNM) is an effective method to simulate pore structure and the mass transfer process in the porous media. The equivalent geometry parameters of pore structure can be extracted by x-ray computer tomography images of a real porous media, and PNMs are then built based on these equivalent geometry parameters to calculate the flow field and mass transfer process in the pore structure efficiently(Hu et al., 2018). Recently, PNMs began to be applied to simulate the transport and retention of colloids in porous media. PNMs can upscale the mass transfer process from pore- to macro-scale, which can help us to understand the relative importance of different mechanisms. However, the size-related mechanisms(effective porosity reduction and size exclusion) are not included in previous PNMs and the study about the influence of particle size and velocity in PNMs is still very limited.

This study aims to investigate the transport and retention of colloids in porous media by PNMs. Two size-related mechanisms(a) effective porosity reduction and(b) size exclusion are included in the model. The influence of particle size and velocity on the breakthrough curves and retention profiles of the model are discussed.

2 Method

2.1 Geometry and flow field

In the PNM, the pore structure is combined by pore bodies and pore throats. The sphere pore bodies are aligned on the point of a $10 \times 10 \times 20$ grid and connected by cylindrical pore throats. The geometry of the model is shown in Fig. 1.

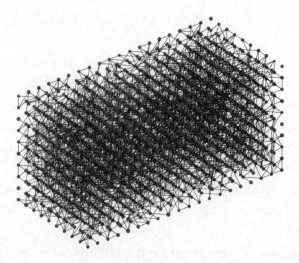

Figure 1. The geometry of the PNM

The radius of pore bodies obey normal distribution(Zhang et al., 2017) with a mean value equals to 30 μm and the standard deviation equals 5 μm. The radius of the pore throats is determined by the BACON equation (Acharya et al., 2004) with the curvature number equals 0.5, of which the distribution is shown in Fig. 2. One pore body in the PNM can be connected to at most 26 pore throats (Raoof and Hassanizadeh, 2009) and the coordination number is set as 8. The porosity of the PNM is 0.3 and the length of the element of the grid is 82.6 μm, so the length of this model is 0.156 cm. The pore volume(PV) of the PNM is 3.2×10^{-4} cm³.

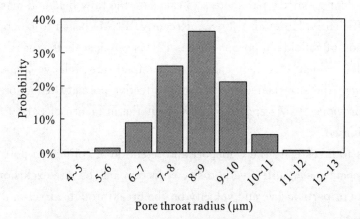

Figure 2. The pore throat radius distribution in the PNM

The left and right boundaries of the PNM are set as constant pressure boundaries to obtain a steady-state flow field with an approach velocity equals to 10^{-5} m/s. Water is assumed to be incompressible liquid and the water flux is in the range of laminar flow. The flux in the pore throat Q_{ij} can be determined as(Raoof et al., 2012):

$$Q_{ij} = \frac{\pi r_{ij}^4}{8\mu l_{ij}}(p_i - p_j) \tag{1}$$

where p_i and p_j at both ends of the pore throat, r_{ij} is the radius of the pore throat, l_{ij} is the length of the pore throat and μ is the viscosity of water. The flux in the pore body Q_i equals the sum of the flux in the upstream pore throats, which can be calculated as:

$$Q_i = \sum_j^{N_{up}} Q_{ij} \qquad (2)$$

where N_{up} is the number of the upstream pore throats connected to this pore body.

2.2 Conservative solute transport

No adsorption happens in both the pore body and the pore throat for conservative solutes. Based on the mass balance, the transport of a conservative solute in the pore body can be written as

$$V_i \frac{dC_i}{dt} = \sum_j^{N_{up}} Q_{ij} C_{ij} - Q_i C_i \qquad (3)$$

where V_i is the volume of the pore body, C_i is the liquid concentration in the pore body. The transport of a conservative solute in the pore throat can be calculated as

$$V_{ij} \frac{dC_{ij}}{dt} = Q_{ij} C_j - Q_{ij} C_{ij} \qquad (4)$$

where V_{ij} is the volume of the pore throat, C_j is the liquid concentration of the input pore body, C_{ij} is the liquid concentration in the input pore throat.

2.3 Colloid transport and retention

To simulate the transport of colloids in the PNM, the size-related mechanisms (a) effective porosity reduction and (b) size exclusion are included in the model, of which the schematic drawing is shown in Fig. 3.

In the PNM, colloids cannot get into the pore throat of which the size is smaller than that of the particle, as shown in the gray regions in Fig. 3. This fraction of particles will go to other pore throats connected to this pore body. Besides, colloids do not sample the truncated portion of the parabolic velocity profile, as shown in the pink region in Fig. 3 in the pore throat, of which the width equals the radius of the particle. In this way, the effective volume of the pore throat where colloids can transport is smaller than the real volume of the pore throat and the mean flux of the colloids in this pore throat is higher than the mean value of solute.

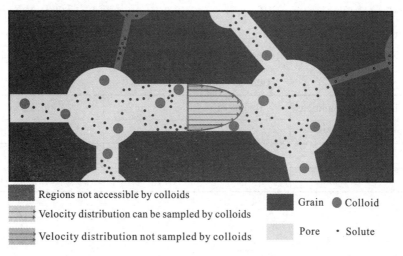

Figure 3. The schematic drawing of the size-related mechanism which can cause the early breakthrough of colloids than tracer in porous media

The surface deposition of the colloids is assumed to only happen in the pore throats. The transport and surface deposition of colloids in thepore throat can be expressed as

$$V'_{ij}\frac{dC_{ij}}{dt} + V'_{ij}\frac{dS_{ij}}{dt} = Q_{ij}C_j - Q'_{ij}C_{ij} \qquad (5)$$

$$\frac{dS_{ij}}{dt} = k_{\text{att},ij}C_{ij} - k_{\text{det},ij}S_{ij} \qquad (6)$$

where S_{ij} is the solid concentration in the pore throat, $k_{\text{att},ij}$ and $k_{\text{det},ij}$ are the attachment and detachment coefficients of the pore throat respectively. Under unfavorable conditions, the value of $k_{\text{att},ij}$ and $k_{\text{det},ij}$ are set to be zero. Under favorable conditions, $k_{\text{att},ij}$ is calculated by the following equation:

$$k_{\text{att},ij} = \frac{\eta_{ij}v_{ij}}{l_{ij}} \qquad (7)$$

where v_{ij} is the mean velocity of the pore throat, η_{ij} is the collection efficiency, which is determined by numerical simulation like Colloid Filtration Theory(Logan et al., 1995) in a cylindrical tube with the particle density set as 1500 kg/m³. The detailed method used to predict η_{ij} can be found in another work(Lin et al., 2021). The value of $k_{\text{det},ij}$ is set as $0.01 \times k_{\text{att},ij}$ according to previous work.

The V'_{ij} is the effective volume of the pore throat for colloids transport, which equals to

$$V'_{ij} = \pi(r_{ij} - r_c)^2 l_{ij} \qquad (8)$$

where r_c is the radius of the colloids. Q'_{ij} is the mean colloid flux, which is can be calculated as

$$Q'_{ij} = Q_{ij} \times \frac{(r_{ij} - r_c)^2}{r_{ij}^2} \times \left[2 - \frac{(r_{ij} - r_c)^2}{r_{ij}^2}\right] \qquad (9)$$

No colloid deposition happens in the pore bodies, so the colloid mass balance in a pore body of colloids is the same as Eq.(4). Based on Eqs.(3)~(8), the concentration distribution of solutes and colloids can be solved in the PNMs.

2.4 Simulation scenarios

In the simulation, colloids with different r_c and conservative solute($C_0 = 10^N \text{c/m}^2$) are injected from the left of the model for 57.9 s and followed by water($C_0 = 0$) under different flow velocity conditions. The whole simulation continues for 250 s. The simulation scenarios are shown in Table 1.

Table 1. Simulation scenarios

No.	Deposition conditions	Radius of particles, r_c (μm)	Approach veolicy, U(m/d)
1	Unfavorable	0.5, 1, 3, 5, 7, 8	0.864
2	Favorable	0.05, 0.5, 1, 2, 3	0.864
3	Favorable	0.5	0.0864, 0.864, 8.64, 86.4

3 Results and Discussion

3.1 Under unfavorable conditions

The influence of colloids size on the BTCs is shown in Fig. 4. Under favorable conditions, $k_{att,ij}$ and $k_{det,ij}$ for each pore throat are assumed to be zero, which means no surface deposition happens in the PNMs. Fig. 4 shows that colloids have an early breakthrough than the conservative solute, which is consistent with the previous experimental observations.

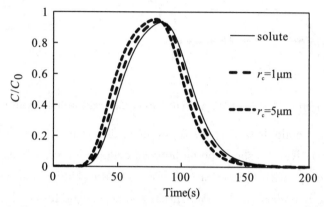

Figure 4. Breakthrough curves of solute and colloids with different particle sizes

To further illustrate this phenomenon, the time it takes for the C/C_0 of the BTCs to reach 0.5, $t_{0.5}$, is compared for colloids with r_c ranges from 1~9.5 μm, which is shown in Fig. 5. The $t_{0.5}$ for solute is marked in Fig. 5 as $r_c=0$. Results show that when $r_c<8$ μm, $t_{0.5}$ decreases with r_c, which means larger colloids have an earlier breakthrough. Results in Fig. 5 also implied that when the colloids injected into a porous media have a wide size distribution, part of the particles can come out much earlier than the other part due to the difference in transport paths. Notice that the length of the model in this study is only 0.156 cm for a high computing speed. In column scale or filed scale, $t_{0.5}$ can be much higher than what is presented in Fig. 5.

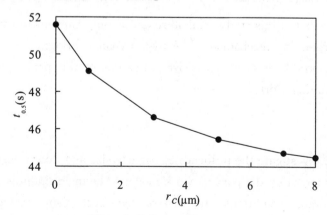

Figure 5. Influence of r_c on $t_{0.5}$

3.2 Under favorable conditions

The breakthrough curves and retention profiles of the model with particles with different r_c when $U=0.864$ m/d are shown in Fig. 6(a) and Fig. 6(b). Results show that the mobility of colloids increases with r_c first and then decreases with r_c. This is due to the coupled influence of diffusion and gravity sedimentation(Tufenkji and Elimelech, 2004). The influence of diffusion is more significant for smaller particles and that of gravity sedimentation is more pronounced for larger particles, therefore there is an optimum r_c with which the colloids have the highest mobility. In this study, this optimum r_c is around 0.5 μm.

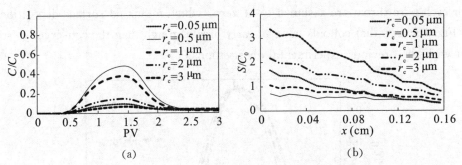

Figure 6. The (a) BTCs and (b) RPs of the model with different r_c

The BTCs and RPs of colloids with $r_c = 0.5$ μm under different flow velocity conditions are shown in Fig. 7(a) and Fig. 7(b). Results show the mobility of colloids increases with flow velocity, which is caused by the decrease of η_{ij} with the increase of flow velocity. Under high-velocity condition, fewer colloids will come to the boundary of the pore throat and this cause less collection.

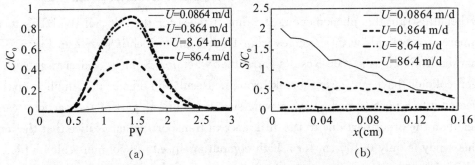

Figure 7. The (a) BTCs and (b) RPs of the model under different velocity conditions

It is worth noticing that in this study, colloid retention only happens in the pore throat by surface sedimentation. Other retention mechanisms like hydrodynamic bridging or straining can cause the retention of particles in pore bodies, which is over the scope of this paper. More discussion on this topic can be found in another work.

4 Conclusions

This study aims to investigate the influence of particle size and flow velocity on colloid transport and retention in porous media by the pore-network model. The main findings of this study include:

(1) Under unfavorable conditions, the size-related mechanisms including effective porosity reduction and size exclusion can cause the early breakthrough of colloids than conservative solute. When the particle size is smaller than the mean size of the pore throat, larger particles break through the model earlier.

(2) Under favorable conditions, the mobility of colloids increases first with the particle size and then decreases with particle size, which is caused by the coupled influence of diffusion and gravity

sedimentation.

(3) Under favorable conditions, higher flow velocity leads to less retention of colloids, due to the decrease of collector efficiency with an increase in velocity.

Acknowledgements

The National Key Research and Development Program of China (2017YFC0403501, 2020YFC1806502) funded this research.

References

[1] Acharya R C, van der Zee S E A T M, Leijnse A. Porosity-permeability properties generated with a new 2-parameter 3D hydraulic pore-network model for consolidated and unconsolidated porous media[J]. Advances in Water Resources, 2004, 27(7):707−723.

[2] Chrysikopoulos C V, Katzourakis V E. Colloid particle size-dependent dispersivity[J]. Water Resources Research, 2015, 51(6):4668−4683.

[3] He F, Zhao D, Paul C. Field assessment of carboxymethyl cellulose stabilized iron nanoparticles for in situ destruction of chlorinated solvents in source zones[J]. Water Research, 2010, 44(7):2360−2370.

[4] Hilpert M, Johnson W P. A binomial modeling approach for upscaling colloid transport under unfavorable attachment conditions: Emergent prediction of non-monotonic retention profiles[J]. Water Resources Research, 2017, 54(1):46−60.

[5] Hu L, Guo H, Zhang P, et al. Pore-Network Model for Geo-Materials. Paper presented at the Geoshanghai International Conference.

[6] Hu L, Xia Z. Application of ozone micro-nano-bubbles to groundwater remediation[J]. Journal of Hazardous Materials, 2018, 342:446−453.

[7] Lin D, Hu L, Bradford S A. Simulation of colloid transport and retention using a pore-network model with roughness and chemical heterogeneity on pore surfaces[J]. Water Resources Research, 2021, 57(2).

[8] Lin D, Hu L, Lo I M C. Size distribution and phosphate removal capacity of nano zero-valent iron(nzvi): Influence of ph and ionic strength[J]. Water, 2020, 12(10).

[9] Logan B E, Jewett D G, Arnold R G. Clarification of clean-bed filtration models[J]. Journal of Environmental Engineering, 1995, 121(12):869−873.

[10] Molnar I L, Johnson W P, Gerhard J I. Predicting colloid transport through saturated porous media: A critical review[J]. Water Resources Research, 2015, 51(9), 6804−6845.

[11] Raoof A, Hassanizadeh S M. A new method for generating pore-network models of porous media[J]. Transport in Porous Media, 2009, 81(3):391−407.

[12] Raoof A, Nick H M, Wolterbeek T K T. Pore-scale modeling of reactive transport in wellbore cement under CO_2 storage conditions[J]. International Journal of Greenhouse Gas Control, 2012, 11,S67-S77.

[13] Sasidharan S, Torkzaban S, Bradford S. Transport and retention of bacteria and viruses in biochar-amended sand[J]. Science of the Total Environment, 2016, 548−549:100−109.

[14] Scheibe T D, Wood B D. A particle-based model of size or anion exclusion with application to microbial transport in porous media[J]. Water Resources Research, 2003, 39(4).

[15] Seetha N, Raoof A, Mohan Kumar M S. Upscaling of nanoparticle transport in porous media under unfavorable conditions: Pore scale to Darcy scale[J]. Journal of Contaminant Hydrology, 2017, 200:1−14.

[16] Torkzaban S, Bradford S A, Vanderzalm J L. Colloid release and clogging in porous media: Effects of solution ionic strength and flow velocity[J]. Journal of Contaminant Hydrology, 2015, 181:161−171.

[17] Tufenkji N, Elimelech M. Correlation equation for predicting single-collector efficiency in physicochemical filtration in saturated porous media[J]. Environmental Science & Technology, 2004, 38(2):529−536.

[18] Yang H T, Balhoff M T. Pore-network modeling of particle retention in porous media[J]. AIChE Journal, 2017, 63(7):3118-3131.

[19] Yu Z, Hu L, Lo I M C. Transport of the arsenic(As)-loaded nano zero-valent iron in groundwater-saturated sand columns: Roles of surface modification and As loading[J]. Chemosphere, 2018, 216:428-436..

[20] Zhang P, Hu L, Meegoda J N. Pore-scale simulation and sensitivity analysis of apparent gas permeability in shale matrix[J]. Materials(Basel), 2017, 10(2).

[21] Zhuang J, Qi J, Jin Y. Retention and transport of amphiphilic colloids under unsaturated flow conditions: effect of particle size and surface property[J]. Environ Sci Technol, 2005, 39(20):7853-7859.

Study on Relationship of Sediment Delivery Ratio with Inflow and Operational Water Level of the Three Gorges Reservior

Ying Zhang[1], Sichen Tong[2], Xiaoping Long[3], Guoxian Huang[4,5]

[1,2,3] School of River and Ocean Engineering, Chongqing Jiaotong University, Chongqing 400074, China

[4] Chinese Research Academy of Environmental Sciences, Beijing 100012, China

[5] State Key Laboratory of Plateau Ecology and Agriculture, Qinghai University, Xining 810016, China

Abstract

The characteristics of water and sediment inflow of the Three Gorges Reservoir (TGR) have greatly changed compared with its design stage after its impoundment in 2003. Sedimentation occurred following the rise of water level due to the decrease of flow velocity and sediment carrying capacity. Deposition rate or sediment delivery ration (SDR) is of key importance for preserve the storage capacity of the TGR for maintaining its comprehensive benefits Based on the measured water and sediment data of the TGR from 2003 to 2018, this paper analyzes the impact of inflow variation and reservoir operation on SDR in flood season, and discusses the response relationship between SDR and its mainly influencing factors, single factor and combination factors. Results show that the SDR of the TGR is positively correlated with the inflow discharge, and negatively correlated with the operational water level and the detention time. With the development of the reservoir operation stage, the response relationship between the SDR and single influencing factor gradually becomes gradually complicated, and the effect of combination factors is enhanced. The flood detention time longer than 16 days should be avoided as far as possible to avoid the minimal SDR.

Keywords: Sediment delivery ratio; Inflow; Water level; Detention time; Three Gorges Reservoir

1 Introduction

The sediment delivery ratio (SDR) is an important index to analyze the sediment deposition status. The larger the SDR is, the better the sediment delivery effect of the reservoir will be, and the smaller the degree of sediment deposition occurs(Dong et al., 2017). In order to reduce the adverse effects of sediment deposition, appropriate sediment delivery operation is needed to be carried out according to the variation of SDR. There are many factors that affecting the ratio of sediment discharge, including water and sediment in and out of the reservoir, reservoir operation mode, river terrain conditions and river sand excavation, soil and water conservation and other factors(Chen et al., 2012; Huang et al., 2013). The characteristics of the SDR of the TGR are constantly changing under the condition of new water and sediment, especially under the condition of rising water leve. The response relationship between the SDR and various factors is also needed to be analyzed. Therefore, from the perspective of controllable engineering operation, this paper mainly analyzes the influence of inflow and water level in front of the dam in flood season on the SDR. Relationship between the SDR of the TGR in flood season and the single or combined factors is obtained.

From 2003 to 2008, the reservoir operates in the mode of 135~156 m water level, and the SDR is greatly affected by the inflow condition. During the operation period of 2009—2012, the variation trend of water and sediment inflow is basically the same, but with the normal operation water level rising to 175m, the sediment discharge efficiency of the reservoir is greatly reduced. During the operation period of 2013—2018, due to the successive impoundment and operation of the Xiangjiaba Reservoir and the Xiluodu Reservoir in the upper reaches of Yangtze River, the inflow and sediment situation of the TGR has greatly changed. The sediment inflow amount has obviously decreased, and the SDR of the TGR has also changed consequently(Hu et al., 2017; Jin et al., 2018; Tang et al., 2019). Therefore, this paper analyzes the influence of various factors on the SDR of the TGR indifferent operational stages.

2 Relationship Between SDR and Single Factor

2.1 Inflow

The annual runoff of the TGR varied very little in recent decades. The inflow is mainly concentrated in the flood season(May to October). The slight decrease of annual runoff after the operation of the reservoir is mainly caused by the decrease of runoff in the main flood season(June to September). Fig.1 shows the relationship between the SDR and the inflow of the TGR in the flood season of 2003—2008, 2009—2012 and 2013—2018 respectively. It can be seen that the SDR is directly proportional to the inflow. With the increase of inflow, the flow velocity in the reservoir increases, which is the reason for the obvious enhancement of sediment discharge effect. Due to the change of inflow sediment and the rise of reservoir water level, the correlation coefficient between SDR and inflow in the three stages is obviously different, and has a decreasing trend. Moreover, the sensitivity of SDR to inflow change after 2013 is less than that before 2013.

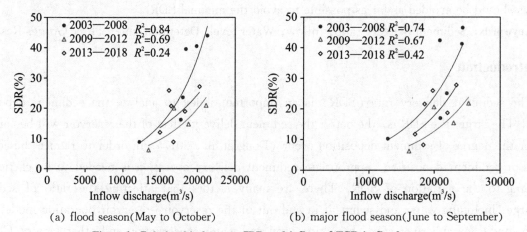

(a) flood season(May to October)　　(b) major flood season(June to September)

Figure 1. Relationship between SDR and inflow of TGR in flood seasons

2.2 Water level of the reservoir

The average water level of the reservoir in the initial impoundment period of the reservoir is comparatively low in the flood season, and the river course is close to the natural river course. When the inflow discharge is large, the flow velocity in the reservoir area is large, the flow power is sufficient and the sediment carried is relatively easy to be discharged out of the reservoir. When the water level rises gradually in the flood season, the difference between the reservoir river and the natural river increases gradually, the flow velocity slows down. In addition, the annual backwater area is silted up the river slope decreases. The sediment is relatively difficult to discharge out of the reservoir, so the SDR decreases(Zhu et al. , 2021). When the average water level of the reservoir in flood season is higher than 145 m, the SDR in flood season is lower than 30%. When the average water level of the reservoir is higher than 155 m, the SDR is almost lower than 25%(Fig. 2). Under the same operation scheme in flood season the sediment inflow volume is greatly reduced after 2013 with the impoundment of upper reservoirs and the sediment grain size is relatively small. The degree of sediment deposition is relatively reduced, so the SDR is increased. At the same time, due to the continuous enrichment of reservoir operation experience, effective measures such as sand peak sediment discharge operation are implemented in the year with large incoming sediment in flood season, and the sediment discharge ratio of the reservoir is increased. As shown in Fig. 3, the SDR in 2013—2018 is slightly higher than that in 2009—2012.

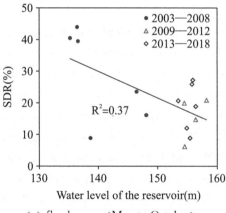 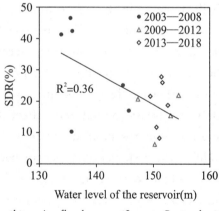

(a) flood season(May to October)   (b) major flood season(June to September)

Figure 2. Relationship between SDR and average water level of the reservoir in flood seasons

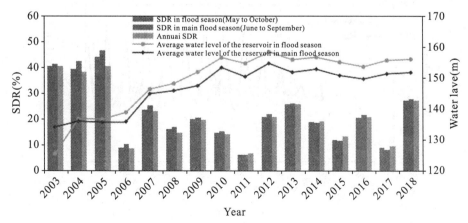

Figure 3. Variation of SDR and average water level of the TGR in flood season

3 Relationship Between SDR and Combination Factors

The SDR is both affected by inflow and average water level in front of dam in flood season. In fact, it is not only affected by single factor but by multiple factors.

The single factor correlation between the SDR and the inflow is slightly better than other factors, but after the water and sediment conditions and dispatching mode changed greatly in 2013, the single correlation between the SDR and the inflow in the reservoir area is obviously weakened. The further increase of water level makes the relationship between the SDR and the influencing factors more complex, so it is necessary to analyze the response relationship between the SDR and the combination factors.

The detention time of flood in the reservoir in flood season is affected by the water level of the reservoir. Before the construction of the reservoir, the time for the flood from Cuntan Hydrological Station to Huanglingmiao Hydrological Station is about 3 days. The sand peak basically lags behind 3 days, and the propagation time of the flood peak and sand peak is basically synchronous. After the completion of the TGR, the detention time of the sand peak in the reservoir area is different due to the different water level of the reservoir. After the operation of 145 m in the flood season, the average time for the sand peak from Cuntan Hydrological Station to Huanglingmiao Hydrological Station is about 7.6 days. When the inflow peak flow is less than 30000 m^3/s, the average time is more than 9 days. Under the same inflow discharge, the higher the water level of the reservoir is, the longer the sand peak will transport in the reservoir(Dong et al., 2014).

The TGR has a lot of comprehensive benefits and plays an important role in flood control. The implementation of flood control operation scheme in flood season affects the SDR. Different water levels of the reservoir correspond to different reservoir capacities. The lower the reservoir water level and greater the inflow is, the greater the flow velocity is, and the better the sediment discharge effect is. The reservoir capacity can not only reflect the water level of the reservoir, but also directly affect the flow velocity in the reservoir area(Hu et al., 2019). Therefore, the reservoir capacity V can be used to establish the response relationship between SDR and V/Q (detention time). The reservoir capacity can be obtained from the water level and capacity curve of the TGR(Zhou et al., 2013) through the average water level. The monthly change of water level and capacity in flood season is shown in Fig. 4.

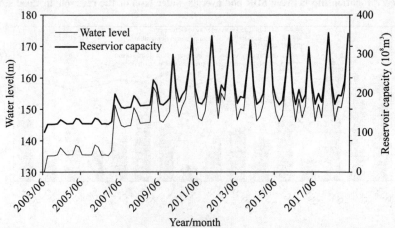

Figure 4. Monthly average water level of the reservoir and corresponding reservoir capacity of the TGR in flood season(2003—2018)

Fig. 5 shows the relationship between SDR and flood detention time of the TGR in flood season. It can be seen that there is an inverse relationship between SDR and flood detention time V/Q. The logarithm function correlation coefficients of 2003—2008, 2009—2012 and 2013—2018 are 0.9, 0.9 and 0.6 respectively. When the detention time is more than 10 days, the sediment delivery ratio is almost less than 30%, and the monthly detention time in the main flood season is generally less than 20 days. The relationship between monthly sediment delivery ratio and flood detention time in flood season from 2003 to 2018 shows a slightly asymmetric "V" shape, especially in 2013—2018 with about 16 days as the dividing line. When the detention time is about 16 days, the minimum sediment delivery ratio of the TGR is about 4%, and the detention time is shortened by 2~3 days. The main flood season does not have "V" shape distribution. It shows that the combined effect of dispatching in May and October is complicated. The key factor affecting the delivery ratio is flood detention time in the main flood season(Liu et al., 2019).

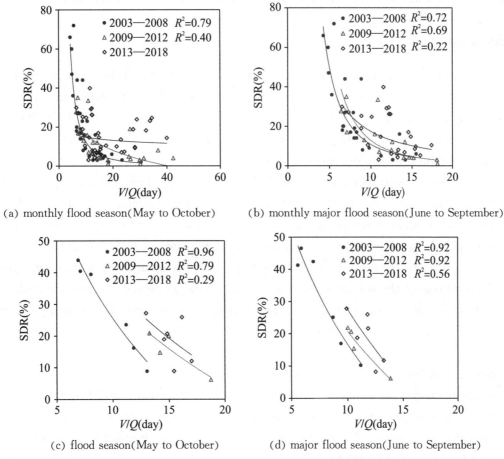

(a) monthly flood season(May to October) (b) monthly major flood season(June to September)

(c) flood season(May to October) (d) major flood season(June to September)

Figure 5. Relationship between SDR and flood detention time in different flood season of the TGR

The relationship between SDR and flood detention time is plotted according to the SDR during the flood season of 2013—2017(Fig.6). When the detention time is less than 16 days, the SDR is negatively correlated with the detention time. This is because the water level of the reservoir does not change obviously in the main flood season, but with the increase of inflow the sediment carrying capacity of the flow increases, resulting in better sediment discharge effect. When the detention time is about 16 days, the turning point appears in the diagram, and the SDR is about 5%. The detention time increases slightly after more than 16 days, which is due to the large difference between the inflow and outflow during the pre-flood fluctuation period and the post flood storage period. In these periods, the SDR is strongly affected by various factors.

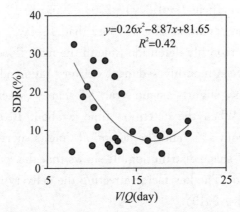

Figure 6. Relationship between SDR of single flood and detention time of the TGR in flood season(2013—2017)

4 Conclusions

Based on the data of the flow and reservoir regulation of the TGR since its operation, from 2003—2008, this paper discussed the relationship between the SDR with inflow and operational water level according to the sediment discharge effect of the TGR. Results show that the SDR of the TGR is positively correlated with the inflow, negatively related to the water level and the flood detention time. When the flood detention time is 16 days, the SDR of the TGR is about 5%, which is unfavorable for the Sediment Ejection of the TGR. In order to reduce the sediment deposition in the reservoir area, the water level should be adjusted reasonably in combination with the inflow condition in order to avoid or shorten the occurrence of this situation. With the development of the reservoir operation stage, the reservoir storage water and sediment conditions may change further. Considering that the sediment from the uncontrolled area is large, and the sediment deposition changes the river boundary conditions of the reservoir area. In addition, the TGR and the cascade reservoirs in the lower reaches of Jinsha River are continuously carrying out joint optimal operation. The SDR of the TGR needs to be further observation and in-depth analysis.

Acknowledgements

This study was partially supported by the National Key Research and Development Program of China(2018YFB1600400) and the Basic and Frontier Research Programs of Chongqing, China (cstc2018jcyjAX0534) and Model Development and Application of Water Environment in the Yangtze River Basin(Grant No. 2019-LHYJ-0102).

References

[1] Chen G Y, Yuan J, Xu Q X. On sediment diversion ratio after the impoundment of the Three Gorges Project[J]. Advances in Water Science, 2012, 23(3): 355—362.

[2] Dong B J, Chen X W, Xu Q X. Investigations and considerations on peak sediment regulation of Three Georges Reservoir[J]. Yangtze River, 2014, 45(19): 1—5.

[3] Dong Z D, Hu H H, Ji Z W. Response of sediment excluding ratio to incoming flow and sediment in the Three Gorges Reservoir[J]. Journal of Sediment Research, 2017, 42(6): 16—21.

[4] Hu C H, Fang C M, Xu Q X. Application and optimization of "storing clean water and discharging muddy flow" in the Three Gorges Reservoir[J]. Journal of Hydraulic Engineering, 2019, 50(1): 2−11.

[5] Hu C H, Fang C M. Research on sediment problem solutions for the Three Gorges Project and it operational effects[J]. Scientia Sinica(Technologica), 2017, 47(8): 832−844.

[6] Huang R Y, Tan G M, Fan B L. Sediment delivery ratio of Three Gorges reservoir in flood events after its impoundment[J]. Journal of Hydroelectric Engineering, 2013, 32(5):129−133, 152.

[7] Jin X P, Xu Q X. Sediment issues in joint dispatch of reservoir group in upper Yangtze River[J]. Yangtze River, 2018,49(3):1−8,31.

[8] Liu S W, Zhang X F, Xu Q X. Estimation of sediment amount from ungauged area and analysis of sediment delivery ratio of Three Gorges Reservoir[J]. Journal of Lake Sciences, 2019, 31(1): 28−38.

[9] Tang X Y, Tong S C, Xu G X. Delayed response of sedimentation in the flood seasons to the pool level of the Three Gorges Reservoir[J]. Advances in Water Science, 2019, 30(4): 528−536.

[10] Zhu L L, Xu Q X, Dong B J. Study on the effect and influencing factors of sand discharge of Xiluodu Reservoir in the Lower Jinsha River[J]. Advances in Water Science, 2021: 1−9.

[11] Zhou J J, Cheng G W, Yuan J. Dynamic storage of Three Gorges reservoir and its application to flood regulations: Sensitivity in regulation[J]. Journal of Hydroelectric Engineering, 2013, 32(1): 165−169.

Study on the Scheme of Diverting Clean Water Operation in the City Proper of Changzhou

Shousheng Mu, Jingxiu Wu, Ziwu Fan, Yang Liu, Guoqing Liu, Chen Xie

Nanjing Hydraulic Research Institute(NHRI),
Nanjing 210029, China

Abstract

Urbanized plain river network features complex river channel system and low fluidity. Influenced by physical factors and human modification, water quality and stream structure in such an area has unique characteristics. Changzhou is a typical plain river network city which encountered poor water mobility and heavy water pollution. On account of the complex river system, numerous water-related projects and poor river water quality in this city, the regulation and control of water environment is a difficult task. The basic strategy to improve the water quality is to reduce the amount of pollutant to the water by applying physical, chemical and biological methods. Although some effects have been made to improve the water quality in this area, several problems still exist. In other words, there is no fundamental change in bring the water pollution under control. In this paper, a new scheme of diverting clean water operation is proposed to improve the water quality in Changzhou city, based on the idea of system governance. This scheme includes the integrated control of guide engineering, river regulation engineering, gates and pumping stations. If necessary, reconstruction of some water-related projects has been taken into consideration. Due to the request of diverting clean water operation, fresh water is obtained from the main stream of Yangtze River. According to the combination of numerical simulation, field observation and numerous methods of diverting clean water operation, the scheme of smooth flow and live water in the city proper of Changzhou is formulated. A smooth flow pattern with three-level water level difference and reasonable flow distribution is formed by scientifically dispatching the existing dams, gates and pumps and other water conservancy projects. The fresh water is transported across the river network in the proper area of Changzhou City form north to the south. The water quality of river network has been significantly improved through the operation of this scheme. Results of the observation during and after this operation indicate that, besides the diverting clean water operation, controlling of the sources of pollution is also needed for the long term improvement of water quality.

Keywords: Plain river network; Water environment; Diverting clean water operation; Changzhou

1 Introduction

Urbanization substantiallychanges the water quality, stream structure and the storage capacity of river channels in lowland plain river networks. Urbanized plain river network features complex river system and low fluidity. Water quality and stream structure in such an area has unique characteristics, with the effects of physical factors and human modification. There are numerous water-related projects in this system and the river water quality is usually poor in this river network. Therefore, the regulation and control of water environment in such an area is a difficult task(Yu et al., 2021). The basic strategy to improve the water quality is to reduce the amount of pollutant to the water by applying physical, chemical and biological methods. Physical methods usually include controlling the sources of pollution, sediment dredging, hydrodynamics controlling and water aeration. Adding algaecide and flocculent settling are the main chemical methods. In addition, microbial degradation, purification by aquatic plants and purification by constructed wetland are typical biological methods. At present, all water environment treatment related technologies are mature. Physical and biological methods are widely applied among these technologies. On the basis of these related technologies, the combination of numerical simulation and joint operation of gates and pumping stations are applied to control the hydrodynamics in plain river network, which could improve the water environment significantly(Jiang et al., 2011; Li et al., 2016; Yuan et al., 2015; Zuo et al., 2015). Practices indicate that application of various methods according to the actual situation, is the key of water environment treatment in urbanized plain river network.

This paper aims to propose a new scheme of diverting clean water operation, which according to the idea of system governance, by focusing on the urbanized plain river network in Changzhou City. This scheme is mainly based on the integrated control of guide engineering, river regulation engineering, gates and pumping stations. Several water-related projects should be reconstructed as necessary. The fresh water is obtained from the main stream of Yangtze River. In addition, this new scheme of diverting clean water operation is proposed to improve the water quality in Changzhou City by combining the numerical simulation and joint operation of gates and pumping stations.

2 Project Site

Changzhou City is located in the south part of Jiangsu province, the downstream region of the Yangtze River. This paper focus on the proper area of Changzhou City, a typical urbanized plain river network. This region strength its border to the Xinlong river and the Huning highway in the north, the Grand Jinghang Canal at the south. The west and east borders are Desheng River and Dingtang port, respectively. Although many effects have been made to improve the water environment in the past few years, some problems still exist. For example, the water quality in some Middle-Small River Channels are still poor since it is difficult to transport the fresh water into these channels. Some block up channels are not well connected in the channel system. In addition, some sources of pollution are not cut off, the related channels are badly polluted.

3 Date Collection Program

General idea of the Scheme of diverting clean water operation includes diverting clean water from the main stream of Yangtze River, construction of two channels and four controlling engineering. In these ways, a smooth flow pattern with three-level water level difference and reasonable flow distribution is formed. This scheme of diverting clean water operation offers a practicable solution to improve the water quality in this urbanized plain river network.

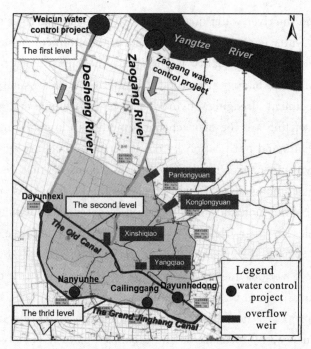

Figure 1. Study sites and water diversion and flow path of this diverting clean water operation

Fig. 1 shows the water diversion and flow path of this diverting clean water operation. The fresh water in the main stream of Yangtze River is drained by two water control projects, Weicun and Zaogang respectively. Then the fresh water is transported though Desheng river and Zaogang river to the proper area of Changzhou City. Four controlling projects (Overflow weirs), Panlongyuan, Konglongyuan, Xinshiqiao and Yangqiao are constructed to form the flow pattern with three-level water level difference. Specifically, the Desheng River and Zaogang River form the first level, the Old Canal is in the second level and the Jinghang Grand Canal is in the third level. The three-level water level difference leads to the fresh water transport across the river network in the Changzhou City form north to the south. The integrated regulation of all water-related projects in this river network is an efficient way to control the allocation of fresh water into each channel. Therefore, every channel in this channel system could keep a reasonable fluidity, the water environment could be improved in the end. The overall arrangement of the diverting clean water operation is shown in Fig. 1.

4 Results

This operation of diverting clean water was executed in 15 days form 28 November to 11 December 2020. The water levels were observed at three points to represent the water level variation at Zaogang River, Old Canal and the Grand Jinghang Canal, respectively. The results of water levels are shown in figure 2. Water levels in these channels indicate the three-level water level difference was formed. This water level difference forms the basis of water diversion and flow path in this plain river network.

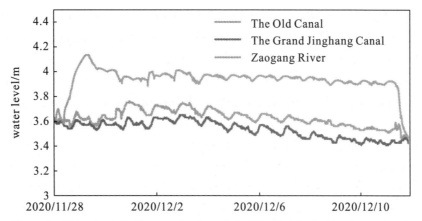

Figure 2. Water levels in three main channels in this urbanized plain river network

Four parameters[Ammonia-nitrogen(A-N), Permanganate Value(PV), Total phosphorus(TN) and Dissolved oxygen(DO)] which represent the water environment are observed during and after the operation. The observation are executed at five key points: 1. Bailongqiao located at Xishi River, Linggangqiao located at Cailinggang, Guangrenqiao located at Baidang River, Jinguqiao located at Nantongzi River, Yunxiangqiao located at Houtang River. In this way, the water environment improvement by the diverting clean water operation are measured. The results of water environment during and after the operation are shown in Fig. 3 and Fig. 4.

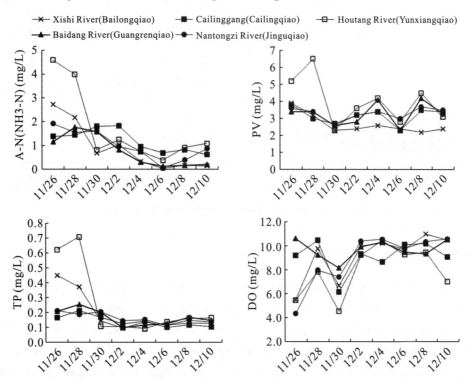

Figure 3. Water environment observed during the operation

During the operation, the concentrations of Ammonia-nitrogen(A-N), Permanganate Value(PV) and Total phosphorus(TN) are decreased. In the meantime the concentration of Dissolved oxygen (DO) is increased. The variation of these parameters indicates that the water environment is improved significantly during the operation. The fluidity in this river network is raised by diverting clean water from the main stream of Yangtze River. As a result, the water self-qurification capacity is improved.

After the end of operation, the observation of water environment continued for 16 days from 11 December to 27 December. In this way, the sustainable effect of diverting clean water operation could be estimated. For Ammonia-nitrogen, the concentration in most points are lower than the initial values even 16 days after the end of operation. Meanwhile, the concentration of Ammonia-nitrogen in Baidang River and Nantongzi River reach and exceed the initial values at about 10 days after the end of operation. For Permanganate Value, the concentration at all points lower than the initial values until about 12 days after the operation. The concentration of Total phosphorus in Xishi River and Houtang River keeps lower than the initial values. For other points, the concentration of Total phosphorus reach to the initial value in 16 days. The concentration of Dissolved oxygen in most points are well kept after the end of operation, except it in Xishi River and Houtang River. In general, the effects of diverting clean water operation could last for at least 10 days. After that, the improved condition at most points could be well kept except some particular points.

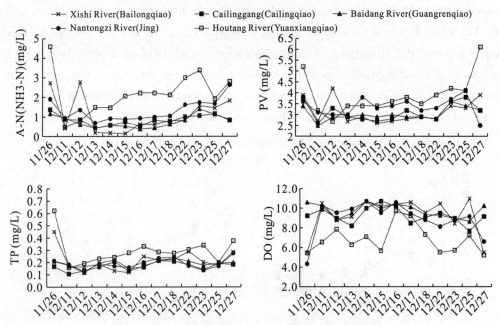

Figure 4. Water environment observed after the operation

5 Conclusions

In this paper, a new scheme of diverting clean water operation is proposed according to the idea of system governance. This scheme is mainly the integrated control of guide engineering, river regulation engineering, gates and pumping stations, and even reconstruction some water-related projects as necessary. Due to the request of diverting clean water operation, fresh water is obtained from the main stream of Yangtze River. According to the combination of numerical simulation and field observation, the scheme of smooth flow and live water in the city proper of Changzhou is formulated. A smooth flow pattern with three-level water level difference and reasonable flow distribution is formed by scientifically dispatching the existing dams, gates and pumps and other water conservancy projects. The water environment of river network has been significantly improved through the operation of this scheme.

Diverting clean water operation is an efficient way to improve the water quality in urbanized plain river networks. Since some block up channels are not well connected in this river system, the diverting clean water operation should be well planned and carried out step by step, according to the actual situation. In this case, the proper area of Changzhou City, a typical urbanized plain river network, are applied to carry out the diverting clean water operation. The controlling of the sources of pollution is necessary for the long term improvement of water quality. In this case, water quality of some channels has been deteriorated after the diverting clean water operation, indicating that some sources of the pollution are not cut off.

References

[1] Yu S, Li Y P, Cheng Y X, et al. The impacts of water diversion on hydrodynamic regulation of plain river network[J]. Journal of Lake Science, 2021, 33(2): 462−473. (in Chinese with English Abstract)

[2] Jiang T, Zhu S L, Zhang Q, et al. Numerical simulation on effects of gate-pump joint operation on water environment in tidal river network[J]. Journal of Hydraulic Engineering, 2011, 42(4): 388−395. (in Chinese with English Abstract)

[3] Li X, Tang H W, Wang L L, et al. Simulation of water environment under joint operation of gates and pumps in plain river network area[J]. Journal of Hohai University(Natural Sciences), 2016, 44(5): 393−399. (in Chinese with English Abstract)

[4] Yuan D, Zhang Y L, Liu J M, et al. Water quantity and quality joint-operation modeling of dams and floodgates in Huai River Basin, China[J]. Journal of Water Resources Planning and Management, 2015, 141(9): 04015005. (in Chinese with English Abstract)

[5] Zuo Q T, Chen H, Dou M, et al. Experimental analysis of the impact of sluice regulation on water quality in the highly polluted Huai River Basin, China[J]. Environmental Monitoring and Assessment, 2015, 187(7): 450.

The Analysis of Flood Limit Water Level in Longyangxia Reservoir During Different Flood Stages Based on Fuzzy Theory

Song Yu[1,2], Kong Dezhi[3], Liu Qiang[3], Li Xinjie[1,2]

[1] College of Water Conservancy and Hydropower Engineering, Hohai University, Nanjing 210024, China
[2] Yellow River Institute of Hydraulic Research, Zhengzhou 450003, China
[3] Hydrological bureau of Yellow River Conservancy Commission, Zhengzhou 450004, China

Abstract

Flood season division and dynamic control of flood limit water level in the reservoir are important methods used in real-time reservoir scheduling for utilizing the water resource of flood safely during the flood season. In this study, the division of flood season and the corresponding changes in flood limit water level in Longyangxia reservoir were analyzed using the fuzzy set theory based on the daily average runoff data collected over 32 years by the Tangnaihai Hydrological Station at the entrance of Longyangxia Reservoir. As shown by the analysis based on the fuzzy set theory, the main flood season of Longyangxia reservoir lasts from June 27th to August 26th. The flood limit water level can be controlled at 2590 m in order to increase the storage capacity of the reservoir during the flood season, to improve the power generation of the reservoir, and to alleviate the conflict between effective flood control and effective reservoir. These findings provide valuable guidance to making full use of the storage capacity and improving the economic benefits of Longyangxia Reservoir.

Keywords: Longyangxia reservoir; Flood limit water level; Flood season division; Fuzzy theory

1 Introduction

The Yellow River is characterized with a high amount of sand sedimentation, which comes from different source areas. Such a feature give rise to serious water scarcity issue along the Yellow River basin. In addition, the flood control infrastructures remain underdeveloped along the Yellow River, which is also manifested by a poor storage capacity of the local reservoirs. The operation of a cascade reservoir often aims to ensure a maximum level of safety. Specifically, the operation policy is formulated according to the flood seasons determined through the analysis of regional hydrological and meteorological characteristics (cause analysis, mathematical statistical analysis). Such an operation strategy is based on the general set theory. By defining different hydrological stages and flood seasons clearly, the reservoir can be regulated to provide a strict control on the water level, thereby ensuring a high level of safety. However, this operation strategy fails to maximize the utilizable benefit of the water resource. The continuous economic development along the Yellow River basin and the lack of water resource have led to an increasingly acute dilemma between effective flood control and effective reservoir scheduling. To resolve this contradiction fundamentally, it is imperative to re-define the traditional concepts of flood season and the non-flood season along with a more objective and dynamic description of the transition period between the flood and non-flood seasons. In other words, the flood season becomes uncertain, which can be classified as a fuzzy set. In this way, the flood season can be described by the degree of membership(membership function) in the fuzzy set(Pan Liu, 2007).

The Longyangxia reservoir, the leading reservoir along the Yellow River, is a multipurpose reservoir capable of multi-year regulation in the upper reaches of the Yellow River. The Longyangxia reservoir is located at the junction of Gonghe County and Guinan County in Qinghai Province on the upper reaches of the Yellow River. Around 131,400 km² of drainage area is regulated by this reservoir, which accounts for 28% of the total drainage area (Guangming Cao, 2007). The Longyangxia reservoir was designed with a normal water level of 2600 m, a total storage capacity of 24.7 billion m³, and a regulated storage capacity of 19.4 billion m³. It is classified as a large-scale level-one reservoir. Apart from water storage, this reservoir has also been used for a number of other purposes such as flood control, power generation, irrigation, and ice jam prevention(Weihua Song, 2020). Therefore, the Longyangxia reservoir plays a particularly important role in flood control in the upper reaches of the Yellow River as well as the optimal allocation of water resources throughout the entire river, as shown in Fig. 1.

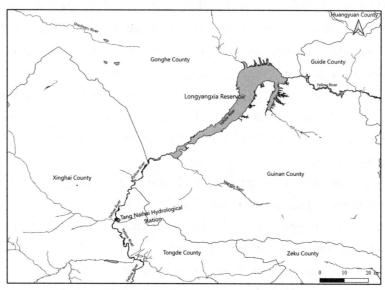

Figure 1. The geological map and location of Longyangxia reservoir

Until 2021, the Longyangxia reservoir has been operating for 35 years. Currently, a single flood limit water level is used for flood regulation during the flood season in Longyangxia reservoir. This dispatch strategy is considered to be relatively conservative, which fails maximize the potential of the reservoir(Shunde He, 2020). To better understand the regulation function of the reservoir, a fuzzy analysis was performed on the daily average runoff data collected over 32 years by the Tangnaihai Hydrological Station at the entrance of Longyangxia Reservoir. Based on the analysis, we re-defined the different flood seasons associated with the reservoir and calculated the corresponding flood limit water level to ensure the safety of the reservoir and the downstream basin as well as to provide higher economic benefits simultaneously. These analyses are expected to help us understand how the flood season and limit water level of the reservoir are affected by the changes in the runoff.

2 Research Method

In this study, the fuzzy set method was used to re-define the flood seasons associated with Longyangxia reservoir(Shouyu Chen, 1998). As indicated by the fuzzy theory clearly, the flood season can be treated as a fuzzy concept. Within a time period(T) of a year, the flood season can be considered as a fuzzy subset A(Bo Gao, 2005). In this way, the opposite event $\bar{A}$ of flood season (i.e., non-flood season) can also be classified as a fuzzy subset. During the transition period between the flood season and the non-flood season, the runoff may exhibit features of both the flood season and the non-flood season. The extent of flood season features can therefore by described by a fuzzy membership function $\mu_A(t)$ while the extent of non-flood season features can be described by $\mu_{\bar{A}}(t)$. Both fuzzy membership functions are defined between 0~1, i.e., $0 \leqslant \mu_A(t) \leqslant 1$ and $0 \leqslant \mu_{\bar{A}}(t) \leqslant 1$. In addition, both functions satisfy the following equality: $\mu_A(t) = 1 - \mu_{\bar{A}}(t)$ (Debo Liu, 2012). Therefore, determining the empirical membership function associated with flood season is the key problem when performing fuzzy set analysis on the flood season(Yuanjie Bi, 2018).

In order to find a proper membership function of the fuzzy set to describe flood and non-flood season, it is important to first define a proper critical value Y_T that distinguishes the flood season from the non-flood season. Subsequently, we can identify the periods of time(starting at t_1 and ending at t_2) during which the flow rate is greater than or equal to the critical value Y_T based on the runoff data collected in the research region over different years. The interval $[t_1, t_2]$ then defines the flood seasons in a year, i.e., the manifestation of the fuzzy set associated with flood season. When analyzing multiple years, the membership of flood season fuzzy set can be determined by identifying the corresponding manifestations in all the years. The detailed calculation procedures are given as follows(Yongsheng Ma, 2008).

(a) Calculate the inflow of Longyangxia reservoir over n years and use it as the total set in this study. The time period over a year is represented by $T=[1, 365]$d.

(b) Analyze and calculate according to the actual scenario. Then select a proper Y_T as the flow criterion to determine the start of the main flood season.

(c) Perform statistical analysis to find out the starting time t_{1i} and ending time t_{2i} during which the flow rate is greater than or equal to Y_T. These time periods then represent the flood season of the year and can be taken as a test result of the fuzzy set A. Subsequently, repeat this procedure to find the flood seasons of all years investigated in this study. The flood seasons are represented by $T_i = [t_{1i}, t_{2i}]$.

(d) If a given time t within a year T was covered by the manifestation interval T_i of the flood season for a certain number of times(m_i), then the membership frequency of which time t belongs to the flood season fuzzy set A is given by $P_A(t) = \dfrac{m_i}{n}$. Therefore, the degree of membership can be obtained as $n \to \infty$, i.e.,

$$\mu_A(t) = \lim_{x \to \infty} P_A(t) = \lim_{x \to \infty}\left(\frac{m_i}{n}\right) \tag{1}$$

The membership function can then be obtained after calculating the degree of membership for all the time frame one by one.

3 Division of the Flood Seasons

Division of the flood season is a clustering problem of high-dimensional time series. Currently, the division of the flood seasons is often handled using the fuzzy set analysis, the fractal method, change-point analysis, and dynamic clustering method.

The inflow data collected from 1987 to 2018($n=32a$) at Tangnaihai hydrological station located on the upstream of Longyangxia reservoir was used as the fuzzy test set A in this study. The precipitation at the upstream of Longyangxia reservoir is distributed unevenly in different spaces and during different times. Most of the precipitation occurs from May to October. Therefore, the pattern of the flood season(i. e. , when the flow rate exceeds $Y_T = 1000$ m³/s) was analyzed in Longyangxia Reservoir from May 1st to November 30th. Within this period time $T = [5.1, 11.30]$, we further identified the starting time(t_{1i}) and ending time(t_{2i}) during which the flow rate is greater than or equal to the critical value $Y_T = 1000$ m³/s. For example, the flood season in 1987 is found to be $T_{1987} = [t_{1,1987}, t_{2,1987}] = [6.18, 8.17]$. This subset represents a test result in the set A, which is also known as a manifestation sample. Therefore, 32 manifestation samples will be obtained during a total of 32 years. The membership value of the fuzzy set A over 32 years can be calculated using equation(1) based on the results shown in Table 1. For example, May 7th was only covered by the flood season $[t_{1i}, t_{2i}]$ for one time(in 1989), then $m_{5.7}=1$ and the corresponding membership value is given by μ_A(May 7th) $=1/32=0.031$. Following this method, we can find the fuzzy membership value of all other time t within a year. The calculation results are shown in Fig. 2.

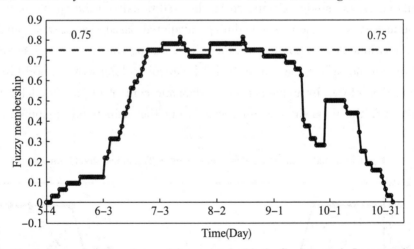

Figure 2. The fuzzy membership value $\mu_A(t)$ associated with the flood season in Longyangxia reservoir

As shown by the fuzzy membership value associated with the flood season in Longyangxia reservoir listed in Table 2, the arrival of the main flood season is marked by a membership value of 0.750. The starting time of the main flood season is given by a_1=June 27th while the ending time is given by a_1=August 26th.

4 Flood Limit Water Level During Different Flood Stages

Maintaining a fixed flood limit water level during the flood season fails to maximize the utilizable benefit of the reservoir. On the other hand, defining different limiting water level during different time periods can resolve the dilemma between an effective flood control and an effective reservoir scheduling for beneficial purpose. However, this approach requires the flood season to be divided into different stages reasonably.

The Longyangxia reservoir has a flood control storage of $V_{control}=4.5$ billion m³ and a flood limit water level of $Z_{limit}=2594$ m during the flood season. The corresponding total storage capacity and water level for check flood are 19.563 billion m³ and 2607 m, respectively. In this study, the limiting water level during different flood stages in Longyangxia reservoir were calculated using the direct method. Specifically, the membership value $\mu_A(t)$ of the flood season was considered as the allocation ratio of the storage capacity during different periods. The detailed calculation procedures are given as follows. For a given $\mu_A(t)$, we first calculate $[1-\mu_A(t)]\times V_{control}$ and use it to define the flood control storage required during different stages of the flood season. In this way, the total storage capacity at different stages of the flood season is given by $[1-\mu_A(t)]\times V_{control}+V_{limit}$. Subsequently, the flood limit water level required at different stages of the flood season can be obtained based on the relationship between Z and V. The calculation results of the limiting water level during different periods of the flood season are summarized in Fig. 2.

As shown by the calculation results, the pre-flood stage lasts from June 16th to June 26th. The corresponding flood limit water level is reduced gradually from 2594 m to 2590 m. The main flood season lasts from June 27th to August 26th and the flood limit water level remains at 2590m. Finally, the post-flood stage lasts from August 27th to September 17th. The corresponding flood limit water level is increased gradually from 2588 m to 2594 m.

Fig. 3 shows the comparison between the conventional flood limit water level and the flood limit water level calculated in this study. Compared to the traditional scheduling of the flood limit water level in Longyangxia reservoir, the new scheduling calculated based the fuzzy set analysis provides a more smooth and continuous variation of the water level during the entire flood season. In addition, the new scheduling will not affect the safety of flood control so long as the flood limit water level is less than 2594 m. Therefore, by implementing a dynamic control of the flood limit water level in Longyangxia reservoir, the flood limit water level during the main flood season can be kept within 2590 m.

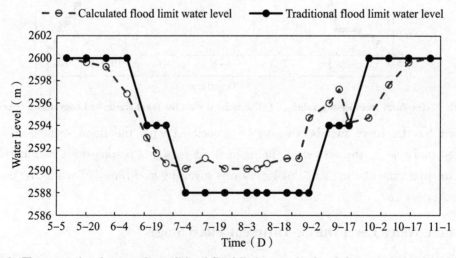

Figure 3. The comparison between the traditional flood limit water level and the one calculated in this study

The results calculated based on fuzzy set analysis were consistent with the starting and ending data of the pre-flood stage used in real operation. The flood limit water level calculated in this study varies gradually over time. On the contrary, a constant flood limit water level was used in actual practice(2588 m). When the main flood season is approaching, keeping a smaller flood limit water level(compared to that used in actual operation) will be more beneficial for flood control. In terms of the main flood season, the starting date calculated in this study is consistent with that used in real operation. However, the ending data of the main flood season calculated in this study is earlier than that used in the real operation. According to the statistics collected over multiple years, the flood limit water level used in Longyangxia reservoir was never able to be kept at the designed value due to the construction of the downstream cascade power stations and the requirement of flood control in downstream rivers. When the flood season is over, the reservoir cannot be filled with water to its maximum storage capacity, thereby failing to maximize its utilizable benefit. Therefore, adopting the main flood season calculated in this study will allow the reservoir to better utilized. In terms of the post-flood stage, the starting data calculated in this study is earlier than that used in real operation. In addition, the flood limit water level increases gradually during the post-flood stage until reaching the target value required for beneficial utilization.

5 Conclusions

In this study, the flood season of Longyangxia reservoir was divided by using the fuzzy set theory. The corresponding flood limit water level during different time periods was further calculated based on the division of different flood stages. As shown by the results obtained based on the real case, the flood season membership value associated with the flood limit water level is different during different months. Furthermore, the flood control storage capacity also varies over different time periods. Therefore, the flood limit water level can be increased to different extent during different flood stages in order to optimize the flood control operation plan of the reservoir. Increasing the flood limit water level is beneficial for intercepting and storing the water from rainfall and flood at the end of the flood season, increasing the storage capacity of the reservoir, reducing the amount of water discarded by the reservoir, and providing greater benefits from the reservoir in general. Such a strategy can alleviate the conflict between effective flood control and effective reservoir scheduling. The results obtained in this study will provide valuable scientific basis to the management of water resources and flood control in Longyangxia reservoir in the future.

Acknowledgements

This research is financially supported by the National Natural Science Fundation of China(Grant No. 51879115).

References

[1] Yuanjie B. Study on flood water level design of Fenhe Reservoir by stages[J]. Water Resources Development and Management, 2018,(12):12,24−27.

[2] Guangming C, Jinhuai X. Some thoughts on the implementation of dynamic flood limit water level in Longyangxia Reservoir[J]. Advances in Power Grid and Hydropower, 2007, 23(7): 28−31.

[3] Shouyu C. Theory and practice of fuzzy set analysis of hydrology and water resources engineering system[M]. Dalian: Dalian University of Technology Press, 1998.

[4] Bo G, Yintang W, Siyi H. Adjustment and application of the limited level of reservoirs during the flood season[J].

Advances in Water Science, 2005(3):326−333.

[5] Shunde H, Weihua S, Peng C. Flood control operation conditions and flood limited water level demonstration of Longyangxia reservoir in 2019[J]. Yellow River, 2020, 42(6): 22−26, 36.

[6] Debo L. Discussion of probability and statistical methods in the design and application of flood control level in reservoirs[J]. South-to-North Water Transfers and Water Science & Technology, 2012, 10(3): 161−164.

[7] Pan L, Shenglian G, Yi X. Study on the optimal reservoir seasonal flood control water level[J]. Journal of Hydroelectric Engineering, 2007, 26(3):5−10.

[8] Yongsheng M, Qiang H, Yimin W. Seasonal flood limited level determination for Shiquan reservoir based on fuzzy set theory[J]. Water Resources and Power, 2008, 26(3):47−49.

[9] Weihua S, Shunde H, Xiaoying X. Study on the scheme of dynamic control of the flood control level of Longyangxia reservoir[J]. Yellow River, 2020, 42(2):18−21.

The Characteristics of Air Flow Driven by Free Surface of Open Channel

Jing Gong, Jun Deng, Wangru Wei, Weiwei Li

State Key Laboratory of Hydraulics and Mountain River Engineering,
Sichuan University, Chengdu 610065, China

Abstract

Spillway tunnel is the key structure of a large-scale water conservancy. The high-head water inlet makes water surface-velocity extremely high, and the air is driven by free surface of water to move downstream. In this paper, air velocity distribution above water was studied by model tests, and assumption that the airflow is a turbulent boundary layer with a rough interface was put forward, and the influence of water depth and water velocity on the air velocity distribution was analyzed and summarized. The results show that air velocity presents an exponential distribution. As the measured position rises, velocity gradually decreases, and gradient decreases. When water depth increases, air velocity distribution moves upward, and distribution form does not change. With water surface-velocity increaes, air velocity of same measuring point increases, variation range near water surface is large, air boundary layer thickness increases slightly, and index coefficient of air velocity distribution function decreases. Through numerical fitting, the calculation formula of air boundary layer thickness of different water surface-velocity is proposed, and the numerical value of index coefficient is given.

Keywords: Open channel; Air velocity distribution; Air boundary layer height; Index coefficient

1 Introduction

With the continuous development and utilization of water resources, many hydraulic engineerings with high head and large flow have been built. Spillway is a key structure of large-scale hydraulic engineering(Novak, 2003), used for various discharge needs of dams, including controlling water level of reservoir, as well as flushing and diversion in emergency situations. The high-head water inlet makes the free surface velocity of water flow in spillway reach $30 \sim 50$ ms^{-1}. The high-speed water flow drags the air, and drives the air above the free surface of water flow in the spillway to move downstream together. High-speed water-gas two-phase flow in open channel(Chanson, 2004; Shi, 2007; Wu, 1989) appeared with the emergence of high-head hydraulic engineering, and is a hydraulic problem that needs to be solved.

The movement law of high-speed water-gas two-phase flow in open channel is very complicated. Prototype observations have obvious constraints and cannot form systematic and structured results and laws. Model tests(Yalin, 1971) or numerical simulations(Bombardelli, 2011; Ma, 2011; Meireles, 2014; Valero, 2016; Wei, 2020) are often used for research. When calculating the high-speed water-gas two-phase flow through numerical simulation, the free surface of water flow is simplified into a plane without fluctuations, which is different from the actual water surface environment. Under the action of water turbulence, water point jumps near water surface. As the water flow speed increases to a certain critical value, not only water point jumps, but also many vortices jump out of water surface to form a water column.

From the research results of previous scholars(Falvey, 1987; Straub & Anderson, 1959; Killen, 1968), in the process of water flow, the water free surface takes on wave-like shape, and there is water-gas mixing area between water flow and air. this water surface deformation is called water surface roughness. Wilhelms & Gulliver(2007) studied the water-air mixing area and proposed the concepts of waves, which are widely used in the numerical simulation of open channel flow.

The velocity distribution in air boundary layer above water free surface is often calculated by following exponential distribution:

$$\frac{u_{fs} - v_a}{u_{fs}} = \left(\frac{y - \bar{h}}{\delta_a}\right)^{1/m} \quad (1)$$

where m is the exponential coefficient of air velocity distribution law, u_{fs} is the water surface velocity, v_a is the air velocity, y is the height from bottom plate, δ_a is the air boundary layer height, $\bar{h}$ is the average water depth. Falvi(1984) considers that the air above water free surface as boundary layer with increasing height, which is 0 at gate and reaches the maximum value at the end of channel, and proposed the exponential coefficient $m = 7$ in velocity distribution function in air boundary layer. Valero and Bung(2016) calculated the air boundary layer height through physical model experiments, and obtained the exponential coefficient $m = 5.15$ by fitting experimental data distribution, and pointed out that the pressure gradient of air flow is similar to the pressure gradient in water flow. Chanson (2009) compared actual engineering and model tests, found that the air-water interface area and turbulence level under prototype conditions may increase dramatically. When self-aeration occurs, the resistance at air-water interface area increases, resulting in higher velocity and greater momentum transfer from water to airflow(Ishii & Hibiki, 2010).

In this paper, through physical model tests, the characteristics of air flow above water free surface are explored. the air velocity under different flow conditions(speed, depth) in open channel is measured, and the data is analyzed. The air boundary layer height and the exponential coefficient of air velocity distribution function under different water flow velocity are calculated. The air velocity distribution law and the changes along the way, as well as the influence relationship between water velocity and water depth are obtained.

2 Experimental Setup

This physical test model is built at the test site of the State Key Laboratory of Hydraulics and Mountain River Engineering of Sichuan University. For different water flow conditions in open channel, the air velocity and water velocity at different section in channel are measured.

From upstream to downstream, the open channel model can be divided into horn-shaped inlet, pressurized section, opening adjustment section, and open channel section. In order to clearly observe the flow state of water and facilitate the measurement of experimental data, the model material mainly uses transparent plexiglass. The inlet pipe is steel pipe connected with round variable square pipe, and rectangular open channel is placed on a rectangular pipe platform. The layout of the open channel model facilities is shown in Fig. 1.

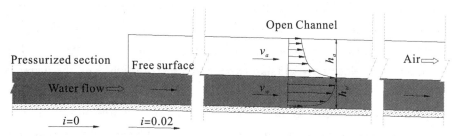

Figure 1. Model sketch with relevant parameters of the open channel

The horn-shaped inlet is built in the front of the pressurized pipeline section, to make water flow smoothly into rectangular open channel and reduce the head loss. The pressurized pipe is composed of a round steel pipe and a round variable square pipe. The diameter of round steel pipe is 30 cm and the length is 100 cm. The round variable square pipe is made of plexiglass, the length is 50 cm, the bottom is horizontal, connected with the round steel pipe, and the end cross-sectional shape is rectangle with width of 25 cm and height of 20 cm. The length of the opening adjustment section is 50 cm, and the opening change range is 10~20 cm. The section of the open channel is rectangle with width of 25 cm and height of 45 cm, with total length of 14.40 m. The bottom slope is 0.02, and the outlet is free flow. The highest measured water level is 3.00 m higher than the open channel floor.

In order to clearly observe the change law of air velocity distribution above water free surface along flow direction of the open channel, a total of 8 measurement sections were set up for the open channel model test. When water level of the reservoir is high, the water velocity at gate is high, and the flow pattern is jet-like, accompanied by water points flying out of water surface. Therefore, the first measurement section is arranged at 3.5 m downstream of the gate, and the second section is 2.0 m away from the first section, Thereafter, the spacing between each section is 1.0 m.

The test points are arranged on the vertical central axis of each section. The air velocity test of open channel mainly includes the measurement of water velocity and air velocity above the water free surface. The water velocity is measured by LGY-II intelligent flow meter(Nanjing Hydraulic Research Institute). The air velocity is measured by KANOMAX 6006−0C thermal anemometer (Kano, Japan). The measurement range of air velocity is 0.01~20 m/s, and the measurement accuracy is ±5%. All test measurements are carried out in accordance with the instrument use regulations.

3 Analysis of Data

3.1 Air velocity distribution law

There is no cover plate or other obstacles above water flow of open channel, and the air moves downstream with water flow under drag action. Both water and air can be regarded as fluids with active surfaces. The bottom plate has blocking effect on water flow opposite to the direction of movement, so that water velocity at bottom plate is low. The air near water surface is moved downstream by the drag force of water free surface. According to the transmission of force, the air above water moves downstream together. Since the friction of air is small, the movement of air will stop when it develops to a certain height.

The maximum measurement height is 0.45 m above the floor. The air velocity at the highest measurement point is greater than 0, and less than the air boundary layer height at the corresponding position. The air velocity at the highest point changes with the change of section position. There are random jumps of water points near water free surface, and the accuracy of air velocity measured value is poor. It is not taken into consideration when analyzing the law of air velocity distribution. As shown in Fig. 2, the water velocity presents an exponential distribution, and the water velocity at the bottom plate is less than the surface velocity. The air velocity presents an exponential distribution, and the air velocity at water free surface is the largest. As the measuring position increases, the air velocity gradually decreases and the change gradient decreases. The air velocity that is infinitely close to water free surface should be equal to the water velocity, that is, the velocity of the water-air interface is consistent.

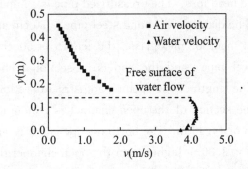

Figure 2. Typical air-water velocity distribution of open channel (h_w =15 cm, u_{fs} =4.1 m/s, Section 5)

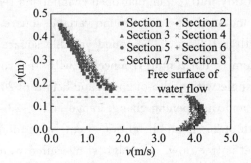

Figure 3. Air-water velocity distribution of different section of open channel (h_w =15 cm, u_{fs} =4.1 m/s)

The water flow at gate is jet-like, so that the water velocity of section 1 presents a distribution pattern of large in the middle and small at the ends. As shown in Fig. 3, as the measurement section moves along the direction of water flow, the water velocity distribution gradually tends to be uniform. Because the water surface is broken, the measured value is slightly smaller than middle area. By comparing the test results of 8 measuring sections, the water surface velocity changes very little.

The measured value at the highest air velocity measurement point on section 1 is greater than 0, indicating that the air boundary layer height at this location is greater than the measured height. There is no obvious change in air velocity near the highest measuring point along the way, indicating that the air boundary layer has developed to stable condition. By comparing the test results of 8 measurement sections, along the direction of water flow, the air velocity increases slightly, and the distribution pattern is the same.

3.2 The influence of water flow depth on air velocity distribution

As the water flow depth increases, the position of water free surface increases. the air velocity distribution pattern remains unchanged, the value of the same measuring point becomes larger, the measuring point of the same air velocity value increases. the entire air velocity distribution moves upward. As the water velocity increases, the influence of water flow depth on air velocity distribution remains unchanged. The air velocity distribution at different water flow depths is shown in Fig. 4.

The air is only in contact with the water free surface, and has not related to the water depth below water free surface. In order to prove that the water depth has no effect on the air velocity distribution, the above air velocity distribution is processed in dimensionless manner, as shown in Fig. 5. Where y_{a13} is the air area height at water depth of 13 cm, y_{a13} =32 cm. The air velocity distribution at different water depths is converted to the distribution at water depth of 13 cm. It can be seen from the figure that the air velocity distribution at different water depths coincides with the part near water free surface of air distribution at water depth of 13 cm. When the water surface velocity and the height of air velocity measuring point from water surface are constant, the air velocity at different water depths is equal, indicating that the air velocity distribution is not affected by water depth, but is related to water surface velocity and the height of air velocity measuring point from water surface.

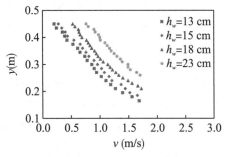
Figure 4. Air velocity distribution at different water flow depths (u_{fs} =4.1 m/s, Section 5)

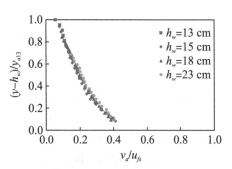
Figure 5. Dimensionless diagram of air velocity distribution (u_{fs} =4.1 m/s, Section 5)

3.3 The influence of water velocity on air velocity distribution

As shown in Fig. 6, as the water surface velocity increases, the air velocity at each measurement point increases, and the entire air velocity distribution moves right. The air velocity at the highest measurement point changes slightly, and near water surface changes greatly. The drag force of water surface on air decreases with the increase of air measuring point.

In order to analysis the degree of influence of water velocity on air velocity distribution, the abovementioned air velocity distribution is processed in dimensionless manner, as shown in Fig. 7. As the water surface velocity increases, the air velocity distribution curve becomes slower. The change gradient of air velocity along elevation under high water velocity is greater than low water velocity. As the elevation decreases, the change gradient increases. When the water velocity is high, the water surface fluctuation will increase, and the resistance at the water-air interface will increase, making the momentum transfer from water to air increase, and the air velocity near water surface is closer to the water surface velocity.

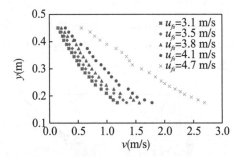
Figure 6. Air velocity distribution at different water velocity (h_w =15 cm, Section 5)

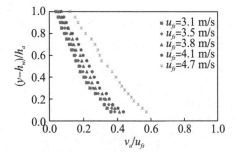
Figure 7. Dimensionless diagram of air velocity distribution (h_w =15 cm, Section 5)

3.4 Calculation results of air boundary layer height and exponential coefficient

Assuming that the position where the air velocity decreases to 10% of maximum air velocity (water surface velocity) is the upper edge of air boundary layer, the height of air boundary layer is the distance between the position at $v_a = 0.1u_{fs}$ and water free surface. According to the changes in air velocity distribution of different sections, the air boundary layer at sections 1~8 has developed to a stable condition. Therefore, the section 5 is selected as calculation section. By determining the position of upper edge of air boundary layer, the height of air boundary layer δ_a has been calculated, shown in Fig. 8.

The air moves downstream under the action of drag force of water surface, which can be regarded as the turbulent boundary layer caused by rough boundary. According to the formation mechanism and influencing factors of the turbulent boundary layer, the relationship function of air boundary layer height is as follows:

$$\delta_a = f(u_{fs}, d_w, \upsilon) \tag{2}$$

where δ_a is the air boundary layer height, d_w is the roughness of water surface, υ is the air kinematic viscosity, $\upsilon = 15.06 \times 10^{-6}$ m²/s (standard atmospheric pressure of 101.33 kPa and temperature of 20℃). The water surface roughness is considered to be the height of water-air mixing area formed by water turbulence, and is calculated by the following formula(Jiang, 2020):

$$\frac{d_w}{y_{90}} = \eta \cdot \left(\frac{x}{h_0}\right)^{0.5} \tag{3}$$

where y_{90} is the height of water free surface of water-air mixing area from bottom plate, η is the correlation coefficient, taking $\eta = 0.016$, x is the distance between measure section and the location where self-entrainment occurs, h_0 is the pressure outlet height.

According to data distribution in Fig. 8, the air boundary layer height is directly proportional to water surface velocity. When the water surface velocity(u_{fs}) is 0, the boundary layer height(δ_a) is 0. The air boundary layer height is positively related to the drag force of water flow, and inversely related to the air kinematic viscosity. The influence of water drag force on air boundary layer depends on the rough boundary height and the transmission efficiency. It is expressed by the height of water-air mixing area(d_w^n). The action object in mixing area is water droplet, for characterizing the shape and transmission efficiency of acting object, taking $n=3$. According to the distribution of data points, the formula for calculating the air boundary layer height is obtained by fitting:

$$\delta_a = \alpha \cdot \left(\frac{u_{fs} \cdot d_w^3}{\upsilon}\right)^{0.5} \tag{4}$$

The experimental measurement value is fitted to obtain the empirical coefficient $\alpha = 0.259$. The correlation coefficient(R^2) is 0.9201. The relative error between measured value and calculated value of the air boundary layer height is within ±10%. The comparison is shown in Fig. 9.

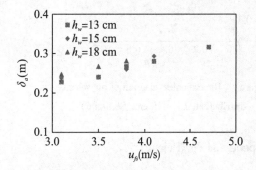

Figure 8. The air boundary layer height under different water velocity(Section 5)

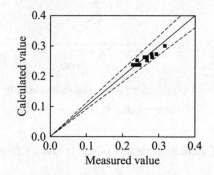

Figure 9. Comparison of measured value to calculated value of air boundary layer height

According to the air velocity distribution, the exponential coefficient is fitted and calculated through air velocity distribution function, and the fitted value of exponential coefficient is shown in Fig. 10. When the water surface velocity changes from 3 m/s to 5 m/s, the corresponding exponential coefficient (m) of air velocity distribution function changes from 3.7 to 2.

From Fig. 9 and Fig. 10, when the water surface velocity increases, the air boundary layer height increases slightly, and the exponential coefficient decreases. When the water velocity is large, the exponential coefficient changes greatly. With the increase of water velocity, the fluctuation of water surface increases, and the water-air interface changes from a relatively smooth wall to a relatively rough wall. Therefore, the interface roughness increases, the air velocity gradient increases, the velocity distribution curve becomes slower, and the exponential coefficient decrease. When the water velocity is high, the fluctuation of water surface increases rapidly, resulting in a large change in the exponential coefficient.

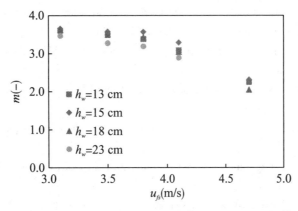

Figure 10. **The exponential coefficient under different water velocity(Section 5)**

4 Conclusions

In this paper, the air velocity above water free surface under different water flow conditions (speed, depth) of the open channel was measured, and the data was analyzed, and the air boundary layer height and the exponential coefficient of air velocity distribution function under different water velocities were calculated.

The air velocity decreases as the elevation increases, showing an exponential distribution. As the water surface velocity increases, the air velocity increases, and the variation range of the highest measuring point is small, and near water surface is large. When the water depth increases, the air velocity distribution does not change. Assuming that the position where air velocity is reduced to 10% of the maximum air velocity(water surface velocity) is the upper edge of air boundary layer, the air boundary layer height and the exponential coefficient of air velocity distribution function, as well as the empirical formula of the air boundary layer height are obtained. When the water surface velocity increases, the air boundary layer height increases slightly, and the exponential coefficient decreases.

Acknowledgements

This research report is financially supported by the Sichuan Science and Technology Program(Grants No. 2020YJ0320) and the National Natural Science Foundation of China(Grants No. 51939007).

References

[1] Novak P, Moffat A I B, Nalluri C, et al. Hydraulic Structures[M]. Taylor & Francis e-library, 2003.

[2] Chanson H. Air-water flows in water engineering and hydraulic structures. Basic Processes and Metrology[M]. London: Taylor & Francis Group, 2004.

[3] Shi Q S. High-speed water-air two-phase flow[M]. Beijing: China Water Resources and Hydropower Press, 2007.

[4] Wu C G. Two-phase flow of water and gas in an open channel[J]. Chengdu: Chengdu University of Science and Technology, 1989.

[5] Yalin M S. Theory of hydraulic models[M]. London: Macmillan, 1971.

[6] Bombardelli F A, Meireles I, Matos J. Laboratory measurements and multi-block numerical simulations of the mean flow and turbulence in the non-aerated skimming flow region of steep stepped spillways[J]. Environmental Fluid Mechanics, 2011, 11(3): 263−288.

[7] Ma J, Oberai A A, Drew D A, et al. A comprehensive sub-grid air entrainment model for RaNS modeling of free-surface bubbly flows[J]. The Journal of Computational Multiphase Flows, 2011, 3(1): 41−56.

[8] Meireles I C, Bombardelli F A, Matos J. Air entrainment onset in skimming flows on steep stepped spillways: An analysis[J]. Journal of Hydraulic Research, 2014, 52(3): 375−385.

[9] Valero D, García-Bartual R. Calibration of an air entrainment model for CFD spillway applications[J]. Advances in hydroinformatics, 2016:571−582.

[10] Wei W R, Deng J, Xu W L. Numerical investigation of air demand by the free surface tunnel flows[J]. Journal of Hydraulic Research, 2020, 59: 185−165.

[11] Ervine D, Falvey H. Behaviour of turbulent water jets in the atmosphere and in plunge pools[J]. Ice Proceeding, 1987, 2: 295−314.

[12] Straub L G, Anderson A G. Experiments on Self-Aerated Flow in Open Channels[J]. Journal of the Hydraulics Division, 1959, 85: 119−121.

[13] Killen J M. The surface characteristics of self-aerated flow in steep channels[J]. Minneapolis: University of Minnesota, 1968.

[14] Wilhelms S C, Gulliver J S. Bubbles and waves description of self-aerated spillway flow[J]. Journal of Hydraulic Research, 2007, 45(1): 142−144.

[15] Falvi H T. Aerated water flow in hydraulic structures[M]. Beijing: Water Conservancy and Electric Power Press, 1984.

[16] Valero D, Bung D B. Development of the interfacial air layer in the non-aerated region of high-velocity spillway flows. Instabilities growth, entrapped air and influence on the self-aeration onset[J]. International Journal of Multiphase Flow, 2016, 84: 66−74.

[17] Chanson H. Turbulent air-water flows in hydraulic structures: dynamic similarity and scale effects[J]. Environmental Fluid Mechanics, 2009, 9(2): 125−142.

[18] Ishii M, Hibiki T. Thermo-Fluid Dynamics of Two-Phase Flow[M]. Springer US, 2010.

[19] Jiang F, Xu W, Deng J, et al. Flow structures of the air-water layer in the free surface region of high-speed open channel flows[J]. Mathematical Problems in Engineering, 2020.

The Impact of Climate and Land-use Changes on Freshwater Ecosystem Services Flows in Lianshui River Basin, China

Yang Zou, Dehua Mao

Hunan Normal University, Changsha 410081, China

Abstract

In order to deal with the shortage of water resources caused by climate and land-use changes, it is necessary to conduct a comprehensive analysis of water ecosystem services. Based on the InVEST-DEM coupling model and scenario analysis method from the perspective of ecosystem service flow, this study analyzed the spatiotemporal evolution of the freshwater supply and demand in the Lianshui River Basin from 2000 to 2020 under climate and land-use changes. The results showed that from 2000 to 2020, the water yield of the river basin increased first and then decreased. From 2000 to 2010, the water yield of the Lianshui River Basin increased by 1.26 billion m^3, and from 2010 to 2020, it decreased by 1.20 billion m^3. The high value areas of water production were mainly distributed in the upper reaches of the basin. Water demand has gradually increased, with a total increase of 1.92 billion m^3. Compared with the impact of land use change, the impact of precipitation on the flow of freshwater ecosystem services is more significant. From 2000 to 2010, the contribution rates of precipitation and land-use changes to freshwater supply were 0.73% and 99.27%, and the contribution rates to service flow were 99.98% and 0.02%, respectively. From 2000 to 2020, the contribution rate to supply were 81.45% and 18.55%, and the contribution rate to service flow were 80.88% and 19.12%. The research revealed the spatiotemporal characteristics of the water production ecosystem service flow in the Lianshui River Basin and its response mechanism to precipitation and land-use changes. It provides a scientific basis for the management of land-use, redistribution of water resources and the sustainable development of water ecosystems in the basin.

Keywords: Ecosystem service flows; Freshwater; Climate; Land-use change; Lianshui River Basin

1 Introduction

Ecosystem services refer to all the benefits that humans obtain from the ecosystem. The United Nations Millennium Ecosystem Assessment(MEA) believes that ecosystem services can be divided into supply services, regulation services, cultural services, and support services(MEA 2005; Peng et al., 2015). In recent years, ecosystem services research has become one of the global research hotspots(Hayley et al., 2021; P et al., 2017). Many scholars have paid attention to the supply and demand of ecosystem services and their balance. However, since the supply and demand dynamics of ecosystem services have not been combined, there is a spatial mismatch between the supply and demand of ecosystem services(Bagstad et al., 2013). Ecosystem service flow can scientifically explain the whole process of ecosystem services from production, delivery to use, dynamically coupling natural ecosystems and human society, clarifying the balance of regional ecosystem service supply and demand, and actively and effectively solve the problems of time-space mismatch between supply and demand, double counting of services, etc.

Freshwater ecosystem service is one of the most important ecosystem services. It can not only provide people with freshwater resources, meet the needs of production and life, and provide basic raw materials and power energy such as aquatic products and inland shipping, but also can store floods, regulate regional climate, improve people's quality of life and other functions. Since the reform and opening up, China's economy has developed rapidly. As people's demand for water continues to increase, ecological problems such as shortage of water resources and reduction of water surface area have gradually emerged, severely restricting regional socio-economic development(Jun and Qiting, 2018; Shunze et al., 2020).

Climate and land use changes are important factors affecting ecosystem services(Zou et al., 2020). Climate change affects the intensity and distribution of meteorological factors such as precipitation and evapotranspiration in the watershed, which in turn affects the supply of freshwater ecosystem services(Alessandro et al., 2021; Wang et al., 2021). Land-use changes can affect the supply of freshwater ecosystems by changing the geographic environment and ecosystem structure of the watershed. Many attempts have been made to explore the impact of climate and land-use changes on the supply of fresh water, but none of them have involved their impact on the demand and flow of ecosystem service flows(Hao et al., 2020; Yang et al., 2021). For the sustainable development of water resources management, it is necessary to understand the widespread impact of climate and land-use changes on freshwater ecosystem service flows.

Most researches on ecosystems are achieved through modeling. Currently, the Soil and Water Assessment Tool(SWAT) model(TJ, NM, 2013), the Integrated Valuation of Ecosystem Services and Tradeoffs(InVEST) model(Richard et al. 2015) and the Artifcial Intelligence for Ecosystem Services(ARIES) model(Bagstad, Johnson, Voigt and Villa, 2013)are the main model used. Many scholars have studied the adaptability and parameter sensitivity of the InVEST model. The InVEST model has the advantages of few parameters, low data requirements and wide adaptability, so it is widely used in ecosystem service evaluation research. The research of ecosystem service flow focuses on the flow process. The Service Path Attribution Networks(SPANs) model can use artificial intelligence to capture the spatial dynamic characteristics of ecosystem service flow, but it is not yet open to the outside world, so it is not widely used. The flow direction is extracted according to the digital elevation model, which can be used to identify the flow path of the freshwater ecosystem service flow.

In this paper, we selected the Lianshui River Basin above the Xiangxiang Hydrological Station as the research area to analyze the impact of climate and land-use changes on freshwater ecosystem service flows. We built a coupled model framework. The purpose of this research were to:(1) map land-use change; (2)quantify the supply of freshwater ecosystem services by the InVEST model;(3) use a multi-index method to calculate water demand in various industries;(4) determine the path by using Digital Elevation Models(DEM), and calculate service flow by supply and demand;(5) quantify the impact of climate and land-use changes on freshwater ecosystem service flows, and introduce contribution rates to distinguish the degree of impact of precipitation and land use changes on freshwater ecosystem service flows. The results can provide a scientific basis for water resources management and ecological compensation in the Lianshui River Basin and other similar ecosystems.

2 Study Area and Methodology

2.1 Study area

The Lianshui River Basin is located in the central part of Hunan Province, at N27°17′~28°04′and E111°31′—112°53′. It is a primary tributary on the left bank of the lower reaches of the Xiangjiang River. The catchment area of the basin above the Xiangxiang Hydrological Station is 5919.03 km^2 (Fig. 1). It is bordered by Weishui in the north, distilled water and trickling water in the south, and the watershed between the Xiangjiang River Basin and the Zishui River Basin in the west. The average elevation of the basin is about 202.9 m, and the terrain is higher in the west and lower in the east. It has a subtropical monsoon climate with sufficient rainfall and rain and heat in the same period. Rainfall is unevenly distributed in time and space, mostly concentrated in April-September, with annual average rainfall of 1350~1450 mm. With the rapid development of social economy, the imbalance between supply and demand of water resources in the river basin has become more and more serious.

2.2 Data sources

The data included in this study was a DEM, land-use data, average annual precipitation, annual mean potential evapotranspiration, soil data and water demand in 2000, 2010, and 2020. ArcGIS software was used to preprocess the DEM, define the outlet location of the watershed and divide the watershed into 35 sub-watersheds. Spatial interpolation of the environmental data (specific data types and data sources in Table 1) was conducted using the inverse distance weighting method.

Table 1. The data type and data sources

Type	Data sources
Digital Elevation Model(30 m resolution)	Geospatial Data Cloud(https://www.gscloud.cn/)
Land use/cover data	GlobeLand30(http://www.globallandcover.com/)
Meteorological data(precipitation, temperature.)	Chinese Academy of Sciences Resource and Environmental Science Data Center(http://www.resdc.cn/)
Soil data	the Food and Agriculture Organization of the United Nations (FAO)(http://www.fao.org)
Hydrological data(runoff.)	The Hydrological Bureau of Hunan Province
Water demand data	Water Resources Bulletin of Hunan province
Socioeconomic data(county scale)	Statistical Yearbook

2.3 Analytical methodology

Following, we analyzed the freshwater supply ecosystem services flow in six steps (Fig. 1). We constructed hydrological models, water demand models, freshwater ecosystem service flow paths and flow models to realize the visualization of freshwater ecosystem services from supply to demand flow.

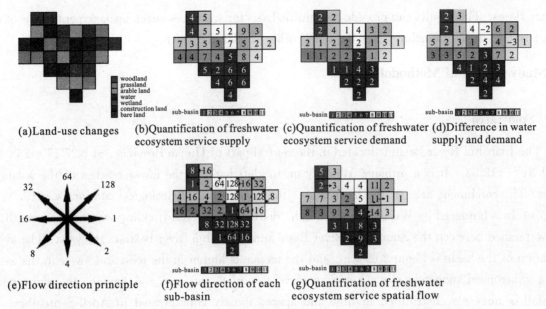

(a) Land-use changes (b) Quantification of freshwater ecosystem service supply (c) Quantification of freshwater ecosystem service demand (d) Difference in water supply and demand

(e) Flow direction principle (f) Flow direction of each sub-basin (g) Quantification of freshwater ecosystem service spatial flow

Figure 1. The Research framework of freshwater ecosystem service flow

2.2.1 *The InVEST model*

The InVEST model was used to was jointly developed in 2007 by Stanford University, the Nature Conservancy (TNC), and the World Wide Fund for Nature (WWF) (Maria et al., 2012). It was used to simulate and quantify the supply of freshwater ecosystem services in this paper. The water production module was based on the water balance formula, ignoring the interactive flow between the surface and groundwater, and calculated the water production through parameters such as precipitation, plant transpiration, surface evaporation, root depth, and soil depth (Han and Dong 2017). The main algorithm was as follows:

$$Y_{xj} = \left(1 - \frac{AET_{xj}}{P_x}\right) P_x \tag{1}$$

where, Y_{xj} is the annual water production of the j-th land use type in the grid x, mm; AET_{xj} is the actual annual evapotranspiration of the j-th land use type in the grid x, mm; P_x is the average annual precipitation in the grid x, mm; $\frac{AET_{xj}}{P_x}$ is an approximation of the Budyko curve estimated by (RJ et al., 2012) as follows:

$$\frac{AET_{xj}}{P_x} = \frac{1 + \omega_x R_{xj}}{1 + \omega_x R_{xj} + \frac{1}{R_{xj}}} \tag{2}$$

$$\omega_x = z \cdot \frac{AWC_x}{P_x} + 1.25 \tag{3}$$

where, ω_x is an unrealistic parameter that describes the soil properties under natural climatic conditions; R_{xj} is the aridity index of the *j-th* type of land use type in the grid x, defined as the ratio of potential evaporation to precipitation; AWC_x is the available water content of the vegetation the in grid x, mm, which is used to determine the amount of water provided by the soil for plant growth; Z is the Zhang coefficient (Zhang et al., 2001), and the more rainfall in the study area, the greater the Zhang coefficient.

2.2.2 Multi-index fusion method

Based on the definition of ecosystem service by the Millennium Ecosystem Assessment (MEA 2005), we regarded actual water consumption as the demand for freshwater ecosystems, including agricultural water, industrial water, and domestic water (urban resident' domestic water and rural resident' domestic water). The formula was as follow:

$$w_{dem} = w_{agr} + w_{ind} + w_{dom} \tag{4}$$

where w_{agr}, w_{ind}, and w_{dom} is agricultural, industrial, and domestic demand for water, respectively. C_{agr} represents water requirement per unit quality of crop products, L_{agr} represents crop yield, C_{ind} is water indicator per 10,000 yuan of GDP, L_{ind} is the GDP, C_{dom_u} stands for the domestic water quota of urban residents, L_{dom_u} stands the number of urban residents, C_{dom_r} is domestic water quota for rural residents, L_{dom_r} is the number of rural residents. Table 2 shows detail for these indices of water demand in the study area, which data comes from the Hunan Provincial Water Resources Bulletin.

Table 2. Average annual water consumption of agriculture, residents, and industry

Year	City	C_{agr} (m³/ha)	C_{ind} (m³/10⁴GDP)	C_{domu} (L/d·person)	C_{domr} (L/d·person)
2000	Shaoyang	9015	483	156	114
	Loudi	8209	326	163	118
	Xiangtan	8955	295	163	120
2010	Shaoyang	8611	228	154	106
	Loudi	8160	182	152	108
	Xiangtan	8534	182	152	112
2018	Shaoyang	8044	158	153	92
	Loudi	8044	96	150	95
	Xiangtan	7910	96	150	101

2.2.3 Spatial flow model

In this study, we ignored some human factors (such as water lifting and diversion), and based on DEM, used the direction of water flow as the path of freshwater ecosystem service flow. This model only flows in adjacent grids, and there was no flow inside the grid. 8 numbers (1, 2, 4, 8, 16, 32, 64, 128) were used to indicate the flow direction of 8 adjacent grids [Fig. 1(e)]. Ensure that each unit will flow to the outlet of the basin. The service flow was the remaining amount of water resources excluding the water demand of this unit. All units with a positive surplus were transferred according to the flow direction until the surplus was negative or to the outlet of the basin.

2.3 Climate change and land-use change scenarios

Based on the simulation study of freshwater ecosystem service flow, precipitation and land-use types are the main factors that affect the simulation results of the model. In order to further explore the impact of precipitation and land use changes on freshwater ecosystem service flows, this study designed three scenarios (Table 3). Scenario 1: Actual situation in 2000 in the base year, using actual climate and land-use data in 2000. Scenario 2: Keep the 2000 land-use data unchanged, and replaced the 2000 precipitation data with the 2010 and 2020 precipitation data to account for the impact of precipitation changes on freshwater ecosystem service flows. Scenario 3: Keep the 2000 precipitation data unchanged, and replaced the 2000 land use data with the 2010 and 2020 land-use data respectively to determine the impact of land use changes on freshwater ecosystem service flows.

Table 3. Scenario design for analyzing the impacts of climate and land-use change on the freshwater service flows

Scenario	2000	2010	2020
1	Precipitation 2000 Land-use 2000	Precipitation 2000 Land-use 2000	Precipitation 2000 Land-use 2000
2	Precipitation 2000 Land-use 2000	Precipitation 2010 Land-use 2000	Precipitation 2020 Land-use 2000
3	Precipitation 2000 Land-use 2000	Precipitation 2000 Land-use 2010	Precipitation 2000 Land-use 2020

In order to distinguish the degree of influence of changes in precipitation and land use on changes in freshwater ecosystem service flows, we introduce the concept of contribution rate to quantify the degree of influence(Yang, Xie, Zhang and Tao 2021). The contribution rate can be calculated by the following formula:

$$R_p = \frac{\Delta p}{|\Delta p| + |\Delta l|} \times 100\%$$
$$R_l = \frac{\Delta l}{|\Delta p| + |\Delta l|} \times 100\%$$
(5)

where R_p and R_l are the contribution rates of precipitation change and land-use change in freshwater ecosystem service flows, respectively. If R_p or R_l is greater than 0, the effect is a positive effect, otherwise it is a negative effect; Δp and Δl represent the changes in precipitation change and land-use change scenarios respectively.

3 Results

3.1 Land-use change

Land use types are classifed into 7 land types: arable land, woodland, grassland, wetland, water, build-up land and bare land. Arable land and woodland are the main land use types in the Lianshui River Basin(Fig. 2), which are widely distributed throughout the whole basin. The second is grassland, which is mainly distributed in the middle of the basin. The build-up land is concentrated in the centers of counties and cities, such as sub-catchments 11 and 14 corresponding to Louxing District of Loudi City, and sub-catchments 17, 20, 25 correspond to the urban area of Lianyuan City, etc.

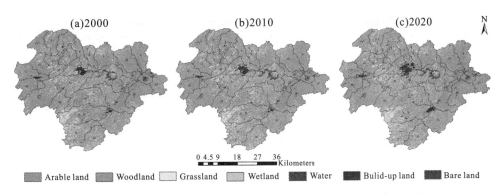

Figure 2. Land-use map of the Lianshui River Basin in 2000(a), 2010(b) and 2020(c)

Table 4 shows that the main land-use types are arable land and woodland, accounting for more than 78% of the watershed area, followed by grassland, which accounts for about 17%. Wetlands, water areas, and build-up land are less, and they occupy less than 5% of the watershed area. By 2020, a very small portion of bare land appeared in the basin. Between 2000 and 2020, different land use types have changed. The area of arable land decreased during 2000—2010, but increased during 2010—2020. In general, the area of arable land decreased by 29.02 km² during 2000—2020. The change trend of forest land increased during 2000—2010, and increased in 2010—2020. During the study period, the area of woodland decreased by 69.75 km². Grassland showed a trend of first increase and then decrease. Wetland area continued to decrease, a total decrease of 2.57 km². The build-up land has been increasing by 101.53 km², which proportion was not large, but the increase was as high as 151%. Bare land appeared in 2020, with an area of 4.09 km².

Table 4. Temporal changes of land-use in the Lianshui River Basin

Land-use types	2000		2010		2020	
	Area(km²)	Percentage(%)	Area(km²)	Percentage(%)	Area(km²)	Percentage(%)
Arable land	2288.27	38.66	2256.11	38.11	2259.25	38.17
Woodland	2478.95	41.88	2487.77	42.03	2409.20	40.7
Grassland	1010.94	17.08	1014.51	17.14	988.41	16.71
Wetland	2.67	0.05	0.50	0.02	0.10	0.002
Water	71.12	1.2	82.51	1.39	89.36	1.5
Build-up land	67.08	1.13	77.62	1.31	168.61	2.85
Bare land	0.00	0	0.00	0	4.09	0.068

3.2 Simulation verification

According to the measured data of Xiangxiang Hydrological Station, the total water resources in the basin are 3.97 billion m³, 5.41 billion m³ and 4.08 billion m³ in 2000, 2010, and 2020. Simulation error was calculated by the measured runoff data and mean relative error(MRE). It turned out that simulation errors were 4.53%, -0.02% and 3.19% respectively(Table 5). Therefore, the InVEST model has been proved to have good adaptability in the simulation of water yield in the Lianshui River Basin, and it can be used for the study of freshwater ecosystem service flows in the basin.

Table 5. Model simulation water yield verification results

Year	Water yield($10^9 m^3$)		MRE(%)
	Observation	simulation	
2000	3.97	4.15	4.53
2010	5.411	5.41	-0.02
2020	4.08	4.21	3.19

3.3 Freshwater ecosystem service flows change

3.3.1 *Spatio-temporal evolution of water yield*

In general, from 2000 to 2020, the water output of the Lianshui Basin increased first and then decreased. The water production in the basin was the highest in 2010, which was 5.41×10^9 m³. Compared with 2000, the water production in the basin increased slightly in 2020, about 0.06×10^9 m³, with an increase of 1.4%. The spatial difference of water production in different periods was relatively small(Figure 4), and the overall performance was higher in the southern region and lower in the northern region. From 2000 to 2010, the water yield of all sub-basins within the basin increased. The most significant increase was in Shuangfeng County in the south, where the average water production of the sub-basins increased by 0.07 billion m³. From 2010 to 2020, although most of the sub-basin water yield increased, but the increase was not high, the water yield of the remaining sub-basins all decreased, the decrease was slightly larger, resulting in a final decrease in the water production of the entire watershed during 2010—2020.

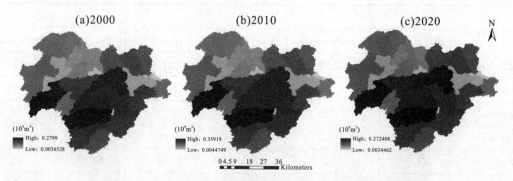

Figure 4. Spatial distribution of water yield in Lianshui River Basin

3.3.2 *Spatio-temporal evolution of water demand*

The total water demand of Lianshui River Basin in 2000, 2010 and 2020 was 5.44×10^9 m³, 6.37×10^9 m³, and 7.36×10^9 m³ respectively, showing a significant upward trend. Water demand was affected by factors such as farmland irrigation and population. From 2000 to 2020, the water demand in the basin was dominated by agricultural irrigation water and industrial water, accounting for 62.4%~64.66% and 21.9%~24.76%, respectively, and the proportion of domestic water demand was relatively small. The areas with low water demand were mainly located in the upper reaches of the Lianshui River Basin (Xinshao County and Shaodong County), where have a small population and insufficient industrial and agricultural development. The area with high water demand was located in the south of the middle reaches of the basin, which was closely related to the wide distribution of arable land in this area and the low utilization coefficient of agricultural irrigation water.

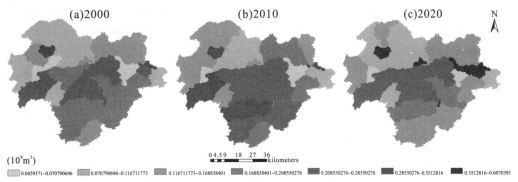

Figure 5. Spatial distribution of water demand in Lianshui River Basin

3.3.3 *Freshwater ecosystem service flows*

In this study, we defined a sub-catchment that cannot meet the actual water demand by its water supply and needs to be supplemented by an upstream sub-catchment as the beneficiary area; conversely called the supply area (Fig. 6). In 2000 and 2010, the beneficiary areas of the basin were both sub-basins 5, 11, 12, 15, 19, and 28. In 2020, the beneficiary areas were sub-basins 5, 11, 12, 15, 16, 18, 19 and 28. From 2000 to 2020, the area of beneficiary areas in the basin had a trend of expansion, reflecting the weakening of the freshwater ecosystem services in the basin. The service flow path of the freshwater ecosystem was consistent with the flow direction of the water system. The service flow was 0.304×10^9 m^3, 0.472×10^9 m^3 and 0.292×10^9 m^3, respectively. The flow increased by 0.168×10^9 m^3 from 2000 to 2010, and decreased by 0.181×10^9 m^3 from 2010 to 2020. In 2000—2020, the freshwater ecosystem service flow showed a decreasing trend, decreased about 0.013×10^9 m^3. Changes in freshwater ecosystem service flows were mainly due to changes in the basin climate and land-use.

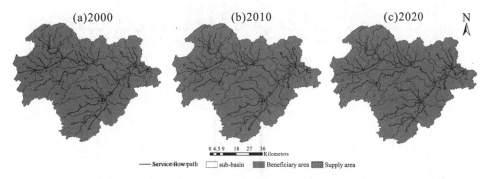

Figure 6. Spatial distribution of freshwater ecosystem service flows in Lianshui River Basin

3.5 Effects of climate and land-use change on freshwater ecosystem service flows

From 2000 to 2020, the precipitation in the Lianshui River Basin was 1510~1780 mm/a, showing a trend of first increasing and then decreasing. Rainfall was greatly affected by topography. The average precipitation in the western part of the basin was far more than 1780 mm, and precipitation in some areas was as high as 1946.48 mm. In contrast, in Shuangfeng County of the southeast, precipitation in some areas was less than 1190 mm. The precipitation in the basin was unevenly distributed in time and space.

Under the scenario of constant land-use and changes in rainfall, the water yield of the whole watershed in 2010 and 2020 of scenario 2 were 6.02 billion m³ and 4.2 billion m³, respectively. Compared with the real scenario in 2000, the water yield of the watershed increased by 1.86 billion m³ and 0.05 billion m³, respectively. And the flow increased 4.36 billion m³ and 0.25 billion m³ [Fig. 7(a)]. Under the change of precipitation from 2000 to 2010, the service flow of all sub-basins increased, with a significant increase, and some of the beneficiary areas in 2000 were transformed into freshwater ecosystem service supply areas (sub-basins 12, 15, 19, 28). Under the change of precipitation from 2000 to 2020, the service flow of each sub-basin will change. Most of the sub-basin service flows have increased, but the increase was small and not obvious compared to 2000—2010.

Under the scenario of constant precipitation and changes in land-use, the water production in the watershed in 2010 and 2020 in Scenario 3 were 4.157 billion m³ and 4.164 billion m³, respectively. Compared with the actual situation in 2000, the water production in the watershed has increased slightly by 0.014 billion m³ and 0.01 billion m³, respectively. Service flow decreased by 0.001 billion m³ in 2010 and increased by 0.06 billion m³ in 2020[Fig. 7(b)]. Under the change of land-use from 2000 to 2010, the water yield and service flow of all sub-basins changed, and the spatial distribution was uneven. The sub-basins 15 and 19 near the outlet of the watershed changed most significantly. Under the change of land-use from 2000 to 2020, most of the sub-basin water production and service flow have increased, and the increase rate was relatively large, which increased the overall service flow.

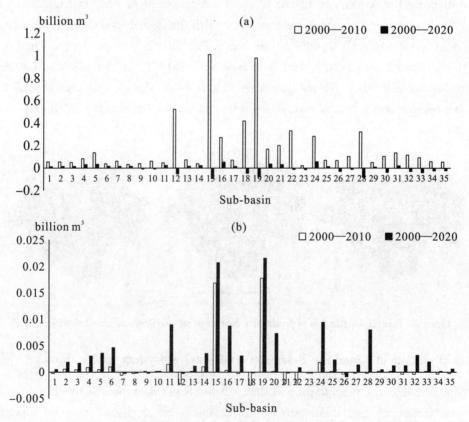

Figure 7. The change in service flow of each sub-basin compared with the actual situation in 2000;
Scenario 2 changes in precipitation(a), Scenario 3 changes in land-use(b)

From 2000 to 2010, the contribution rates of precipitation changes and land use changes to water production changes were 99.27% and 0.73%, respectively, and their contribution rates to service flow were 99.98% and 0.02%, respectively. From 2000 to 2020, the contribution rates of precipitation changes and land use cover changes to water production changes were 81.45% and 18.55%, and their contribution rates to service flow changes were 80.88% and 19.12%, respectively. Obviously, changes in precipitation have a more significant impact on freshwater ecosystem service flows, and changes in land use have less impact on them.

4 Discussion

4.1 The impact mechanism of precipitation and land-use change on freshwater ecosystem service flow

The impact of precipitation and land-use changes on the service flow of freshwater ecosystems is mainly reflected in their impact on the water supply. Precipitation replenishes groundwater through infiltration, and the rest collects on the surface to form runoff, which directly affects water production (Bagher, Arash, Ali, Bahram, Hossein, Ziga and Peter 2020). The increase in precipitation can directly lead to an increase in water output, thereby changing the freshwater ecosystem services flow.

The impact process of land-use change on water production is more complicated. Changes in land-use first changed soil conditions, biodiversity, etc., and then changed the underlying surface, thereby affecting surface runoff. Different land-use types have different effects on water yield. Usually, the water yield of woodland is very low, because vegetation roots and forest canopy can effectively intercept precipitation, forest vegetation can absorb precipitation trapped in litter and precipitation that penetrates into the soil layer, and trees also have a strong transpiration effect. Both will lead to a decrease in surface runoff. Therefore, the increase in the area of woodland will reduce the water production in this area. Arable land, grassland and wetland have similar effects on precipitation as woodland. The difference in plant density and root depth will lead to different effects on surface runoff. When precipitation directly reaches the water surface, it is most likely to form runoff, but the evaporation of the water area is strong. Some studies have shown that increasing the water area will reduce water production. The impervious rate of build-up land and bare land is relatively high. The decrease in infiltration and the increase of surface runoff will increase the water production of the area. Therefore, when the area of build-up land and bare land increases, the water yield will increase significantly.

The mutual transfer between land use types will also lead to changes in water yield. This mutual transfer often has a trade-off effect. For example, under the condition of constant precipitation, an increase in construction land will increase water production, while an increase in forest land will reduce water production. This leads to insignificant changes in total water production. The increase in construction land and bare land can increase the water production capacity of the river basin, but it will increase the risk of soil erosion and reduce the water storage capacity. Therefore, it is necessary to further quantify the impact of land-use types on ecosystem services. In the process of urbanization, while protecting the ecology, rational development and management of land resources are needed to achieve sustainable regional development.

4.2 Limitations and uncertainties

It is worth noting that our research relied on the application of models, and any model has uncertainty(Zhang et al., 2021; Zhang et al., 2015). Both land-use classification and hydrological process modeling bring uncertainty into the analysis of freshwater ecosystem service flows. The

InVEST model had a wide range of applications in ecosystem assessment research, but the model itself still has certain uncertainties. It cannot fully reflect the complex topography, nor can it describe the groundwater balance process. In the simulation process, the input of parameters also brings uncertainty to the simulation. The water yield module of the InVEST model uses several parameters in the application to simulate water production. Among them, the potential evapotranspiration was calculated by the Penman formula revised by FAO. The soil depth was derived from the World Soil Database established by FAO, and it was used to replace the root depth of vegetation. At the same time, according to the percentage content of the soil texture, we calculated the actual evapotranspiration of the reference crop in the SPAW software using the empirical formula of soil effective water content. These would not affect the basic water production model of the basin, but would affect the accuracy of the model simulation to a certain extent.

Since the precipitation data observed by meteorological stations were limited and they were all point records, we adopted the inverse distance weighting method for spatial interpolation. The location and density of meteorological stations have an impact on the accuracy of the data. Although we calibrated the sensitivity parameters of the model with the data of measured hydrological stations, there were still errors. The water demand data was obtained according to the statistics of the water resources bulletin of the city and county, and the accuracy was poor. The water demand data corresponding to each sub-basin cannot truly reflect the actual situation, which limits the identification and calculation of the freshwater ecosystem service flow.

5 Conclusions

Precipitation and land-use changes affect the flow of freshwater ecosystem services by changing the relationship between supply and demand. We found that from 2000 to 2020, the freshwater ecosystems service flow in the Lianshui River Basin changed in time and space. During the study period, the supply of freshwater increased first and then decreased, while the demand continued to increase. The service flow path was consistent, but the flow increased first and then decreased. Changes in freshwater ecosystem service flows were mainly affected by precipitation changes, and the land-use change have relatively small effects on them. Analyzing the widespread impact of precipitation and land use changes on the flow of freshwater ecosystem services will help support the management of water resources in the basin, and is essential to promote the sustainable development of freshwater ecosystem services.

Acknowledgements

This study was supported by the Joint Fund for Regional Innovation and Development of NSFC (U19A2051), the Research Project of Hunan Provincial Water Resources Department (XSKJ 2018179-09), the Key R&D Project of Hunan Province (2017SK2301), and the Construction Program for First-Class Disciplines(Geography) of Hunan Province, China(5010002).

References

[1] Alessandro F, F M S, Scott M J. The direct and indirect effects of extreme climate events on insects[J]. The Science of the total environment, 2021:769.

[2] Brendan F, Kerry T R. Ecosystem services: classification for valuation[J]. Biological Conservation, 2008, 141(5):

1167−1169.

[3] Bagher S, Arash M, Ali S, et al. Impacts of future climate and land use change on water yield in a semiarid basin in Iran[J]. Land Degradation & Development, 2020, 31(10): 1252−1264.

[4] Bagstad K J, Johnson G W, Voigt B, et al. Spatial dynamics of ecosystem service flows: A comprehensive approach to quantifying actual services[J]. Ecosystem Services, 2013, 4:117−125.

[5] Cheng Z, Jing L, Zixiang Z, et al. Application of ecosystem service flows model in water security assessment: A case study in Weihe River Basin, China[J]. Ecological Indicators, 2021, 120.

[6] Gao J, Jiang Y, Wang H, et al. Identification of Dominant Factors Affecting Soil Erosion and Water Yield within Ecological Red Line Areas[J]. Remote Sensing, 2020, 12(3).

[7] Gao X, Huang B, Hou Y, et al. Using Ecosystem Service Flows to Inform Ecological Compensation: Theory & Application[J]. Int J Environ Res Public Health, 2020, 17(9).

[8] Han H, Dong Y. Spatio-temporal variation of water supply in Guizhou Province, China[J]. Water Policy, 2017, 19(1):181−195.

[9] Hao L, Zheng W, Guangxing J, et al. Quantifying the impacts of climate change and human activities on runoff in the lancang river basin based on the budyko hypothesis[J]. Water, 2020, 12(12).

[10] Hayley S, Graeme A, Iris A J, et al. The Practical Fit of Concepts: Ecosystem Services and the Value of Nature[J]. Global Environmental Politics, 2021, 21(2).

[11] Jeong P Y, Jin L H, Kwan S J, et al. Water quality and structure of aquatic ecosystem in water source, lake Gachang[J]. Korean Journal of Environmental Biology, 2011, 29(4).

[12] Jun X, Qiting Z. The utilization and protection of water resource in China (1978−2018) [J]. Urban and Environmental Studies, 2018,(02):18−32.

[13] Ke X, Wang L, Ma Y, et al. Impacts of strict cropland protection on water yield: a case study of Wuhan, China[J]. Sustainability, 2019, 11(1).

[14] Mononen L, Auvinen A P, Ahokumpu A L, et al. National ecosystem service indicators Measures of social-ecological sustainability[J]. Ecological Indicators, 2016, 61:27−37.

[15] Lin J, Huang J, Prell C. Changes in supply and demand mediate the effects of land-use change on freshwater ecosystem services flows[J]. Science of the Total Environment, 2021,1(763):143012.

[16] Maria S C, Alfredo LB, Ana P, et al. Sensitivity analysis of ecosystem service valuation in a Mediterranean watershed[J]. Science of the Total Environment, 2012, 440:140−153.

[17] MEA. Millenium ecosystem assessment: ecosystems and human well-being[M]. Synthesis Island Press, 2005.

[18] Hejnowicz P A, Rudd A M, Haberl H. The Value Landscape in Ecosystem Services: Value, Value Wherefore Art Thou Value? [J]. Sustainability, 2017, 9(5).

[19] Peng H, Qiao W, Shenwenming, J. Progress of integrated ecosystem assessment: Concept, framework and challenges[J]. Geographical Research, 2015, 34(10):1809−1823.

[20] Richard S, Rebecca C K, Spencer W. InVEST 3.2.0 User's Guide[M].2015

[21] Donohue R J, Roderick M L, Mcvicar T R. Roots, storms and soil pores: incorporating key ecohydrological processes into Budyko's hydrological model[J]. Journal of Hydrology, 2012, 436:35−50.

[22] Schirpke U, Tappeiner U, Tasser E. A transnational perspective of global and regional ecosystem service flows from and to mountain regions[J]. Scientific Reports, 2019, 9(4).

[23] Shunze W, Lingjie Z, Yu S. Review on the research of Xi Jinping's Thought on ecological civilization [J]. Environment and sustainable development, 2020, 45(6):37−42.

[24] Sun Y, Liu S, Dong Y, et al. Spatio-temporal evolution scenarios and the coupling analysis of ecosystem services with land use change in China[J]. Science of the Total Environment, 2019, 681.

[25] TJ B, NM S. Using the Soil and Water Assessment Tool(SWAT) to assess land use impact on water resources in an East African watershed[J]. Journal of Hydrology, 2013, 485:100−111.

[26] Wang J L, Zhou W Q. Ecosystem service flows: Recent progress and future perspectives[J]. Acta Ecologica Sinica,

2019, 39(12):4213-4222.

[27] Wang Y, Li X, Liu S, et al. Climate services for water resource management in China: the case study of Danjiangkou Reservoir[J]. Journal of Meteorological Research, 2021, 35(1):87-100.

[28] Wenjie Y, Yue Z, Kangping Z, et al. Evaluation on the Ecosystem services value of the Xin'anjiang river in Huangshan[J]. China Environmental Management, 2018, 10(4):100-106.

[29] Xiao Y, Ouyang Z. Spatial-temporal patterns and driving forces of water retention service in China[J]. Chinese Geographical Science, 2019, 29(1):100-111.

[30] Xu J, Xiao Y, Xie G, et al. Computing payments for wind erosion prevention service incorporating ecosystem services flow and regional disparity in Yanchi County[J]. Sci Total Environ, 2019, 674:563-579.

[31] Yang J, Xie B, Zhang D. Climate and land use change impacts on water yield ecosystem service in the Yellow River Basin, China[J]. Environmental Earth Sciences, 2021, 80(3).

[32] Zhang C, Li J, Zhou Z, et al. Application of ecosystem service flows model in water security assessment: a case study in Weihe River Basin, China[J]. Ecological Indicators, 2021, 120.

[33] Zhang L, Dawes W R, Walker G R. Response of mean annual evapotranspiration to vegetation changes at catchment scale[J]. Water Resources Research, 2001, 37(3):701-708.

[34] Zhang X K, Fan J H, Cheng G W. Modelling the effects of land-use change on runoff and sediment yield in the Weicheng River watershed, Southwest China[J]. Journal of Mountain Science, 2015, 12(2):434-445.

[35] Zou M, Kang S, Niu J, et al. Untangling the effects of future climate change and human activity on evapotranspiration in the Heihe agricultural region, Northwest China[J]. Journal of Hydrology, 2020(585).

The Influence of Atmospheric Pressure Decrease on Cavity Characteristics of the Drop-step Aeration Facilities

Yameng Wang, Jun Deng, Wangru Wei

State Key Laboratory of Hydraulics and Mountain River Engineering,
Sichuan University, Chengdu 610065, China

Abstract

Cavitation is often accompanied by flood discharge in hydropower projects. More and more water conservancy projects are built in high altitude areas, and cavitation is more likely to occur during flood discharge. At the same time, there are also problems of low atmospheric pressure and thin air concentration. Practice has proved that aeration is one of the effective measures to solve cavitation damage. Cavity characteristics are one of the important factors that influence the flow aeration. In order to study the influence of cavity characteristics on atmospheric pressure change, the comparative analysis method of decompression test and atmospheric pressure test was used to study the change of bottom cavity length and cavity pressure after the drop-step aeration facility. The experimental results show that the change of atmospheric pressure has little effect on the length of the bottom cavity. When the atmospheric pressure changes, the cavity length is basically unchanged, and the cavity length mainly depends on the incoming flow conditions and the shape of the drop-step. The reduction of atmospheric pressure is beneficial to improve the backwater in the bottom cavity. The lower the atmospheric pressure, the less the backwater in the cavity and the increase in the effective area of the bottom cavity. The decrease in atmospheric pressure causes the pressure difference between the inside and outside of the bottom cavity to gradually decrease and eventually tend to zero.

Keywords: Cavitation; Aeration; Cavity length; Cavity pressure

1 Introduction

Hydropower is a kind of zero-emission, clean and renewable energy. As the hydropower development blooms in China, many hydraulic projects with high head and large discharge are under construction or to be constructed in the future. In hydroelectric flood discharge projects, high flow velocity and high altitude make the cavitation number on the wall surface lower, and cavitation erosion is easy to occur. With the development of water resources, more and more dams are being built at altitudes of 3,000 m or more. At high altitudes, the air is thin and the atmospheric pressure is lower, which has a great impact on the aeration characteristics of the water flow. At present, it is considered that aeration is one of the most effective measures to solve the cavitation problem. Most engineering model tests are carried out according to the gravity similarity criterion at normal pressure, and atmospheric pressure is not similar. Although some major flood discharge projects will use decompression tests to check whether there is a problem with water flow aeration in the flood discharge project. However, these experiments did not pay attention to the effect of atmospheric pressure reduction on the variation of flow aeration characteristics.

The existing research shows that aeration can effectively solve the cavitation problem. The jet box behind the aerator step is one of the important factors affecting the aerated concentration. The research results of cavity characteristics at atmospheric pressure are very comprehensive. Xu Yimin (Xu Yimin et al., 2004) and Wu Jianhua(Wu Jianhua et al., 2008; Wu Jianhua et al., 2013) conducted in-depth research on the calculation of jet length, and proposed the corresponding theoretical calculation formula and the method of improving parameters. Pfister and Hager(Pfister et al., 2010; Pfister et al., 2010; Pfister et al., 2011; Pfister et al. 2011) studied the distribution characteristics of aeration concentration near and far away from the aerator step in 2010 and 2011. However, there is no systematic research and analysis on the influence of atmospheric pressure reduction on the cavity characteristics. In this paper, the length and pressure of the cavity are measured by physical model test under different atmospheric pressure, and the influence of atmospheric pressure reduction on the cavity is analyzed.

2 Experimental Setup and Methodology

Experiments were conducted in a 0.3 m-wide and 6 m-long sectional chute model at the State Key Laboratory of Hydraulics and Mountain River Engineering, Sichuan University, China. The test model is made of plexiglass. The model consists of an upstream channel, a step, and a downstream channel(Fig. 1). The width of the channel is 30 cm, the depth of the incoming water is 10 cm, the height of the step is 5 cm, the air holes are located on both sides of the step, 0.05 m long and 0.03 m high. In the test, the size of the air hole meets the requirements, and there is no insufficient air supplement. The gradient of downstream slop section was $i = 0.12$, and the length was 5.0 m. The water level in front of the upstream gate is set to $h_0 = 1.0$ m, 1.5 m, 2.0 m, 2.4 m respectively. Flows of variable approach flow velocity V_0 and Froude numbers $F_0 = V_0/(gh)^{0.5}$ were generated with a bottom cavity, with V_0 = approach flow velocity and g = gravitational acceleration. The parameters h and F_0 relate to unaerated black-water flow. The range of Froude number $F_0 = 4.47 \sim 6.93$.

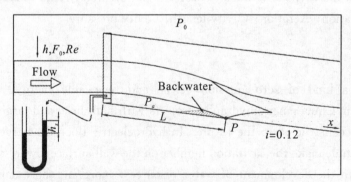

Figure 1. Definition sketch with relevant parameters

The experimental model is placed in a decompression box, which can adjust the atmospheric pressure value, and the maximum vacuum degree can reach 97%. The vacuum degree refers to the degree of gas rarefaction in a vacuum state. $\eta_m = (P_0 - P_a)/P_0$, η_m represent the vacuum degree; P_0 represent the atmospheric pressure outside the decompression box, kPa; P_a represent the atmospheric pressure inside the decompression box, kPa. During the test, the water temperature was 11℃ and the atmospheric pressure outside the decompression box was 96 kPa. Each 10 kPa reduction of non-pressurized pressure is set as a group of test conditions, a total of 10 groups. In this test, the variation range of vacuum degree is as follows, 0%(96 kPa), 10.4%(86 kPa), 20.8%(76 kPa), 31.3%(66 kPa), 41.7%(56 kPa), 52.1%(46 kPa), 62.5%(36 kPa), 72.9%(26 kPa), 83.3%(16 kPa), 93.8%(6 kPa). Pressure holes are arranged in the bottom plate of the downstream flume. A total of 34 measuring points are arranged on the bottom of the downstream channel to measure the pressure change inside the bottom cavity and near the water flow impact area. The minimum distance between the pressure points is 5 cm. The pressure head h_s in the air cavity with an U-shaped manometer. ΔP is the pressure difference between inside and outside the cavity. $\Delta P = P_0 - P_a = \rho g h_s$. The contour of $C = 0.9$ is taken as the surface line of the lower edge of the water tongue, and the jet length was determined by combining the measurement of aeration concentration with observation.

The distance from the pressure $P = 0.15 P_{max}$ measuring point to the aerator step is defined as the jet length L, $L = L_{0.15 P_{max}}$, P_{max} represents the maximum pressure generated by water impacting the bottom. Ten series of four test were conducted in which only P_0 varied, the other parameters were kept constant. Ten of these test series were conducted according to Table 1, including different hydraulic conditions.

Table 1. Experimental conditions and experimental parameters

Sreies(kPa)	test	h_0(m)	Fr	Re
96~6(0%~93.8%)	1	1.0	4.47	265,580.91
	2	1.5	5.48	325,268.86
	3	2.0	6.32	375,588.12
	4	2.4	6.93	411,436.17

3 The Bottom Cavity

Aeration is an effective measure to prevent cavitation damage of discharge structure. Cavity characteristics are one of the important factors affecting the effect of aeration. Jet length and backwater are two important indexes to reflect cavity characteristics. Luo Yongqin et al. (2008) divided the factors affecting the backwater of the bottom cavity into three categories: inflow conditions, boundary conditions and water quality and temperature. The backwater will significantly reduce the length of the bottom cavity. The existing research results show that the jet length is related to the size of the aerator and the hydraulic conditions(Mohaghegh et al., 2009). These results were carried out under non-pressurized pressure. The influence of atmospheric pressure reduction on backwater, cavity pressure ΔP and jet length will be further analyzed in the following sections.

3.1 The backwater

When $P_0 = 96$ kPa, $h = 1.0$ m, $h = 1.5$ m, backwater appeared in the bottom cavity. It can be clearly observed that when the atmospheric pressure decreases gradually, the backwater depth in the bottom cavity decreases gradually. As shown in Fig. 2. When $h_0 = 1.0$ m, the bottom cavity was basically blocked by backwater under non-pressurized pressure. The backwater depth fluctuates with time. When the backwater depth was the maximum, the air holes will be completely blocked, when the backwater depth was the lowest, some of the air holes were still blocked. With the decrease of atmospheric pressure, the backwater depth in the bottom cavity decreases gradually. When $P_0 = 46$ kPa, the backwater depth nearest to the step was less than 1cm, and the influence on the air hole was greatly reduced. When $P_0 = 6$ kPa, the backwater in the bottom cavity was greatly reduced, and the nearest backwater position will not contact the air holes, the jet length was greatly increased. When $h_0 = 1.5$ m, there was less backwater in the bottom cavity at non-pressurized pressure, and only a thin layer of backwater at the bottom of the cavity. With the decrease of atmospheric pressure, the backwater in the bottom cavity gradually disappears.

The pressure measurement results on the channel floor are shown in Fig. 3. The data before the maximum pressure is the floor pressure in the bottom cavity. Fig. 3(a), when $h_0 = 1.0$ m, the vacuum degree was less than 20.8% and the atmospheric pressure was greater than 76 kPa, the floor pressure was greater than zero, which indicates that the backwater was serious and basically filled the bottom cavity. With the decrease of atmospheric pressure and the increase of vacuum, the bottom plate pressure decreases gradually, and the negative pressure appears at the first point. It indicates that the backwater position was gradually away from the air holes. When $P_0 \leqslant 76$ kPa, $\eta_m \geqslant 20.8\%$, the minimum distance from the backwater to the step was $x_{min} > 5$ cm. When $h_0 = 1.0$ m, the backwater was little and only occupied part of the bottom cavity. The minimum distance from the backwater to the step was $x_{min} > 35$ cm, and a part of negative pressure appeared in the bottom cavity. With the decrease of atmospheric pressure, backwater is less and less.

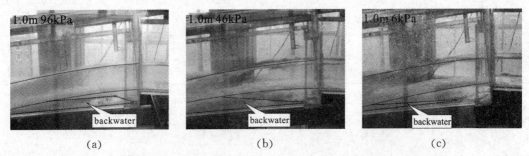

Figure 2. When $h_0 = 1.0$ m, the backwater in the bottom cavity

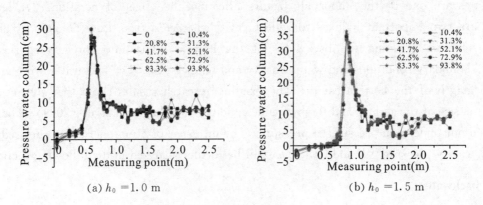

(a) $h_0 = 1.0$ m

(b) $h_0 = 1.5$ m

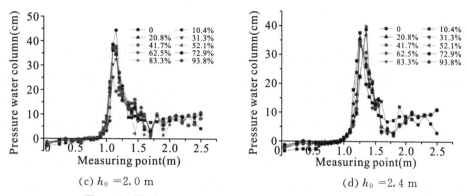

(c) $h_0 = 2.0$ m   (d) $h_0 = 2.4$ m

Figure 3. **The bottom plate pressure at different vacuum degree**

3.2 Cavity pressure ΔP

In Fig. 3, when $h_0 = 2.0$ m, $h_0 = 2.4$ m, the bottom plate pressure was all negative pressure. With the increasing of vacuum degree (the atmospheric pressure decreases), the pressure on the floor increases gradually. In order to measure the pressure change in the bottom cavity more accurately, a U-shaped tube was set in the bottom cavity to measure the change law, and the measurement accuracy reaches 1 mm. The density of water increases with the increase of pressure. In the range of 0~96 kPa, the change of water density is not large and can be ignored. The cavity pressure ΔP affects not only the backwater but also the jet length. The research results of Xu Yimin et al. (2009) show that with the increase of ΔP, the jet length decreases and the backwater depth increases. The negative pressure in the cavity will exert a force on the water flow in the direction of the cavity, making the water flow into the cavity and produce backwater, so as to achieve the momentum balance of the water body. The measurement results of ΔP under different vacuum degrees are shown in Fig. 4. The ΔP decreases with the decrease of atmospheric pressure. When $\eta_m = 0$, the value of ΔP was the largest. With the increasing of vacuum degrees, the atmospheric pressure decreases, ΔP was gradually decreasing. The smaller the h_0 is, the smaller the V_0 is, and the larger the variation range of ΔP is when the atmospheric pressure decreases. In general, when the V_0 is constant, the greater the η_m is, the lower the atmospheric pressure is, and the smaller ΔP is. When the size of the air holes is constant, the decrease of ΔP will lead to the decrease of the wind speed in the air holes, and the ventilation will also decrease accordingly, which is unfavorable to the flow aeration. It is necessary to increase the size of air holes in high altitude area to improve the ventilation capacity.

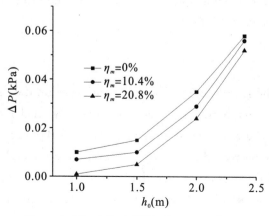

Figure 4. **The ΔP at different vacuum degrees**

3.3 The jet length

The jet length L was optically detected as the distance between the jet takeoff point at $x = 0$ and the reattachment point P of the lower jet trajectory on the chute bottom. There are many formulas for calculating the jet length at non-pressurized pressure, and the calculation accuracy is high. Three theoretical formulas to calculate the jet length and compared the calculated results with the measured results at non-pressurized pressure(Wu Jianhua et al., 2008; Pfister et al., 2010; Rutschmann P et al., 1990), the comparison results are shown in Fig.5(a). It can be seen from Fig.5(a) that the experimental value is slightly lower than the theoretical value only when $h_0 = 1.0$ m due to the influence of backwater. In other conditions, the experimental values are between the theoretical values. This proves that the measurement result of the jet length is reliable.

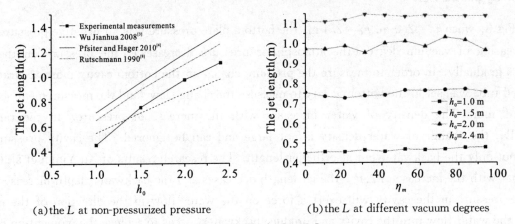

(a) the L at non-pressurized pressure (b) the L at different vacuum degrees

Figure 5. The jet length L

It can be seen from Fig.3 that the position of the maximum pressure point $L_{P_{max}}$ is basically unchanged when the atmospheric pressure decreases. When $h_0 = 2.4$ m, the flow velocity was high, the fluctuation of water impact on the floor was large, and the distance between the maximum pressure points is less than 5 cm ($d_{min} < 5$ cm). The measurement results of the jet length under different atmospheric pressure are shown in Fig.5(b). It can be seen from Fig.5(b) that the jet length is basically unchanged when the atmospheric pressure decreases. When $h_0 = 1.0$ m, the backwater was serious. With the decrease of atmospheric pressure, the backwater weakens, and the jet length increases slightly, but the increase length is small. The test results show that the jet length is mainly related to the flow condition and the shape of the step, and the effect of the atmospheric pressure reduction on the jet length is not obvious.

4 Conclusions

The decrease of atmospheric pressure had a certain influence on the cavity characteristics. The jet length, cavity pressure ΔP and backwater were measured by model test. With the decrease of atmospheric pressure, the ΔP gradually decreases and finally tends to zero. The smaller the ΔP, the less the backwater. With the decreased of atmospheric pressure, the ΔP will gradually decreased and eventually became zero, and the pressure inside the bottom cavity was equal to atmospheric pressure. It is obvious that the decrease of ΔP will weaken the ventilation and reduce the air content in the flow. In the case of low atmospheric pressure, the size of the air holes can be appropriately increased.

The measurement results of the jet length show that the effect of atmospheric pressure reduction on the jet length is not obvious. When the backwater was serious, the effective jet length increased greatly with the decreased of atmospheric pressure, but it did not increase the actual jet length. The jet length mainly depends on the inflow conditions and the shape parameters of the step.

There are still many effects of atmospheric pressure reduction on the aeration characteristics of water flow, such as the change of aeration concentration, the change of bubble size, the change of bubble floating speed and the maximum distance of aeration protection, which need to be further analyzed and studied. In addition, in high altitude areas, how to increase the ventilation rate, increase the aeration concentration and increase the aeration protection distance need to be further studied.

Acknowledgements

This work was supported by the National Science Foundation of China(Grant No. 51939007) and Sichuan Science and Technology Program(Grant No. 2019JDTD0007).

References

[1] Luo Y Q, Zhang S C, Li S S. Research of aeration backwater in knick point[J]. Journal of Southwest University for Nationalities: Natural Science Edition, 2008, 34(1):199−202.

[2] Mohaghegh, Wu J H. Effects of hydraulic and geometric parameters on downstream cavity length of discharge tunnel service gate[J]. Journal of Hydrodynamic, 2009, 21(6): 774−778.

[3] Pfister M, Hager W H. Chute aerators I: air transport characteristics[J]. Journal of Hydraulic Engineering, 2010, 136(6): 352−359.

[4] Pfister M, Hager W H. Chute aerators II: hydraulic design[J]. Journal of Hydraulic Engineering, 2010, 136(6): 360−367.

[5] Pfister M, Hager W H. Chute aerators: pre-aerated approach flow[J]. Journal of Hydraulic Engineering, 2011, 137(11): 1452−1461.

[6] Pfister M, Hager W H. Self-entrainment of air on the stepped spillways[J]. International Journal of Multiphase Flow, 2011, 37(2): 99−107.

[7] Rutschmann P, Hager W H. Air entrainment by spillway aerators[J]. Journal of Hydraulic Engineering, 1990, 116(6): 76−782.

[8] Wu J H, Ruan S P. Cavity length below chute aerators[J]. Science in China series E-technological sciences, 2008, 51(2): 170−178.

Water Quality Management(Sediment and Nutrients Run Off) at Brisbane Catchment

Fangrui Dong[1], Yongping Wei[2], Jinghan Li[2], Hui Li[1], Miao Zhang[1]

[1] China Institute of Water Resources and Hydropower Research, Beijing 100036, China

Research Center on Flood and Drought Disaster Reduction
of the Ministry of Water Resources, Beijing100036, China

[2] The University of Queensland, Queensland, Australia

Abstract

As one of the largest catchments in Queensland, the Brisbane catchment has played a very important role for the development of the Brisbane regions. However, the climate change and human impacts have made the water quality of the Brisbane catchment become worse and this issue has drawn many attentions. In this paper, the hydrological processes, geomorphological characteristics, and ecological processes related to this catchment would be introduced firstly. The whole water flow direction in this catchment is from northwest to southeast, which is originally from Mount Stanley and finally flow to the Moreton Bay. Then, through the data collection by the field trip and online references, the nutrients generation and sediment transportation at the upper catchment is the results of geomorphological characteristics and grazing; while for the middle catchment, agricultural development makes the decrease of native vegetation and the increased use of pesticides and large machinery. In addition, the construction of dams also causes the decrease in water quality and has the problem to increase the risk of flood within the extreme flood events. For the lower catchment, the construction of the residential area, central business district, and factories, and different transports in cities are the main reasons to makes the water quality problems. This paper also analyzes the interests of key stakeholders and the possible improvement points of current policies at the end. The main conclusions are: (1) the water quality management should consider the comprehensive measures, because the sediments and nutrients in this catchment have an accumulation and transport process from upstream to downstream and water pollutions also generate in each part of the catchment; (2) the increase of adaptive management and the maximum degree of participation for stakeholders could also help to improve the water quality management; (3) the catchment management should also make sure the unified standards and highly cooperation between the different local departments or suburb councils within this region.

Keywords: Water quality management; The Brisbane Catchment; The Southeast Queensland; Sediment and nutrients transport

1 Introduction

With the rapid increase of population, not only for Brisbane River, other catchments in the southeast Queensland region but also experience great pressure(Gleeson & Steele, 2010). These environmental pressures are shown in the decrease in water quality (the increase of pollutant and turbidity), and the increase of sediment and bank erosion (Garzon-Garcia et al., 2015). In order to have a critical analysis and give valuable water quality management suggestions of SEQ catchment, this report will introduce the hydrological, geomorphological and ecological features of this catchment; the environmental impacts brought by climate change and human; and management challenges related to stakeholders and social development.

2 General Information of Catchments in Brisbane(SEQ)

The Southeast Queensland (SEQ) Catchment consists of 15 major catchments and it occupies around 23,000 km^2. According to Bunn et al. (2010), the western area of this catchment is the major resource of water supply for both urban people and agricultural landowners; and the downstream estuaries (such as Moreton Bay) provide the source of fishery and local people also benefit from tourism (Pascoe et al., 2014). Brisbane Catchment accounts for the large percentage in SEQ Catchment, which is nearly 13,570 km^2.

The headwaters of Brisbane Catchment are originally from Mount Stanley which is around 121 kilometers distance from the center of Brisbane city, and the final discharge estuary is Moreton Bay. These catchment areas are subtropical climate which experiences wet and dry season within a year (Department of Infrastructure and Planning, 2009), so the feature of water flow in this area is defined as highly variable. For the geomorphological features of this catchment, its direction is from northwest to southeast. This is because that plateau (elevation more than 300 m) is mainly gathered at the margins of northern and most parts of southern and western areas, while eastern areas consist of coastal plains and foothills (Hodgkinson et al., 2007). In addition, the land use types along this catchment has experienced the huge changes. According to Cottingham et al. (2010), the initial Southeast Queensland is a densely forested area with high coverage along the coastal zone and open forests in inland areas. However, with the development of the SEQ area, the vegetation coverage along this catchment is limited.

The brief introduction of field trip sites

The four sites in this field trip are included in Brisbane Catchment areas. From Fig. 1, it is clear to see that Townson locality on Laidley Creek and Mulgowie farm lie on the west of Southeast Catchment within Lockyer Valley. According to the flow direction of SEQ catchment (from northwest to southeast), these two sites can be seen as the upstream. Wivenhoe Dam is clearly to be identified as the middle part of Brisbane Catchment, while Myrtletown is the lower Brisbane area from this figure. Both Wivenhoe Dam and Myrtletown experience the changes of land use types. The more information on these four sites will be given in the next section.

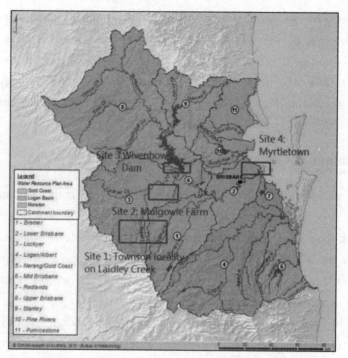

Figure 1. The location of four sites in SEQ catchment

3 Climate Change and Human Activity Impacts on Catchment

The conditions of catchments are founded closely related to climate, because the influences of climate change on catchment include the changes in precipitation, evapotranspiration and groundwater recharge, and precipitation is the most direct factor related to climate variables. According to Cottingham et al. (2010), the Inter-Decadal Pacific Oscillation(IPO) and El Niño-Southern Oscillation (ENSO) are the major factors that influence the precipitation in the SEQ region across years, which means rainfall in dry years is just the half amount in wet years. However, with the increase of global warming level, the precipitation in SEQ catchment tends to have high intensity and frequency influenced by Tropical Cyclones(Lin et al., 2015). According to Wetz and Yoskowitz(2013), extreme precipitation can cause an increased risk of flood and further lead to a decrease in water quality. SEQ region has already experienced four times extreme flood, which are in 1893, 1974, 2011 and 2013 and the water quality problem becomes one of the most troublesome challenges in this area. In addition, human activities can also make significant influences on catchment, and the specific effects on upstream, middle stream and downstream will be discussed respectively.

3.1 Upstream of brisbane catchment

The conditions of catchments are founded closely related to climate. According to EHP(2015), upstream of the Brisbane Catchment covers around 5493 square kilometres, the boundaries are defined by the surrounding low hills and mountain ranges, which is east to D'Aguilar Range, west to the Great Dividing Range, and north to the Brisbane and Jimna Ranges. The flow direction of this catchment is toward to south and final flow into Wivenhoe Dam. The field trip site 1, Townson locality on Laidley Creek, locate in the Lockyer Valley. Croke et al. (2013) state that the width of the upstream in this valley is around 20 m which is limited by narrow bedrock, but the downstream parts are around 2~13 km width and many alluvial cutoffs gathered there. Because the low fertile soil and narrow entrance, the high elevation areas in this Valley are the public national park and less impacted by human activity. However, the lower regions of upstream have high vegetation coverage and are relatively easy to reach, so grazing activities and dairy industry(such as Mulgowie Farm) are located there. According to Apan et al. (2002), riparian forests and vegetation play a significant role in managing the transport of sediments, nutrients and water between upstream and downstream, and can filter the toxins includes dissolved nitrogen and phosphorous. However, the development of grazing and agriculture make the decrease of native riparian vegetation, and the pollutants which source from the livestock waste and the use of fertilizer and pesticide make the increase of water quality severity. In addition, the steep slope makes the fast velocity of water flow, and each of extreme flood will also widen the channel and erosion banks(Lisenby & Fryirs, 2016). This process will make more sediment and nutrients transport. Therefore, in order to manage water quality, the changes of vegetation and catchment channel width all need focus on.

3.2 Middle stream of brisbane catchment

The Mid Brisbane catchment is located at the west of Brisbane which occupies around 522 km^2 and the length is around 61 km, the range of this catchment is from Wivenhoe Dam to Mount Crosby Weir. According to Aryal et al. (2016), the land use types in this catchment area include native forest (27%), intensive agriculture and rural residential(29%), and grazing land(47%). The development of intensive agriculture leads to the increase of nitrogen and phosphorous and many areas become compacted land which decreases the infiltration rate and makes more runoff in this area(Alaoui et al., 2018). In addition, because the intensive agriculture, much amount of water is used for irrigation and also causes excessive runoff. Furthermore, in order to reduce the weed and maintain the ecological balance of grazing land, burning of pastures is very common in this area and leads to the increase of organic debris and dissolved organic carbon gathered in runoff or banks(Ghadiri et al., 2011). Water in Wivenhoe Dam always flow to the Mount Crosby treatment plant before it is supplied to residents as the drinking water, but because the velocity of water flow in the middle catchment is gradually decreased, it always takes a long time for water to flow between these two sites and during the flow process, many sediments and salinity will be included in water and cause the decrease of water quality. Therefore, the decrease of water quality in the middle reaches is not only influenced by sediment accumulations from the upstream but also includes pollution caused by local land types and watershed characteristics.

3.3 Downstream of brisbane catchment

The total area of the lower Brisbane Catchment is around 1,195 km^2 and the total length of stream network is nearly 2,475 km, the source of water is from the Mid Brisbane River and the final flow of lower catchment is Moreton Bay. The most areas of the Brisbane Catchment downstream are the metropolitan region, so the residential areas, commercial districts and industrial estates(such as Myrtletown) are very common, but less natural conservation is still reserved(EHP, 2015). However, because of the construction of Wivenhoe and Somerset Dams at the middle Brisbane Catchment, the bedload transport is interrupted, so the upstream catchment (Lockyer Creek) become the major sediment source flow to Morton Bay at the normal conditions(Kemp et al., 2015). Lovelock et al.

(2011) state that high sediment supply and surface accumulation has made the increase of the elevation in Moreton Bay in recent years. Because of the urbanization at Lower Brisbane Catchment, the decrease of native vegetation and forests and the increase of impervious cover have impacts on water recycle in urban areas and contribute to the generation of runoff (Miller et al., 2014) and the non-point pollution in these urban areas has increased heavy metal contamination into the water flow. In addition, when the extreme weather occurs, excessive stormwater excesses the capability of the urban drainage system and make more untreated water directly flow to Moreton Bay. In addition, under the extreme precipitation condition, in order to protect many rural water reservoirs and weirs at the middle stream of Brisbane Catchment from deconstruction, the 'dam release flood' occurs on the downstream Brisbane Catchment areas (Van den Honert & McAneney, 2011). According to Warner and Catchments (2013), in the 2011 flood, the Wivenhoe Dam also had a flood discharge, because the flood caused a large amount of soil to be washed into the reservoir and thus reducing the storage capacity. Furthermore, the increase of annual temperature, the frequency of high intensity precipitation and the slow flow rate of lower Brisbane Catchment make the growth rate of bacteria's accumulation as well (Cottingham et al., 2010). Therefore, all the above conditions make the downstream part of Brisbane catchment has the worst water quality.

4 Key Stakeholders and Potential Conflicts

Different parts of catchment have various land use types and different sources of water pollution; therefore, the following table will show the different members of stakeholders in each part of the catchment and their interests (Department of Natural Resources, Mines and Energy, 2018; Head, 2014).

Table 1. The interests of Stakeholders

Stakeholders	Scale	Their interests
The Government	In upper, middle and lower catchment	Development and implementation of strategies that incorporate commitments from all levels of stakeholders
Research and Academic Institutions	In upper, middle and lower catchment	Focus on targeted scientific research and give suggestions of appropriate management action
Community (herdsman)	In upper catchment	Consider local resident benefits, such as the areas of water quality management land and their grazing and agricultural land
Community	In middle catchment	Same as to the community in the upstream but also focus on domestic and irrigation water supply from dams
Industry	In middle and lower catchment	Consider company profit; consider water supply and water treatment related to them
Community (Moreton bay and urban areas)	In lower catchment	Consider local resident benefits; Some community organization also consider sustainable environment issues
Urban residents	In lower catchment	Water supply and quality (fully depend on water from large reservoirs at upstream); the problems of city runoff

In this table, research and academic institutions are responsible for the education, but they may not know local conditions well. According to Brisbane Catchment Network (2016), the stakeholders which include in Research and Academic Institutions are universities, SEQ Catchments and Healthy

Waterways. Communities in the upstream always have their own management strategies and lack interest in government management plan; Community in the middle catchment always have their own management strategies but are also interested in government management plan if they can have some shared information and data of outcomes, and local landowner and residents are included in this group; while the communities in the lower catchment include many local catchment groups, such as Bulimba Creek Catchment Coordinating Committee and Oxley Creek Catchment Association. Government in the whole catchment areas includes state and local government, Department of Environment and Heritage Protection and Brisbane City Council. In addition, industry stakeholders at the middle and lower catchment include sewage treatment plants, and dam operator, Queensland Urban Utilities, Port of Brisbane Pty Ltd and Brisbane Airports Corporation.

Potential conflicts

Water quality problems exist in the whole catchment and always interdependent without clear boundaries, so sometimes potential conflicts will occur between different stakeholders in various parts of catchment or between government agencies, professional water management organizations or local communities. For example, Ravnborg and Westermann (2002) state that the development of agriculture along the riverbanks or at the bottom of the valley will cause bank erosion, water shortage and pollution downstream. However, at the middle streams, the land ownership is always controlled by different local users and many water managements measures need land to implement, but land ownership regard these as the waste(Ravnborg & Westermann, 2002). Therefore, sometimes they refuse to participate in management actions and think that they have enough practical knowledge to solve problems. For water utility, they usually think that water pollution is caused by many reasons and needs a lot of responsible people, but often they are the final organizations that need spend a lot of money on pollution elimination(Brisbane Catchment Network, 2016). In addition, potential conflicts will also erupt between different local government, such as different departments or suburb councils. This is because that during water management period, some water quality problems will involve the issues of jurisdictions and shared information(Smith & Porter, 2010). Therefore, the integrated reasons make water quality management difficult to reach the best outcomes.

5 Critical Analysis of Current Policies and Management Plans

According to Cottingham et al. (2010), the current water quality management plan is mainly focused on non-urban systems and urban systems which include the rural, agricultural, forest, industry and commercial areas. Among these areas, erosive soils caused by cropping land and grazing which occupy the largest areas at the upstream and middle stream and account for the largest percentage of sediment(79%) for the whole catchment system; while the high sediment loads in urban areas are caused by the runoff flow on the surface of the impermeable road(Cottingham et al., 2010). Therefore, water quality management has priority in the upstream but also focus on downstream. From Healthy Waters Management Plan Guideline(2009) and the State Planning Policy-water quality (2016), the current management strategies include vegetation and riparian zones restoration at upstream in order to slow run-off flow between hillslopes and gullies; changing the practices and crop types of agriculture, and reducing the possibility that livestock enters the waterways and erosive riparian zones at the middle stream; and improving the management of point source pollution(such as industrial discharge and septic systems) and non-point source pollution(use Water Sensitive Urban

Design) at urban and peri-urban areas. These two plans are very detailed and all of them provide specific management priority of catchment parts and practically effective strategy action. However, both these two plans do not consider the willingness of stakeholder and potential conflicts. For example, in the State Planning Policy-water quality(2016), it only states that there is no limit from policy to prevent local government and professional staff to explore more detailed information in order to improve water quality management outcomes. Therefore, current management plans are not strict enough.

6 Recommendation and Conclusion

Comprehensive water quality management plan should include the parts of adaptive management and the maximum degree of participation for stakeholders. Adaptive management is always used when there is high uncertainty occurs and natural resources managements have this feature(Dutra et al., 2014). As mentioned above, water quality is very related to human activities and climate change. High precipitation and temperature will increase the bacteria content in the water and increase the transportation of sediments throughout the basin, and human activities will exacerbate these factors and make worse impacts. Therefore, adding adaptive management into water quality management plan is my first suggestion. In addition, Prell and Reed (2009) state that, many natural resources management strategies are failed because the plan designer does not include all the interests and types of stakeholders. Therefore, the second suggestion I give is the maximum degree of participation for stakeholders. Although the current water quality management plans have already clarified the management scope and specific measures, it often encounters a situation where stakeholders do not cooperate in the implementation. The reason for this problem is usually that people always lack the correct understanding and self-positioning on management plans(Ravnborg & Westermann, 2002). For example, water resource is generally thought as the public good, and management plan always makes the upstream and middle stream as the key sites to build dams, set buffer zones and restore vegetation(Smith & Porter, 2010). However, Smith and Porter(2010) also state that this strategy makes people live in these areas feel unfair, because people live in the whole catchment areas receive the benefits of the improvement of water quality in principle, but only they need to contribute to fertile agricultural lands while people live in downstream need do nothing. Therefore, these facts make parts of stakeholders do not want to participate in water quality management plan. However, if the government can realize this situation and give some compensation and education, maybe the management outcomes could be better.

Therefore, there are three major improvement suggestions are given, which are the increase of adaptive management, the maximum degree of participation for stakeholders, and make sure the unified standards and high cooperation between the different local departments or suburb councils within this region. Water quality problems are closely linked to climate change and human activities, so in future management, these factors and three improvements suggestions all need to be considered by the government.

References

[1] Alaoui A, Rogger M, Peth S. Does soil compaction increase floods?[J]. Journal of Hydrology, 2018, 557(557): 631−642.

[2] Alexandra J. Australia's landscapes in a changing climate-caution, hope, inspiration, and transformation[J]. Crop & Pasture Science, 2012, 63(3): 215−231.

[3] Apan A A, Raine S R, Paterson M S. Mapping and analysis of changes in the riparian landscape structure of the Lockyer Valley catchment, Queensland, Australia[J]. Landscape and Urban Planning, 2002, 59(1): 43−57.

[4] Aryal R, Grinham A, Beecham S. Insight into dissolved organic matter fractions in Lake Wivenhoe during and after a major flood[J]. Environmental Monitoring and Assessment, 2016, 188(3): 134.

[5] Brisbane Catchments Network. Reducing urban nutrients and pollution in moreton bay[EB/OL]. [2016−05−18]. https://brisbanecatchments.org.au/projects/urban-nutrient-pollution-moreton-bay/.

[6] Bunn S E, Abal E G, Smith M J. Integration of science and monitoring of river ecosystem health to guide investments in catchment protection and rehabilitation[J]. Freshwater Biology, 2010, 55: 223−240.

[7] Croke J, Todd P, Thompson C. The use of multi temporal LiDAR to assess basin-scale erosion and deposition following the catastrophic January 2011 Lockyer flood, SE Queensland, Australia[J]. Geomorphology, 2013, 184: 111−126.

[8] Cottingham R, Delfau K F, Garde P. Managing diffuse water pollution in South East Queensland. 2010.

[9] Daley J S, Cohen T J. Climatically-controlled River terraces in eastern Australia[J]. Quaternary, 2018, 1(3): 23.

[10] Department of environment and heritage protection. healthy waters management plan guideline[EB/OL]. [2009−11−21]. https://environment.des.qld.gov.au/_data/assets/pdf_file/0020/87122/healthy-water-mp-guideline.pdf.

[11] Department of Infrastructure and Planning. South East Queensland Regional Plan 2009—2031[EB/OL]. [2009−12−02]. http://www.dlgrma.qld.gov.au/resources/plan/seq/regional-plan-2009/seq-regional-plan-2009.pdf.

[12] Department of Infrastructure, Local Government and Planning. State Planning Policy-state interest guideline water quality[EB/OL]. [2016−08−10]. http://www.dlgrma.qld.gov.au/resources/guideline/spp/spp-guideline-water-quality.pdf.

[13] Dutra L X, Ellis N, Perez P. Drivers influencing adaptive management: a retrospective evaluation of water quality decisions in South East Queensland(Australia)[J]. Ambio, 2014, 43(8).

[14] EHP. An Aquatic Conservation Assessment for the riverine and non-riverine wetlands of Southeast Queensland catchments[M]. Brisbane: Department of Environment and Heritage Protection, Queensland Government, 2015.

[15] Garzon-Garcia A, Olley J M, Bunn S E. (2015). Controls on carbon and nitrogen export in an eroding catchment of south-eastern Queensland, Australia[J]. Hydrological Processes, 2015, 29(5): 739−751.

[16] Ghadiri H, Hussein J, Rose C W. Effect of pasture buffer length and pasture type on runoff water quality following prescribed burning in the Wivenhoe Catchment[J]. Soil research, 2011, 49(6): 513−522.

[17] Gleeson B, W Steele, editors. A climate for growth: planning South-east Queensland[M]. Brisbane: University of Queensland Press, 2010.

[18] Head B W. Managing urban water crises: adaptive policy responses to drought and flood in Southeast Queensland, Australia[J]. Ecology and Society, 2014, 19(2).

[19] Hodgkinson J H, McLoughlin S, Cox M E. Drainage patterns in southeast Queensland: the key to concealed geological structures?[J]. Australian Journal of Earth Sciences, 2007, 54(8): 1137−1150.

[20] Hoegh-Guldberg O, Bruno J F. The impact of climate change on the world's marine ecosystems[J]. Science, 2010, 328(5985): 1523−1528.

[21] Kemp J, Olley J M, Ellison T. River response to European settlement in the subtropical Brisbane River, Australia[J]. Anthropocene, 2015, 11: 48−60.

[22] Lin Y, Zhao M, Zhang M. Tropical cyclone rainfall area controlled by relative sea surface temperature[J]. Nature communications, 2015, 6(1): 1−7.

[23] Lisenby P E, Fryirs K A. Catchment-and reach-scale controls on the distribution and expectation of geomorphic channel adjustment[J]. Water Resources Research, 2016, 52(5): 3408−3427.

[24] Lovelock C E, Bennion V, Grinham A. The role of surface and subsurface processes in keeping pace with sea level

rise in intertidal wetlands of Moreton Bay, Queensland, Australia[J]. Ecosystems, 2011, 14(5): 745−757.

[25] Miller J D, Kim H, Kjeldsen T R. Assessing the impact of urbanization on storm runoff in a peri-urban catchment using historical change in impervious cover[J]. Journal of Hydrology, 2014, 515: 59−70.

[26] Pascoe S, Doshi A, Dell Q. Economic value of recreational fishing in Moreton Bay and the potential impact of the marine park rezoning[J]. Tourism Management, 2014, 41: 53−63.

[27] Prell C, Hubacek K, Reed M. Stakeholder analysis and social network analysis in natural resource management[J]. Society and Natural Resources, 2009, 22(6): 501−518.

[28] Smith L E, Porter K S. Management of catchments for the protection of water resources: drawing on the New York City watershed experience[J]. Regional Environmental Change, 2010, 10(4): 311−326.

[29] Wagener T, Sivapalan M, Troch P. Catchment classification and hydrologic similarity[J]. Geography Compass, 2007, 1(4): 901−931.

[30] Warner S, Catchments S E Q. Maintenance of water infrastructure assets[EB/OL]. [2013−02−15]. https://www.parliament.qld.gov.au/documents/committees/SDIIC/2013/12-WaterInfraAssets/submissions/008.pdf.

[31] Wetz M S, Yoskowitz D W. An 'extreme' future for estuaries? Effects of extreme climatic events on estuarine water quality and ecology[J]. Marine Pollution Bulletin, 2013, 69(1−2): 7−18.

A Particle-based Model for Simulation of Nonpoint Source Pollutant Dynamics Induced by Rainfall

Qiuhua Liang[1], Jinghua Jiang[1], Xue Tong[2]

[1] School of Architecture, Building and Civil Engineering, Loughborough University, Loughborough, UK
[2] College of Harbour, Coastal and Offshore Engineering, Hohai University, Nanjing 210098, China

Abstract

Nonpoint source (NPS) particulate pollutants are a major source of urban surface pollution, which may be washed off during a rainfall event and degrade water quality in receiving water. In the last two decades, nature-based solutions such as Sustainable Drainage Systems (SuDS) have been promoted to manage urban flood risk and NPS pollutants in the wider context of urban water management. Different stormwater quality models have been developed and used to predict the dynamic process of NPS particulate pollutants to inform the design and operation of SuDS to achieve optimised performance. However, most of the existing stormwater quality models adopt hydrological models or simplified hydraulic models to simulate urban water flow processes, which are not able to capture the detailed flow dynamics to subsequently predict the dynamic behaviours of pollutants. When modelling the pollutant dynamics, grid-based Eulerian models are commonly used, which are incapable of depicting the sources and transport pathways of pollutants. Therefore, most of the existing stormwater quality modelling approaches may not be able to provide sufficiently detailed information to support the planning and design of a 'management train' in which SuDS devices of different types and sizes are integrated to achieve optimised performance in mitigating flood risk, improving water quality and creating other benefits. This work presents a novel coupled hydrodynamic and Lagrangian particle-based model to simulate the full-process dynamics of NPS particulate pollutants from particle detachment, transport to deposition driven by rainfall-runoff and overland flows. In this new modelling framework, the overland flow and flooding processes are simulated using a GPU-accelerated 2D hydrodynamic model, producing the necessary flow fields to drive the simulation of pollutant transport using a random-walk particle-tracking model to directly trace out the trajectories of individual particles and hence identify the pathways of pollutants. A particle-based approach is also implemented to represent the physical processes of pollutant detachment and deposition, as shown in Fig. 1. The new coupled hydrodynamic and particle-based stormwater quality model is validated against analytical and experimental test cases to demonstrate its capability in accurately simulating the full-process dynamics of NPS particulate pollutants in urban areas.

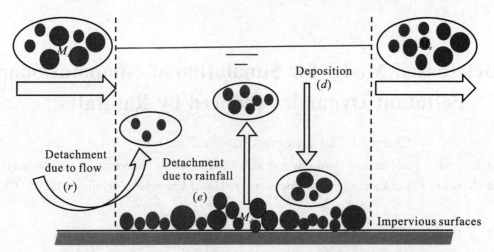

Figure 1. Particle-based Hairsine-Rose model for pollutant wash-off and deposition

An Experimental Study of Turbulent Structures in a Combined Fishway of Central Orifice and Vertical Slot

Zhiyong Dong, Zeyang Yan, Zhou Huang, Junpeng Yu, Jianli Tong

College of Civil Engineering, Zhejiang University of Technology, Hangzhou 310023, China

Abstract

There are a variety of migratory fish species in the seas and rivers in China. It is difficult for single fishway (vertical slot, overflow weir, and orifice) to meet the requirements of different fish swimming behaviors. Turbulent structures such as flow characteristics, main stream trajectory, three-dimensional flow velocity, turbulence intensity, Reynolds stress, frequency-spectrum characteristics, correlation function, turbulence scales (micro scale and integral scale) were experimentally investigated by Acoustic Doppler Velocimetry (ADV) in a combined fishway of central orifice and vertical slot. The staggered and identical vertical slot layouts were considered in the test. For each layout, turbulent structures on different horizontal planes, longitudinal profiles, and cross-sections were respectively studied, and a comparison between the two layouts was made. The experimental results showed that in the case of staggered layout, orifice jet and vertical slot wall jet interacted to combine into one flow, accompanying a clockwise large vortex, whereas in the identical layout the orifice jet and the slot wall jet weakly interacted, and almost exhibited parallel flows; both the maximum turbulence intensity and Reynolds stress occurred in the mixing region between orifice jet and vertical slot wall jet; the auto-correlation coefficient of velocity fluctuation gradually decayed with time, and fluctuated within a small range of amplitude; for the staggered layout, energy spectrum of vertical velocity fluctuation exhibited an extreme value in the orifice zone, and the spectrum of longitudinal velocity fluctuation reached the extreme value in the vertical slot zone; however, for the identical layout, the maximum energy spectrum referred to the vertical velocity fluctuation; the integral and micro time scales showed eddy structures were related to flow zones in the combined fishway.

Keywords: central orifice; vertical slot; combined fishway; turbulent structures

Assessing the Hydraulics Safety of Existing Dams —A View to Dam Rehabilitation

Arturo Marcano

Universidad Católica Andrés Bello, Puerto Ordaz, Venezuela

Abstract

Present Dam Statistics (ICOLD, 2021) indicate a worldwide register of 58 713 large dams, 50% of which are at least 50 years old. Further, more than 45% of the dam failures documented, have occurred by hydraulic related reasons (Donnelly R, 2019). Hydraulic Safety of Dams encompasses all hydraulic works such as the reservoir and its rim, the dam and the appurtenant works, spillways, outlets and all control structures located close to the dam. Thus, Hydraulic Safety of the dam and the appurtenant works include retaining their physical integrity, stability and resistance withstanding all the forces acting under all technically feasible conditions of loading. Properly safe, well maintained and engineered dams may have an expected useful life of 100 years, that can be extended, depending of the level of care owners can provide. On the other hand, without proper maintenance and repairs, an unsafe dam may not be able to serve its intended purpose and could be at risk of failure, increasing the probability of disasters which can otherwise be prevented by timely actions. Presently, Dam Rehabilitation is receiving much attention by Dam owners and regulators (World Bank, 2020). Rehabilitation of a dam is the act of restoring the distressed dam not only to its original state but of improving it to meet added requirements. Dam Rehabilitation is required to counter for various deficiencies which develop along the time line of the dam operation. Deficiencies are caused primarily by dam ageing, underestimated spillway capacity (Fig. 1), degradation caused by weathering, wear and tear of equipment due to normal use or misuse, loss of serviceability due to long time operation, damage from natural events including normal and extraordinary events like flood, earthquakes, slides, ice jams (Fig. 2), vandalism or even war, among others (CWC, 2020). As the ageing of dams in the world increases, there will be increasing demand for dam rehabilitation. This paper presents the discussion, on iconic case studies of dam failure/incidents, to emphasize defects, hydraulics safety and rehabilitation measures. A summary will be presented in Matrix Format, showing Indicator/Cause/Response/Measures as applied to the different elements of a dam spillway/outlet structure from the reservoir to the exit channel, in an attempt to compile useful information for dam Engineers involved in the assessment of Hydraulics Safety in Large Dam Rehabilitation Projects.

Keywords: Dams; Hydraulic Dam Safety; Dam rehabilitation; Ageing; Floods

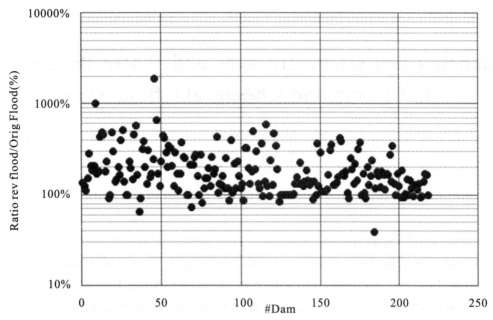

Figure 1. Ratio "Revised Flood/Original Flood" for 223 large dam reservoirs in India (CWC, 2019)

Figure 2. Spencer Dam Failure

Note: Gates were not able to open because some were frozen shut. (https://starherald.com/news/regional)

References

[1] ICOLD[EB/OL]. [2021-06-30]. https://www.icold-cigb.org.

[2] Donnelly R. (2017), Question 97, ICOLD Congress, Spillways, XXV Congress of Large Dams, Stavanger, 2015.

[3] Central Water Commission (CWC, 2019), Manual for Rehabilitation of Large Dams, under Dam Rehabilitation and Improvement Project, DRIP, India.

[4] World Bank, Dam Rehabilitation and Improvement Project. DRIP, India, 2020.

Balancing Ecosystem Recovery and Water Demands in California and Chesapeake Bay, USA

Peter Goodwin
F. ASCE, F. ICE

Abstract

Many of the world's river basins are severely stressed due to population growth, water quality and quantity problems, emerging contaminants, vulnerability to flood and drought, and the loss of native species and cultural resources. Consequences of climate change further increase uncertainties about the future. There has been an increasing interest in balancing water supply reliability, protecting communities and existing infrastructure, fostering economic development while ensuring a healthy earth system. Managers of large river basins face considerable uncertainty in making decisions to strike the balance between ensuring a reliable water supply, economic growth and the recovery or preservation of ecosystem function. Success requires an adaptive management strategy that can be responsive to natural perturbations such as floods and extended droughts, work across multiple agencies and engage the local communities most influenced by the management actions that will determine the environmental conditions for future generations.

The presentation will explore two large ecosystems in the US. The first will consider the management of water in California and the link between water diversions and the riverine and estuarine conditions in San Francisco Bay watershed. The primary driver is the quantity of water as both a scarce resource and the potential for devastating floods. The second example will review the progress being made in the recovery of Chesapeake Bay where significant investments have been made to decrease nutrient and sediment loading with a goal of restoring key ecosystem services. In both cases, the dramatic effects of climate change already being experienced will be highlighted.

Characteristics of Tsunami Damping with Different Friction Factors

Hitoshi Tanaka, Nguyen Xuan Tinh

Tohoku University, Japan

Abstract

In most of tsunami numerical models, steady flow friction factor, such as Manning's n, has commonly been applied for accessing the tsunami-induced bottom shear stress. According to authors' recent investigations, however, it has theoretically been proved that the wave friction law is valid, rather than the steady flow friction law in almost the entire computational domain from tsunami source to shallow area (Tinh and Tanaka, 2019, Tanaka and Tinh 2020).

In the present study, we carried out a numerical study on tsunami damping due to bottom friction using the following governing equation, considering both wave shoaling and energy loss due to bottom friction.

$$\frac{dH}{dx} = -\frac{H}{4h} \cdot \frac{dh}{dx} - \frac{fH^2}{3\pi h^2} \quad (1)$$

in which various friction factors are applied, as summarised in Table 1. Among these expressions, Method 7 is the most reliable approach with considering flow regime transition in a wave boundary layer, as well as depth-limited behavior in the shallow water region. In addition, Table 2 indicates the input boundary condition at the tsunami source area.

Table 1. Friction factors

Method	Used friction factor	
1	$f = 0$ (Green's law)	(2)
2	$f = 0.002$	(3)
3	$f = 0.02$	(4)
4	$f = \dfrac{2k^2}{\left[\ln\left(\dfrac{h}{z_0}\right) - 1\right]^2}$	(5)
5	$f = \dfrac{2gn^2}{h^{1/3}}$	(6)
6	• If $r < 1$ (non-depth-limited), $f = f_w$ (wave friction factor) for rough turbulent regime by Tanaka (1992) $$f_w = \exp\left\{-7.53 + 8.07\left(\frac{30a_m}{k_s}\right)^{-0.100}\right\}$$ • If $r \geq 1$ (depth-limited), Equation (5).	(7)

(To be continued)

(Continue)

Method	Used friction factor
7	• If $r<1$ (non-depth-limited), $f=f_w$ (wave friction factor) from the full-range equation by Tanaka and Thu (1994) $$f_w = f_2[f_1 f_{w(L)} + (1-f_1)f_{w(S)}] + (1-f_2)f_{w(R)} \quad (8)$$ • If $r \geqslant 1$ (depth-limited), Equation (5)

Table 2. Tsunami input conditions

Bottom slope	1/100
Tsunami source depth	$h_0 = 4000$ m
Source tsunami height	$H_0 = 1$ m
Wave period	$T = 15$ min
Diameter of bed material	$d = 0.3$ mm

Fig. 1 indicates computation results in the shallow sea region. In the deeper region, the computed wave height is independent on the friction coefficient, as seen in Fig. 1(b). Using Method 1 without the friction term in Eq. (1), the wave height near the shoreline becomes infinite, following Green's low. Except the frictionless method, the constant friction coefficients, Methods 2 and 3, gives the maximum and the minimum wave height. It is interesting to note that Methods 4, 6 and 7 yield almost the same wave height during the shoaling process. However, the friction factor values are different each other as illustrated in Fig. 1(c). Hence, in numerical simulations of sediment transport and resultant morphology change, it is necessary to consider spatial variation of the friction coefficient due to flow regime transition as observed in Fig. 1(d).

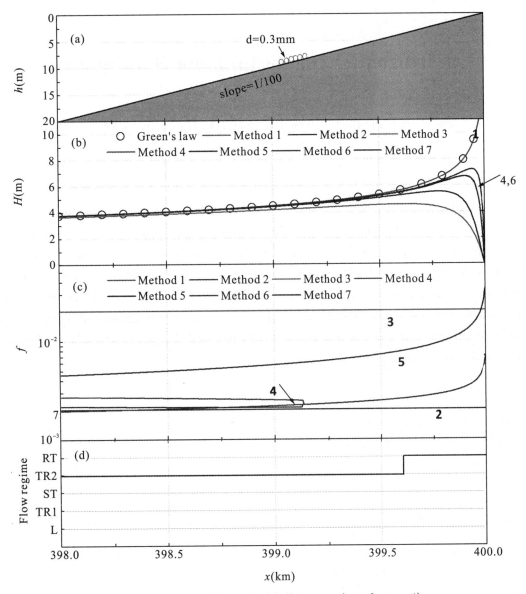

Figure 1. Computation results (shallow sea region enlargement)

Note: (a) Bathymetry; (b) Wave height; (c) Friction coefficient; (d) Flow regime.

Eco-hydraulics for Sustainable Hydropower

David Z. Zhu

Department of Civil and Environmental Engineering, University of Alberta, Canada

Abstract

In this talk, I will discuss recent research development on eco-hydraulics related to sustainable hydropower. I will first talk about our research work on assessing fish entrainment risk related to hydropower intake operations for run-of-the river dams as well as deep dams. The effect of temperature stratifications on fish entrainment will be discussed. This will be followed by our current research on predicting and mitigating the generation of supersaturated total dissolved gases caused by dam spill events. The upstream migration of fish through vertical slot fishways and fishway hydraulics will then be discussed. Our recent study on the hydraulics and design of nature-like fishways using rock ramps will then be summarized.

Effects of Data Temporal Resolution on Simulating Water Flux Extremes at the Field Scale

Lianhai Wu

Department of Sustainable Agriculture Sciences, Rothamsted Research,
North Wyke EX20 2SB, UK

Abstract

Projected changes to rainfall patterns may exacerbate existing risks posed by flooding. Furthermore, increased surface runoff from agricultural land increases pollution through nutrient losses. Agricultural systems are complex because they are managed in individual fields, and it is impractical to provide resources to monitor their water fluxes. In this respect, modelling provides an inexpensive tool for simulating fluxes. At the field-scale, a daily time-step is used routinely, however, it was hypothesized that a finer time-step will provide more accurate identification of peak fluxes. To investigate this, a 15-minute water flux dataset from April 2013 to February 2016 from a pasture within a monitored grassland research farm was up-scaled to hourly, 6-hourly and daily data; and a daily time-step process-based model was adapted to provide corresponding down-scaled simulations at 15-minute, hourly and 6-hourly resolution (in addition to its usual daily output). Analyses were conducted with respect to model performance for: (a) each of the four data resolutions, separately (15-minute measured versus 15-minute simulated; hourly measured versus hourly simulated; etc.); and (b) at the daily resolution only, where 15-minute, hourly and 6-hourly simulations were each aggregated to the daily scale. Comparison between measured and simulated fluxes at the four resolutions (unaggregated approach) revealed that hourly simulations provided the smallest missclassification rate for identifying water flux peaks. Conversely, aggregating to the daily scale using either 15-minute or hourly simulations increased accuracy, both in prediction of general trends and identification of peak fluxes. The improved identification of extremes resulted in 9 out of 11 peak flow events being correctly identified with only 2 false positives, compared with 5 peaks being identified with 4 false positives of the usual daily simulations. Increased peak flow detection accuracy has the potential to provide clear field management benefits in reducing nutrient losses to water.

Field and ExperimentalModeling of Scour Induced by Turbulent Bores around Structures

Ioan Nistor, Razieh Mehrzad, Colin Rennie

Dept. of Civil Engineering, University of Ottawa, Canada

Abstract

Post-tsunami field survey evidence indicate that destructive tsunamis cause substantial coastal sediment mobilization. Measurements collected following several tsunami events, such as the 1992 Nicaragua Tsunami, the 2004 Indian Ocean Tsunami, and the 2011 Tohoku Tsunami provided substantial evidence of scour around damaged buildings and bridges. Tsunami-induced coastal inundation is characterized by high overland flow velocities, both during the inland flow phase as well as during the drawdown. These high flow velocities produce high bed shear stresses and large amounts of sediment movement over extensive areas, resulting in substantial beach erosion and significant scour around many of the structures located in the inundation zone.

The paper willpresent results of several tsunami forensic engineering campaigns as well as results of a comprehensive new experimental program dealing with bore-induced local scour around structures. The hydrodynamic forcing conditions for the experimental program was due to a dam-break wave generated in a hydraulic flume located in the Hydraulics Laboratory at the University of Ottawa, Canada. This highly turbulent bore was generated by the rapid release of water impounded behind a rapidly-opening swing gate equipped with a lock and sudden release mechanism. A video system was used to record the evolution the of scour and vortex characteristics using video imaging obtained from inside of the structure model which was built from transparent PVC. Image processing allowed tracking the time and spatial evolution of the scour around the structure. Finally, new provisions for scour around structures stemming from this research which will be included in the ASCE7 Chapter 6 Tsunami Loads and Effects will be presented.

Flood Control Ability of the Three Gorges Reservoir and the Upstream Cascade Reservoirs in a Catastrophic Flood

Shanghong Zhang, Zhu Jing, Wenjie An, Rongqi Zhang

School of Water Resources and Hydropower Engineering,
North China Electric Power University, Beijing 102206, China

Abstract

A large number of artificial cascade reservoirs upstream of the Three Gorges Reservoir (TGR) has significantly changed the actual flow conditions of the TGR to change considerably from the original/design baseline and further caused substantial impacts on the flood management in the downstream Jingjiang area. In this study, a stochastic model of inflow floods and a dynamic capacity-based river regulation model for the TGR were established. Based on the TGR's flood control compensation rules for Chenglingji area, the flood control risk in the Jingjiang area under joint operation of the TGR and the upstream Wudongde, Baihetan, Xiluodu, and Xiangjiaba cascade reservoirs was fully assessed in terms of (ⅰ) the forms of the inflow floods, (ⅱ) the flood propagation model; (ⅲ) and the flood regulation rules. The results show that only relying on the TGR alone to intercept a 1000-year flood, the maximum discharge within a safety control margin is 70,160~81,160 m^3/s, and the flood diversion capacity of Jingjiang area is 7.295~10.208 billion m^3, which exceeds the threshold of the effective flood diversion capacity of 7.276 billion m^3 specified in the Jingjiang area. Obviously, it is difficult to ensure the safety if still implementing the current flood diversion programme when confronted with large, unprecedent flood events. In order to avoid the huge losses caused by solely using the Jingjiang flood diversion drainage, the upstream Wudongde, Baihetan, Xiluodu, and Xiangjiaba cascade reservoirs are required to reserve a minimum storage capacity of 7.211~9.474 billion m^3 under the conditions of different inflow flood forms and the flood control compensation rules for Chenglingji area. These results can be used as a scientific reference for the joint flood control operation of the cascade reservoirs in the upper reaches of the Yangtze River.

Flow and Sediment Resuspension in Vegetated Channels

Chao Liu[1], Yuqi Shan[2], Kejun Yang[1], Xingnian Liu[1]

[1] State Key Laboratory of Hydraulics and Mountain River Engineering,
Sichuan University, Chengdu 610065, China

[2] Institute for Disaster Management and Reconstruction, Sichuan University, Chengdu 610065, China

Abstract

Aquatic vegetation is widely distributed in rivers and marshes and can protect floodplains from erosion and stabilize beds by suppressing sediment resuspension. Understanding flow and sediment resuspension are important for understanding the evolution of vegetated landscapes. This study answered three questions: (1) What is the connection between turbulence and sediment resuspension? (2) In nature, plants have different stem diameters. How does stem size impact turbulence and resuspension? and (3) Inside an emergent canopy, how to predict longitudinal evolution of velocities in the canopy and for the bare beds? First, based on flume experiments, the threshold for vegetation-generated turbulence is determined, $Re_c = 120$. The Vegetation-vegetated turbulence is present and absent for $Re_c > 120$ and $Re_c < 120$, respectively. The criteria for enhanced deposition inside emergent canopies are determined. Second, for the same solid volume fraction, the stem diameter has a negligible impact on turbulence magnitude and sediment resuspension. The critical velocity for sediment resuspension has no dependence on stem diameter. Third, a model for predicting velocity evolution inside an emergent canopy is proposed, which is combining with an existing model that predicts the critical velocity for resuspension. The minimum length of a canopy for suppressing sediment resuspension can be estimated. For a sparser and/or finer sediment, a longer canopy is required to suppress resuspension.

Keywords: Vegetation; Flow velocity; Turbulence; Sediment resuspension

Future Changes in Storm Surge due to Climate Change in the Western Pacific

Nobuhito Mori

Disaster Prevention Research Institute, Kyoto University, Japan

Swansea University, UK

Abstract

Extreme tropical cyclones (TCs) and their related storm surges have been observed worldwide over the last few decades (e.g., Mori and Takemi, 2016). Some disastrous TCs have produced catastrophic storm surges, including Hurricane Katrina in 2005, Cyclone Nargis in 2008, Typhoon Morakot in 2009, Hurricane Sandy in 2012, Typhoon Haiyan in 2013, Cyclone Pam in 2015, Hurricanes Irma and Maria in 2017, Typhoon Jebi in 2018 and Typhoon Hagibis in 2019 (e.g., Mori et al., 2014). To simulate the storm surge for a particular TC, the accuracy of the dynamical model-based (e.g., nonlinear shallow water equation) computation is high, but the performance of long-term storm surge projection over hundred years is insufficient due to small number of TC and storm surge events at particular location. The computational load is also a major problem to handle such a large computation cost due to fine resolution of storm surge modeling with large ensemble.

Based on this background, weconducted impact assessment of climate change in storm surge in the Western Pacific by using the global circulation model, downscaling, the maximum potential storm surge height (MPS) theory (Mori et al., 2021) and the other methods. The comprehensive study shows that the future changes in the storm surge height by the TCs will be increased in the western Pacific significantly.

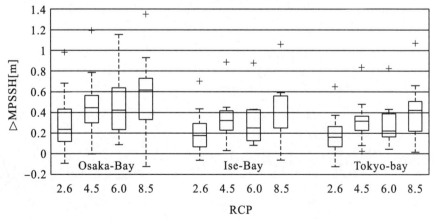

Figure 1. Future changes in maximum potential storm surge height (MPS) for the three major bays in September, for each RCP scenario

References

[1] Emanuel K A. The maximum intensity of hurricanes[J]. Journal of the Atmospheric Sciences, 1997, 45: 1143−1155.

[2] Mori N. Local amplification of storm surge by super typhoon Haiyan in Leyte Gulf[J]. Geophysical Research Letters, 2014, 41: 5106−5113.

[3] Mori N, Takemi T. Impact assessment of coastal hazards due to future changes of tropical cyclones in the North Pacific Ocean[J]. Weather and Climate Extremes, 2016, 11: 53−69.

[4] Mori N, Ariyoshi N, Shimura T, et al. Future projection of maximum potential storm surge height at three major bays in Japan using the maximum potential intensity of a tropical cyclone[J]. Climatic Change, 2021, 164: 25.

Grade Controlin Degrading Channels

Ruihua Nie, Lu Wang, Xingnian Liu

State Key Laboratory of Hydraulics and Mountain River Engineering,
Sichuan University, Chengdu 610065, China

Abstract

In the past decades, the increasing human activities such as sand mining and massive construction of river barriers (e.g., dams and weirs) have significantly reduced the sediment flux in rivers, leading to severe bed degradation. The bed degradation can cause general scour at instream infrastructures (e.g., bridge piers, buried pipelines or crossing-river tunnels), causing structural damages and failure. To protect the river bed from being excessively degraded, grade control structures (e.g., weirs, bed sills, and check dams) have been extensively built in degrading rivers. However, up to the present, the impacts of grade control structures (GCSs) on the bed morpho-dynamic process of a degrading channel is still unknown; and there is still a lack of a proper design method for GCSs.

The present study includes four parts. Part 1 reports a case of GCS failure in Shi-ting River, China, illustrating the failure mechanism of GCS and providing design suggestions for grade control structures in wide degrading rivers. Part 2 studied the impacts of grade control structures on the bed morpho-dynamic process of a degrading channel based on flume experiments, evaluating the effects of GCS height, separation distance on the bed morpho-dynamic process in a degrading channel. Based on experimental and field data, a semi-theoretical predictor for the equilibrium bed slope upstream of GCSs is proposed. Part 3 experimentally studied the local scour at a submerged impermeable GCS (termed submerged weir), evaluating the quantitative impacts of flow, sediment and structural geometry parameters on the local scour depth at submerged weirs. Based on those evaluated relations, a scour design method for submerged weirs is proposed. Part 4 conducted flume experiments to study the dislodgement mechanism of rock-made permeable GCSs (termed rock weir) in sediment bed channels, evaluating the impacts of flow, sediment and structural parameters on the critical condition ofrock weir dislodgment. Based on theoretical derivation and experimental data, a size selecting method for rock weirs in alluvial rivers is proposed.

Granular Column Collapses: Implications for Studies of Landslides and Other Geo-Hazards

Ling Li

School of Engineering, Westlake University, Hangzhou 310024, China

Abstract

Granular column collapses result in an array of flow phenomena and deposition morphologies, the understanding of which brings insights into studying granular flows in connection to various geo-hazards such as landslides. Guided by experiments, we carried out computational studies with theDiscrete Element Method (DEM) to identify fundamental links between the macro-scale behaviour and micro-scale properties of granular columns. A dimensionless number combining particle and bulk properties of the column, a_{eff}, was found from dimensional analysis to determine three collapse regimes (quasistatic, inertial, and liquid-like), revealing universal trends of flow regimes and deposition morphologies under different conditions. This dimensionless number may represent physically the competing inertial and frictional effects that govern the behaviour of the granular column collapse, including energy conversion and dissipation. Furthermore, a finite-size scaling analysis, which is inspired by a phase transition of granular column collapses around an inflection point, was performed to obtain a general scaling equation with critical exponents for the run-out distance and the energy consumption of the granular column collapses. We further formalized a correlation length scale which exponentially scales with the effective aspect ratio. Such a scaling solution shows similarities with that of the percolation problem of two-dimensional random networks and can be extended to other similar natural and engineering systems. These findings are important for understanding quantitatively the flow of granular materials and their deposits.

How Does the Free Jump Dissipate Energy Efficiently?

Sung-Uk Choi

Department of Civil and Environmental Engineering, Yonsei University, Seoul, Korea

Abstract

The flows over an embankment-type weir show four flow types depending on the tailwater level. They change from swept-out jump, optimum jump, submerged jump, and washed-out jump as the tailwater level rises. In practical engineering, the free jump, referring to both swept-out jump and optimum jump, and submerged jump are of interest due to their capacity of energy dissipation. The submerged jump dissipates kinetic energy over an extended length of roller, by creating strong backward flows. The pressure fluctuations at the bed by the submerged jump in the roller region are known to be much weaker than those for the free jump, which enables the downstream apron much safer. However, previous studies that investigated the kinetic energy dissipation of the submerged jump are very rare. This study explores the longitudinal changes of the mean flow and turbulence statistics for the free jump and submerged jump and identify which causes the difference in the longitudinal changes. For this, numerical simulations are carried out to compute free jump and submerged jumps using the open source platform OpenFOAM. The 2D URANS equations are solved with the k-ω SST model. For a fixed unit discharge, a free jump and various submerged jumps with three degrees of submergence are reproduced by varying the tailwater levels.

The figure below shows the longitudinal decay of the peak streamwise mean velocity, peak turbulence intensity (streamwise component), and peak Reynolds stress for both free jump and submerged jumps. The streamwise mean velocity and second-order statistics are normalized using the bulk velocity, and the horizontal axis by the characteristic length, the longitudinal distance where the peak streamwise mean velocity becomes the half of the bulk velocity. Measured data for submerged jumps are also given for comparisons. It can be seen that the decays of the peaks of such vertical structures for free jump are not similar to those for the submerged jumps. This indicates that the decaying patterns of the mean flow and turbulence statistics for the submerged jump are different from those for the free jump.

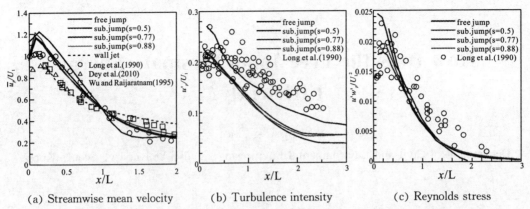

(a) Streamwise mean velocity (b) Turbulence intensity (c) Reynolds stress

Figure 1. Similarity of mean flow and turbulence statistics for free jump and submerged jump

Hydraulics and Energy Dissipation on Stepped Spillways-prototype and Laboratory Experiences

Hubert Chanson

The University of Queensland, School of Civil Engineering, Brisbane QLD 4072, Australia

Abstract

Stepped spillways are used since more than 3,000 years. With the introduction of new construction materials (e.g. RCC, PVC coated gabions), the stepped spillway design has regained some interest. The steps increase significantly the rate of energy dissipation taking place along the chute, thus reducing the size of the required downstream energy dissipation basin. Stepped cascades are used also for in-stream re-aeration and in water treatment plants to enhance the air-water transfer of atmospheric gases and of volatile organic components. Yet the engineering design stepped spillways is not trivial because of the complicated hydrodynamics, with three possible flow regimes, complex two-phase air-water flow motions and massive rate of energy dissipation above the staircase invert.

Altogether, the technical challenges in hydraulic engineering and design of stepped spillways are massive for the 21st century hydraulic engineers. This keynote lecture reviews the hydraulic characteristics of stepped channel flows and develops a reflection on nearly 30 years of active research. The speaker aims to share his passion for hydraulic engineering, as well as share some advice for engineering professionals and researchers.

Hydrodynamic Performance of a Bragg Reflection Breakwater

L. F. Chen, S. B. Zhang, D. Z. Ning

State Key Laboratory of Coastal and Offshore Engineering, Dalian University of Technology, Dalian 116024, China

Abstract

It iswell recognized that in certain circumstances, very few small structures can reflect the incident wave energy substantially, i. e. Bragg resonance occurred. Hence, a concept of Bragg reflection breakwater consisting of several periodic small bars has been proposed for shore-protection. However, the feasibility and effectiveness of Bragg reflection breakwater remain unclear. In this work, Bragg resonances, resulting from nonlinear wave evolution and wave-structure interaction as a surface wave propagating over a periodic sinusoidal structure, are investigated using a Higher Order Boundary Element Model (HOBEM) within a framework of potential flow theory. The model is firstly verified by comparing with benchmarking published experimental results and the analytical solutions before being extended to a wider range of wave conditions and breakwater sizes. Two parameters, the wave-reflection coefficient and the reflection bandwidth (within which the wave-reflection coefficient is larger than a certain value, e.g., 0.5), are used in assessing the hydrodynamic performance of the Bragg breakwater concept; the larger of both, the more effective of the breakwater is. It is found that the former increases with the increasing breakwater dimensions, characterized by the bar number and the bar height, while the opposite trend is true for the latter. Even though, we found that the performance of the Bragg breakwater maybe highly underscored, as multiple resonances at different orders maybe obtained for the same system depending on the size of the bar(s), as shown in the following figure. In addition to Bragg resonance conditions (i.e., class I and III that are studied extensively), resonances at a surface wavenumber close to an integer multiple of half a bottom wavenumber (the integer could be larger than 1) may occur. This is due to the fact that higher free harmonic waves of significant magnitude (i.e., redistribution of wave energy from the fundamental frequency into the first and higher harmonics) are generated over a submerged breakwater. More details will be presented at the workshop.

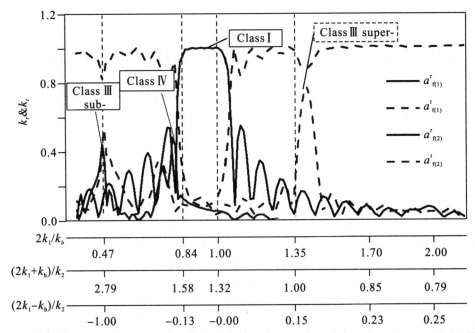

Figure 1. Distribution of reflection and transmission coefficients at various type Bragg resonance conditions

Hyporheic Exchange of Reactive Substances

Guangqiu Jin, Zhongtian Zhang, Hongwu Tang, Qihao Jiang, Wenhui Shao

State Key Laboratory of Hydrology-Water Resources and Hydraulic Engineering,
Hohai University, Nanjing 210098, China

Abstract

Due to natural and human activities, a large number of reactive substances, such as metals and organic matter, enter the hyporheic zone and accumulate in the sediments, which affect the biochemical reaction and pose a threat to the aquatic ecosystem. Although there are many studies on the interface process of reactive substances in the hyporheic zone, most of them focus on the mass exchange flux at the sediment-surface water interface, lack of research on the transport and transformation process of reactive substances in the hyporheic zone. And how reaction and adsorption dominate the transport of such substances is less studied. In this study, the transport and reaction process of reactive substances in the hyporheic zone were studied by flume experiment and numerical simulation with metal ion zinc, benzene, nutrient nitrogen and phosphorus. Due to the adsorption reaction of the bed, the concentration of substances in the overlying water drops quickly and reaches a lower level eventually. Most of the reactive substances stay in the shallow area of the streambed and are difficult to penetrate. When the river is recharging the regional groundwater, the input flux of reactive substances will increase with the change of the head distribution on the surface of bedforms. When regional groundwater is discharged into the river, the input flux of reactive substances is limited as the surface inflow area of bedforms decreases. For the benzene spill accident in the Huaihe River in 2012, we develop a process-based mathematical model to analyze the monitoring data collected shortly after the accident, and explore not only how effective the adopted measures were over the incident but more importantly the mechanisms and critical conditions underlying the effectiveness of these measures. The study of the migration and transformation of reactive substances in the hyporheic zone is helpful to improve people's understanding of the environmental effects of reactive substances in the river system, to provide reference for the restoration of ecological environment in the future.

Keywords: Reactive substances; Hyporheic exchange; Zinc; Benzene; Environmental effect

Methane Emission from Inland Waters and Its Response to Climate Warming

Xinghui Xia, Liwei Zhang, Gongqin Wang

Key Laboratory of Water and Sediment Sciences of Ministry of Education, Beijing Normal University, Beijing 100091, China

Abstract

Inland waters are large sources of methane to the atmosphere. However, large uncertainty exists in estimating emissions of this potent greenhouse gas from global streams and rivers due, in part, to a lack of direct measurements in high-altitude cryosphere and urban inland waters and a poor accounting of ebullition. Here we present methane concentrations and fluxes over three years in four basins on the East Qinghai-Tibet Plateau. Methane ebullition rates decreased exponentially while diffusion declined linearly with increasing stream order. Nonetheless, average ebullition rate (11.9 mmol$CH_4 \cdot m^{-2} \cdot d^{-1}$) from these streams and rivers that have large organic stocks in surrounding permafrost, abundant cold-tolerant methanogens, shallow water depths, and experience low air pressure were 6 times greater than the global average. These results demonstrate the significance of high-elevation rivers on the Qinghai-Tibet Plateau as hotspots of methane to the atmosphere driven by large ebullitive fluxes, which constitute an important fraction of global fluvial methane emissions, and unfolding of a positive feedback of climate warming, permafrost thaw, and methane emissions.

In addition, we measured the concentrations and fluxes of CH_4 and CO_2 in rivers and lakes in the megacity of Beijing, China, between 2018 and 2019. The CH_4 concentration ranged from 0.08 to 70.2 $\mu mol \cdot L^{-1}$ with an average of 2.5 ± 5.9 $\mu mol \cdot L^{-1}$. The average CH_4 ebullition was 11.3 ± 30.4 $mmol \cdot m^{-2} \cdot d^{-1}$ and was approximately 6 times higher than the global average. The high surface water CH_4 concentrations and ebullitive fluxes were caused by high sediment organic carbon/dissolved organic carbon contents, high aquatic primary productivity and shallow water depths in the urban inland waters. The CH_4 emissions accounted for 20% of CO_2 emissions in terms of the carbon release and were 1.7 times higher in terms of CO_2 equivalent emissions from Beijing inland waters. Furthermore, the CH_4 ebullition and its contribution to the total carbon gas emissions increased exponentially with the water temperature, suggesting a positive feedback probably occurs between the greenhouse gas emissions from urban inland waters and climate warming.

Keywords: Methane (CH_4); Ebullition; Carbon dioxide (CO_2); Inland waters; Climate warming

Numerical Simulation of Turbulent Flow and Sediment Transport in Meandering Channel and River

Hefang Jing[1], Chunguang Li[1], Yakun Guo[2]

[1] School of Civil Engineering, Beifang University of Nationalities, Yinchuan 750021, China
[2] Faculty of Engineering & Informatics, University of Bradford, BD7 1DP, UK

Abstract

Flow field and sediment transport in rivers is of importance to the prediction of the river flooding and flooding risk mitigation. Free surface and bed and banks friction effect makes the flow field in open channel and rivers complex. Usually, natural river is meandering and compound in which the secondary current plays an important role, interacts with and affects the main flow structure. This further complicates the flow structures in meandering rivers. To plan river flooding prediction and prevention, accurate prediction of the flow structure and the conveyance capacity in meandering rivers is essential. Due to its practical engineering importance, extensive numerical simulation studies using various numerical models have been carried out in past decades. These studies showed that the standard k-ε model cannot accurately predict the secondary flows (Cokljat & Younis, 1995). Though the nonlinear k-ε model can reproduce the secondary current, such model cannot accurately simulate some turbulence structures in meandering compound channel (Sofialidis & Prinos, 1998). Prediction of the mean flow and turbulence structures in meandering channel using the Algebraic Reynolds stress model (ASM) is also poor (Sugiyama et al. 2006).

In this study, the Reynolds stress equation model (RSM) is adopted to simulate the three dimensional flow structures in the meandering compound channel. The governing equations are discretized using finite volume method and the SIMPLEC algorithm is applied for pressure-velocity coupling. For the simulation of the flow and sediment transport in river with continuous bends, a 2D depth averaged RNG k-ε model, which includes flow model and sediment model, is developed to simulate the flow and sediment transport in the upper meandering reach of the Yellow River. To efficiently simulate the flow and sediment transport, we proposed a semi-coupled scheme, which combines the advantage of the coupled and decoupled schemes. In this semi-coupled procedure, sediment model continues running and the flow model runs intermittently. As such, the proposed scheme is more efficient than the coupled scheme and is more accurate than the decoupled scheme.

Fig. 1 shows the comparison of the calculated and measured flow velocity field. The measured data are from the UK-FCF (Knight & Sellin 1987). It is seen that the good agreement between the numerical simulation and measurements is achieved. Fig. 2 is the plot of the simulated and measured depth averaged velocity distribution at SH5 (b); river bed elevation at SH7 (c) and bed elevation along longitudinal direction (d). Good agreement between the calculated and measured depth averaged flow velocity, cross sectional bed elevation and bed elevation in longitudinal direction indicates that the proposed model can accurately and efficiently simulate the flow and sediment transport in complex meandering river reaches.

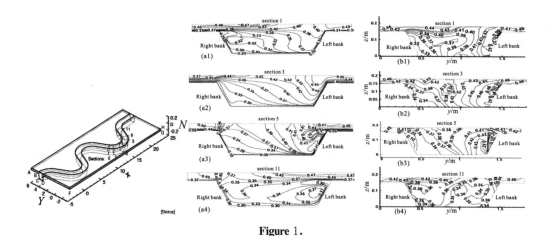

Figure 1.

Note: (a) Sketch of the computational domain (UK-FCF): 26 m, length of one wave of 60° meandering channel. (b) Comparison of simulated (left) and measured (right) flow velocity contours at four sections 1, 3, 5 and 11: H (maximum water depth of main channel)=0.2 m, Q=0.248 m³/s, u=0.378 m/s and h (water depth over flood plain) =0.05 m at the inlet.

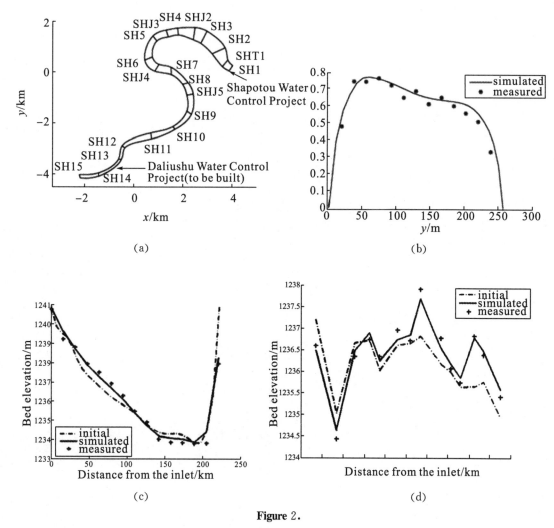

Figure 2.

Note: (a) Sketch of the upper meandering reach of the Yellow River; (b) simulated and measured depth averaged velocity at SH5; (c) simulated and measured bed elevation at SH7; (d) simulated and measured bed elevation along longitudinal direction.

References

[1] Cokljat D, Younis B A. Second-order closure study of open channel flows[J]. Journal of Hydraulic Engineering (ASCE), 1995, 121: 94−107.

[2] Knight D W, Sellin R H J. The SERC flood channel facility[J]. Water and Environment Journal, 1987, 1: 198−204.

[3] Sofialidis D, Prinos P. Compound open-channel flow modeling with nonlinear low-Reynolds k-ε models[J]. Journal of Hydraulic Engineering (ASCE), 1998, 124: 253−262.

[4] Sugiyama H, Hitomi D, Saito T. Numerical analysis of turbulent structure in compound meandering open channel by algebraic Reynolds stress model[J]. International Journal for Numerical Methods in Fluids, 2006, 51: 791−818.

Responsive Strategies of Water Hazard Risk Under Global Change: Case Study in China

Jun Xia

State Key Laboratory of Water Resources & Hydro Power Engineering Sciences, Wuhan University, Wuhan 430072, China

Institute of Geographical Sciences & Natural Resources Research, Chinese Academy of Sciences, Beijing 100101, China

Abstract

This presentation addresses to discusses the severe waterstress that all countries in the world are subjected to, such as water shortage, water pollution, water disaster and ecosystem degradation, in the context of global changes including climate change and frequent human activities. Furthermore, the water hazard events, such as major floods and droughts caused by climate change and human-induced urban flooding in China during the past decades are introduced through actual cases. It is clearly that water hazard tends to continuously increase, and the water hazard risk is significantly enhanced. As a result, an adaptive approach for water resources management is proposed to realize the new vision of water security in China and even the world by reducing water hazard risk caused by global change. This strategy is based on my previous research in climate change and its impact on water cycle, water security, and extreme events such as floods and droughts. It is suggested to improve the capacity for adaptive water management to better response to the future opportunities and challenges from water hazard. Evidently, to effectively defend against water hazard risks, it is urgent for hydrology to play a key role in developing and practicing adaptive strategies for water resources management.

Keywords: Water hazard risk; Global change; Adaptive water management; Water stress; Hydrology

Sediment Transport in Vegetated Open Channel Flow

Wenxin Huai

State Key Laboratory of Water Resources and Hydropower Engineering Science,
Wuhan University, Wuhan 430072, China

Abstract

An understanding of sediment transport in vegetated open channel flow is greatly important for water ecological conservation and restoration. The present study explores sediment transport by investigatingincipient sediment motion, Suspended Sediment Concentration (SSC) and suspended sediment transport capacity in the context of vegetation. Firstly, to quantify incipient sediment motion in vegetated open channel flow, a new formula based on the criterion of critical flow velocity is developed. The derivation of the formula incorporates the influence of vegetation drag that characterizes the effects of mean flow and turbulence on sediment movement with emergent vegetation. The proposed formula is shown to agree with existing experimental data well. Moreover, it is extended to scenarios with submerged vegetation, suggesting that the vegetation drag may be the inherent impact factor for incipient sediment motion in vegetated open channel flow. Secondly, the analytical solution of SSC in both submerged and emergent vegetated open channel flows is obtained by considering the dispersive flux and solving the vertical double-averaging sediment advection-diffusion equation. The dispersive coefficient is assumed as the product of a scale factor and the morphological coefficient and expressed as partitioned linear profiles above or below the half height of vegetation. The analytically predicted SSC agrees well with the experimental data, indicating that the proposed model can be used to accurately predict the SSC in vegetated open channel flow. Results show that the dispersive term can be ignored in the region without vegetation while having a significant effect on the vertical SSC profile within the region of vegetation. Finally, the suspended sediment transport capacity in vegetated open channel flow is predicted with a new formula that is derived by considering the absolute value of the energy loss between the sediment-laden flow and the clear water flow. The prediction of the proposed formula is in good agreement with the collected available experimental data with the appropriate coefficient values adopted. Subsequently, the formula is expressed in a practical form by using the logarithmic matching method.

Keywords: Vegetation; Incipient sediment motion; Suspended sediment concentration; Sediment transport capacity

The Key Technologies and Application of Intelligent Urban Drainage Management

Xiaohui Lei

Institute of Water Resources and Hydropower Research, Beijing 100038, China

Hebei Key Laboratory of Intelligent Water Conservancy, Hebei University of Engineering, Handan 056038, China

Abstract

As you know, China is still enjoying the fast development of urban areas, and has established ten clusters of cities, especially in the Southeast of China. However, many big cities are continuously suffering from the threats of extreme rainstorm events due to rapid urbanization and global climate change. In order to achieve the effective regulation of the urban drainage cycle, we developed an Intelligent Drainage Management Platform including the "eyes" part of a comprehensive monitoring system, the "brain" part of urban drainage operation models and the "hands" part of remote control of sluices and pumps. Relevant technologies have been successfully applied in Fuzhou, providing great help to reduce urban waterlogging.

Theory and Practice of Water-soil Environmental Effects in the Process of Farmland Freezing-thawing Soil in Cold Regions

Qiang Fu

School of Water Conservancy and Civil Engineering, Northeast Agricultural University, Heilongjiang 150030, China

Abstract

Heilongjiang Province is the largest commodity grain base, and also the ballast stone to ensure China's food security. However, due to years of reclamation and soil freeze-thaw erosion, soil organic matter generally decreased, soil structure was seriously damaged, ecological function was degraded, and black soil layer became thinner. In order to maintain the sustainability of black soil resources in the seasonal frozen soil areas and retain the health of farmland soil habitat, we used the method of field experiment and laboratory simulation to intervene and regulate the farmland soil environment with straw and straw biochar. The structural changes of soil caused by freezing-thawing process, the effects of biochar application on soil porosity and water holding capacity, and the migration and transformation of soil carbon and nitrogen elements were discussed. The main research results are as follows: Under the conditions of freeze-thaw cycles, the water-heat-salinity in farmland soil are not independent of each other. The response function of water-heat-salinity can be built to reveal the migration and diffusion of soil water, heat, and salt in cold regions and their interaction mechanisms; Biochar, as an exogenous medium, can undergo certain physical and chemical changes with the soil, increase the soil porosity, improve the soil structure, and enhance the original soil's moisture holding capacity. In addition, it can accelerate the mineralization of soil carbon and nitrogen by acting together with freeze-thaw cycles, improve the urease activity and nutrient availability of farmland soil, and thus improve the soil ecological environment. The research results can provide important theoretical support for the efficient use of farmland freeze-thaw soil water-heat resources and the improvement of farmland water-soil environment in black soil area.

Understanding Hydro-eco-environment Changes in the Yangtze River

Dawen Yang[1], Ai wang[2], Ruijie Shi[3]

[1,3] Tsinghua University, Beijing 100084, China

[2] Tianjin Agricultural University, Tianjin 300350, China

Abstract

Climate change and human activities influence the hydrological and biogeochemical processes and impact on the water resources, water quality and river ecosystem. This study investigated changes in streamflow, flow regime and water pollution in the upper Yangtze River using a comprehensive method including the distributed hydrological model and statistical models. Results showed that annual streamflow decreased in 50 years. Autumn streamflow evidently decreased after the 1980s, which resulted from the decrease in precipitation and water storing by reservoirs. Summer high flow decreased after the 1980s which was also primarily attributed to the decrease in precipitation. Winter streamflow increased in the two most recent decades, which resulted from the reservoir release. Results also showed that the Three Gorges Reservoir (TGR) elevated low flow in the dry season and reduced peak flow in summer since 2003. The simulated results showed that annual total nitrogen loading in the watershed was 1.50 ton/km^2 in average. The amount of nitrogen loading in July and August took more than 65% of the annual total nitrogen loading. The export coefficient was influenced by the nitrogen supply and hydrological processes especially the rainfall-runoff processes. The concentration of total nitrogen was higher in rainy season. The decrease in autumn precipitation since the 1990s, suggests that TGR is facing a serious challenge in maintaining water storage in the reservoir and releasing the water to the downstream ecosystem.

Keywords: Climate change; Human activities; Water resources; Water quality; River ecosystem; Yangtze River

Unified Approach to Extreme Value Analysis of Coastal Storm Waves

Zaijin You

Centre for Ports and Maritime Safety, Dalian Maritime University, Dalian 116026, China

Abstract

Stormwaves can cause severe damage to coastal infrastructure and beachfront property, erode sandy duns and beaches, inundate coastal low-lying areas, and degrade coastal and estuarine ecosystems. On the other hand, accurate estimation of extreme storm waves with different return periods is always required for inshore and offshore structure designs, coastal hazard assessments, coastal planning, and coastal port operations and channel navigations. Even though a large number of studies have been undertaken to develop and improve methodologies for extrapolation of historical wave record to extreme waves with different return periods, there still exists great uncertainty in extreme wave analysis mainly due to the lack of a unified approach to extreme value analysis of coastal storm waves.

In thisstudy, a unified approach is proposed to extract extreme wave data points from historical wave record, simplify a wide range of candidate distribution functions into only two types of simple power and exponential distribution functions, and develop an extended linear regression model to determine the distribution parameters. The historical wave data used in this study were collected with wave rider buoys on the NSW coast of Australia since 1976. Based on the preliminary results of this study, it is found that the uncertainty in extreme wave analysis is caused mainly by short length of wave record, missing storm wave data, improper methods for generation of extreme wave data, and unsuitable candidate distribution function for extrapolation of extreme waves. The unified approach proposed in this study will significantly reduce the uncertainty in extreme wave analysis and improve accuracy in estimation of extreme wave heights with different return periods.

Using Nature-based Solutions for Integrated Water Management and Ecological Restoration: Lessons Learnt from the Netherlands

W. Ellis Penning

Deltares, Delft, the Netherlands

Abstract

The global call for 'working with nature, instead of against it' is getting more and more attention, both on the policy level and the practical implementation level for integrated water management and ecological restoration. Within the Netherlands this has resulted in a large effort to implement Nature-based Solutions as part of the efforts to reduce flood risks and improve ecological status along the coast and large rivers and lakes. Many large-scale pilot projects were carried out over the last decade to demonstrate how to 'build with nature' to improve quality of landscapes and their functioning for the safety of society. The successful implementation of these pilots was greatly helped by the pro-active cooperation between water management authorities, scientific communities, consultancy firms and the dredging industry, who were joined in the pre-competitive EcoShape consortium. This consortium aimed to develop the knowledge needed to actually implement these types of nature based solutions at large scale. By working together under a united umbrella as different stakeholders the consortium provided the scientific underpinning combined with clear awareness of the business-case and practical management aspects of the selected solutions. Proper monitoring and evaluation programs were put into action to follow the development of the pilots for multiple years. The resulting extended knowledge and understanding of the natural and socio-economic aspects has greatly helped the further development of upscaling and new initiatives in which the nature-based solutions were judged in an equal manner as the 'traditional grey' engineering solutions that might otherwise have been selected.

Velocity, Inundation and Run-up Measurements of Laboratory Generated Bores on a Planar Beach

Philip. L.-F. Liu, Lgnacio Barranco

Department of Civil and Environmental Engineering, National University of Singapore, Singapore

Abstract

Decaying and non-decaying bores are generated in the laboratory using a programmable long-stroke wavemaker and a dam-break system. Employing the method of characteristics, the bore strength and length along the wave tank are related to the wavemaker stroke and velocity as well as the reservoir length and the depth ratios. Flow velocities under undulating bores and breaking bore are measured on the constant water depth and near the still water shoreline, using a High-speed Particle Image Velocimetry (HSPIV) system. The ensembled velocity and turbulent flow characteristics are analyzed for both undulating bores and breaking bores. The inundation depth and run-up height are also measured. Empirical relationships between them and the bore length and bore strength at the beach toe are deduced from the laboratory experiments and numerical simulations. Finally, the minimum bore lengths at the beach toe necessary to produce the maximum inundation depths and the flooding plateau.

A GPU-based Two-phase SPH Model for Intense Sediment Transport in Geological-hazard Flows

Huabin Shi

State Key Laboratory of Internet of Things for Smart City and Department of Civil and Environmental Engineering, University of Macau, Macao, China

Abstract

A two-phase SPH model using a continuum description of solid-fluid mixtures is proposed for intense sediment transport in geological-hazard flows, such as submarine landslides and dam-break erosion flows. In the model, the water and the solid particles are treated as two miscible continuous fluids and both phases are controlled by the conservation laws for mass and momentum. A constitutive law based on the rheology of dense granular flows, which consists of a frictional component for the enduring-contact between grains and a collisional component for the collisions among solid particles, is integrated for the inter-grain stresses. In the frictional constitutive law, a theoretical formulation of dilatation/contraction arising from rearrangement of contact force chains under shear deformation is coupled. On the other hand, an inter-phase drag force is formulated to account for the interaction between the water and the solid phase. The Ergun equation is utilized for dense solid-liquid mixtures and the power law is applied to dilute suspensions. A weakly-compressible SPH method is utilized to solve the controlling equations, in which the water is assumed to be weakly compressible with a variant density while the solid phase is incompressible. The solid-liquid mixture is represented by a single set of SPH particles that move with a velocity equal to the water velocity while carry the properties of both phases. The model is numerically solved using GPU-parallel computing. The proposed model is firstly applied to dam-break flows over movable sediment bed. The computed violently-changed water surface and bed profiles as well as the propagation of the dam-break wave are in good agreement with the measured data. The computed vertical profiles of granular velocity and sediment concentration are also well validated. The model is then applied to simulate the collapse of packed granular columns submerged in a liquid. The computed profiles of the granular columns agree very well with the measured data. The significant effect of the packing fraction of columns on the collapse is clearly shown and discussed in-depth. It is noted that the huge difference in the collapse behavior of columns with different initial packing fractions arises from the frictional dilatation/contraction of the granular columns, which is well captured by the integrated theoretical formulation in the present model. In summary, the proposed model in the present study can describe both intense sediment transport by violent water flows and gravity-induced underwater granular flows.

Keywords: Geological-hazard flows; Sediment transport; Granular flows; Two-phase SPH model; Dilatation/contraction

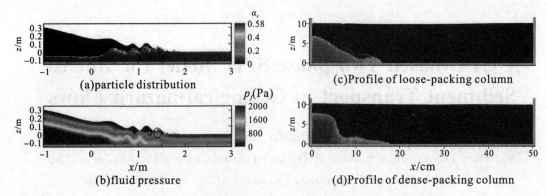

Figure 1. Simulated results of dam-break flows over mobile bed

Note: (a) for sediment concentration and particle distribution; (b) for fluid pressure and underwater granular column collapse; (c) for the initially loosely-packed column; (d) for the densely-packed column.

Accurate Short-term Prediction of the Water Levels Along the Yangtze Estuary by Using the NS_TIDE&AR Model

Yongping Chen[1], Min Gan[2], Shunqi Pan[3]

[1,2] State Key Laboratory of Hydrology-Water Resources & Hydraulic Engineering, Nanjing 210098, China

[1,2] College of Harbor, Coastal, and Offshore Engineering, Hohai University, Nanjing 210098, China

[2,3] Hydro-environmental Research Centre, School of Engineering, Cardiff University, Cardiff, United Kingdom

Abstract

Tides in estuary zones are strongly affected by the factors such as the changes of estuarine width and topography, and upstream river discharge. River discharge can not only increase the mean water levels but also nonlinearly interact with tides and play a frictional effect on tides. Complicated influences of external forces on estuarine tides make the accurate prediction of estuarine water levels challenging. The commonly-used nonstationary tidal harmonic analysis tool (NS_TIDE) achieved a significant improvement in analyzing estuarine water levels relative to the classical harmonic analysis model. However, there is still room for improving the prediction accuracy of the NS_TIDE model, especially in flood season when river discharge is strong. Spectral analysis proves that the prediction errors of the NS_TIDE model mainly source from the inaccurate description of subtidal water levels. Based on this finding, the Auto-regressive (AR) model is used to model the prediction errors of the NS_TIDE and thereby correct the short-term predictions (within 48 h) of the NS_TIDE model. The long-term measurements test in the Yangtze estuary shows that the NS_TIDE&AR model achieved the short-term (24 h) prediction accuracy with the root-mean-square error (RMSE) being in a range of 0.10~0.13 m, which is much smaller relative to the NS_TIDE model (0.22~0.26 m). In the summer of 2020, severe flood events happened in the Yangtze River basin, which made the water levels of several hydrometric stations along the Yangtze estuary break their maximum water level records. During the flood event of 2020, the NS_TIDE&AR model was also applied to actually predict the water levels along the Yangtze estuary. For example, the NS_TIDE&AR model predicted that the maximum water level at the Nanjing station during the 2020 flood is 10.32 m, close to the measured value (10.38 m). The application of the NS_TIDE&AR model in the 2020 flood period proved its reliability in extreme flood conditions.

Keywords: Water level prediction; Estuarine tides; NS_TIDE; Auto-regressive; Yangtze estuary

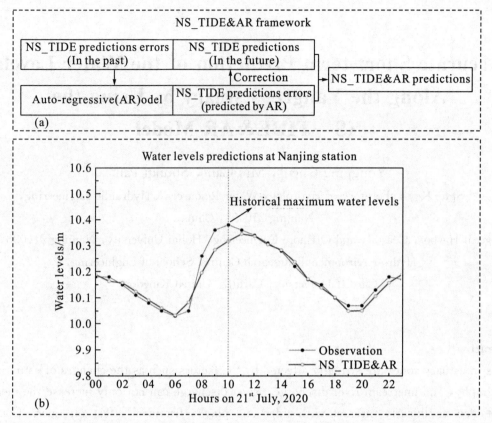

Figure 1. A diagram of the NS_TIDE&AR model (a) and its application in predicting the water levels of Nanjing station during the 2020 flood event of the Yangtze River basin (b)

Analysis of the Recirculation Length Downstream of Spur Dikes by Using a Two-dimensional Depth-averaged Mathematical Model

Bingdong Li

State Key Laboratory of Hydraulics and Mountain River Engineering, Chengdu 610065, China

Abstract

Spur dike is a common channel regulating structure with important functions of adjusting water flow direction and increasing water depth which are benefit to shipping. Besides, spur dikes are found recently that they have a significant ecological function of increasing aquatic habitat. Downstream of spur dikes, recirculation zones will develop. In these areas water flows smoothly and slowly. Sediment will deposit and nutrient substance will enrich in these zones. Enough nutrient substance prompts the production of benthos which causes aquatic community propagates gradually and becomes prosperity. The occurrence of recirculation zones behind spur dikes will greatly improve aquatic ecological environment and their sizes will determine the potential areas of aquatic habitat. In this paper, a two-dimensional depth-averaged mathematical model was applied to predict the lengths of recirculation zones formed behind of spur dikes in the Jinjiang River. The model was firstly calibrated by using field test data of water depth. Then, different influencing factors, such as discharge, water level, and bed slope, were selected and were used as boundary conditions to simulate water flowing around spur dikes in open channels. Based on these data from the simulated water flow, recirculation zones were depicted and the lengths of the recirculation zones were evaluated. The simulated results show that the regulating effect of spur dikes on water flow is obvious and the influencing range of recirculation zones varies greatly for different impacting parameters. Compared with several existed empirical formulas for predicting the length of recirculation zones downstream of spur dikes, the results from the two-dimensional depth-averaged model reveal the recirculation length gained from these formulas might be over-predicted. Systematical analysis of the selected impact factors on the size of recirculation zones downstream of spur dikes presents the recirculation length is closely related to the length of the spur dike. Also the increasing discharge will increase the recirculation length. However, the increasing bed slope will decrease the length of the recirculation. Water level given in our study seems to have a little effect on recirculation zone. According to our study, though the existing empirical formulas can provide a preliminary and satisfied evaluation of the recirculation length, the two-dimensional mathematical model performs well in some complicate situations because it takes more the factors of flow and topography into account. The results reveal that when spur dikes are used as river ecological restoration measures the two-dimensional mathematical model should be a prior alternative to predicting recirculation zones which determine the potential area of fish habitat.

Keywords: Spur dike; Recirculation; Mathematical model; Two-dimensional

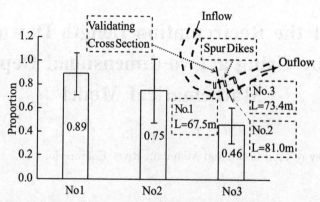

Figure 1. The proportion (Mean±SD) of the recirculation length to that of spur dikes

Note: According to the statistical data from 98 simulated sample cases including different of fluxes, water levels, and bed slopes. For the No. 1 spur dike, in different flow conditions, the average proportion is 0.89, and the standard deviation is 0.18; For the No. 2 spur dike, in different flow conditions, the average proportion is 0.75 and the standard deviation is 0.27; and for the No. 3 spur dike, the average proportion is 0.46 and the standard deviation is 0.15.

Assessing Impact Factors for Geological Hazards in Shaanxi Province Using Machine Learning

Shizhengxiong Liang[1,2], Dong Chen[1,2]

[1] Key Laboratory of Water Cycle and Related Land Surface Processes, Institute of Geographic Sciences and Natural Resources Research, Chinese Academy of Sciences, Beijing 100101, China

[2] College of Resources and Environment, University of Chinese Academy of Sciences, Beijing 100049, China

Abstract

The study interprets the spatial distribution of geological hazards, including collapse (CL), landslide (LS) and debris flows (DF) during 1951—2018 in Shaanxi Province, China. The potential impact factors for those hazards, involving geomorphologic types, distance to river, distance to road, rainfall parameters, permeability ecoefficiency, soil erodibility, relief and population, are analyzed in ArcGIS. Then, we divided Shaanxi Province into two parts, i. e., Southern Shaanxi (SS) and Northern Shaanxi (NS), by Weihe river and Liupan mountain. Random forest (RF), a widely used machine learning model, is applied to identify the main impact factors and develop the geological hazards susceptibility map of SS and NS, respectively. The results indicated that the RF has certain applicability to assess the susceptibility of three geological hazards, especially in SS. For instance, RF has a better performance of CL and LS prediction in SS (AUC=0.8 and 0.917, respectively) than in SN (AUC=0.642 and 0.736, respectively), which may indicate more complex impact factors for CL and LS occurrences in the Loess Plateau. Since the sampling pool for DF in NS is not large enough, we only estimate DF in SS with satisfactory RF performance (AUC=0.826). The dominant factors of different types of geological hazards are quite different in SS and NS. The controlling factors of CL in NS are population and distance from roads, while they are landform types and rainfall CV values in SS. LS in NS are mainly affected by rainfall intensity and soil permeability coefficient, while in SS, the main factors are population and soil permeability. For DF in SS, the permeability ecoefficiency is the key factor, followed by the maximum hourly rainfall intensity. The high-risk region of CL mainly concentrated in the North bank of Weihe river, while DF concentrated in the mountainous area of Ankang. For landslides, both the above two areas are high-risk regions (Fig. 1).

Keywords: Geological hazards; Spatio-temporal distribution; Impact factors; Shaanxi Province; Machine learning

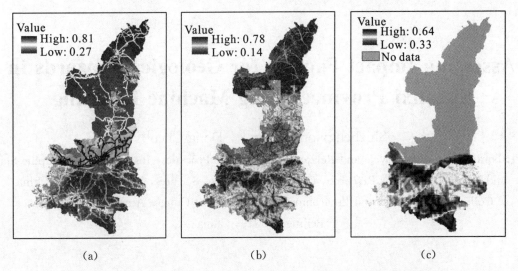

Figure 1. The susceptibility of CL (a), LS (b) and DF (c) in Shaanxi Province

Biofilms as the "Architect": The Microbiological Mediation of Intertidal Sediment Behavior

Xindi Chen[1], Changkuan Zhang[2], Xiping Yu[3]

[1,3] Department of Hydraulic Engineering, Tsinghua University, Beijing 100084, China
[2] College of Harbor, Coastal and Offshore Engineering, Hohai University, Nanjing 100083, China

Abstract

The term "ecosystem engineering" emerged in the 1990s, which commonly refers to the activities of larger organisms like mangroves. However, while people think that bigger organisms generate bigger potential engineering effects, there may be microscale organisms who can result in significant impacts on the ecosystems through their number rather than their size. Currently, cohesive extracellular polymeric substances (EPS) generated by microorganisms have been widely reported to increase the threshold for sediment erosion by flowing water, which is known as "biostabilization". However, we demonstrate that this is not the case under wave action. On the contrary, EPS show a destabilization effect of the system, turning an otherwise stable sedimentary bed into "soup". Our analysis clarifies how neglecting even low content of EPS can result in inaccurate prediction of the bed stability and coastal safety under wave action. The risk of bed liquefaction is expected to pose potential threats to wetlands where microbial communities occupy habitats while the production of EPS is much higher. The misinterpretation of the vulnerability of wetlands when exposed to waves could put the existing ecosystems at risk, considering that these ecosystem services are valued at about US $10, 000 per hectare.

Keywords: Sediment behavior; Biofilms; Intertidal sediment; Bed stability; Wave action

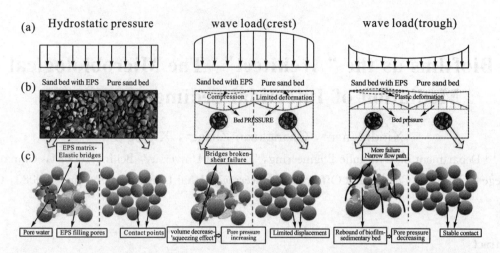

Figure 1. **Schematic of mechanism for EPS (low content) induced liquefaction of a sandy bed. Under load from the hydrostatic condition (left column) and wave forcing (crest, center column and trough, right column, respectively) (a), the bed responds differently for the biological condition (left) and abiotic condition (right) (b). The images in (b) are SEM images. The grain-EPS-grain contact and the excess-pore water response are represented at the μm scale (c)**

Note: This work has recently been published. Full details can be found in Chen X D, Zhang C K, Townend I, et al., 2021. Biological cohesion as the architect of bed movement under wave action. Geophysical Research Letters, 48: e2020G-e92137G.

Changes in the Floodplain Channel Resistance in the Middle Yangtze River and Influencing Factors after the Impoundment of the Three Gorges Reservoir

Yong Hu, Yitian Li, Jinyun Deng

State Key Laboratory of Water Resources and Hydropower Engineering Science,
Wuhan University, Wuhan 430072, China

Abstract

The river resistance notably changed after large-scale reservoirs were impounded in the upper Yangtze River, the water and sediment regimes were altered, and the riverbed was adjusted, which significantly affects the flood control, aquatic ecology security, and water resource utilization. To investigate the resistance of the middle reaches of the Yangtze River after the impoundment of the Three Gorges Reservoir (TGR), the spatiotemporal variations of the comprehensive roughness, main channel, and floodplain roughness were analyzed based on hydrology and topography data from 2004 to 2015. The reason for the resistance change was also investigated. The results show that all the comprehensive roughness and main channel and floodplain roughness values have increased since the start of the operation of the TGR, especially after the change of reservoir operation. The floodplain roughness is generally larger than that of the main channel. The main channel roughness correlates with the flow discharge, whereas the floodplain roughness changes nonmontonously. The longitudinal variation of the roughness in the middle reaches of the Yangtze River did not change before and after the impoundment of the TGR, but the roughness decreased from Yichang to Hukou. The change in the longitudinal profile of the middle reaches in the Yangtze River due to the continuous erosion after the impoundment of the TGR and changing hydrological rhythm is a key factor affecting the resistance alteration and vegetation growth on the floodplain. The reduction of the overbank-flood probability is another important factor. Moreover, the size of floodplain roughness and its proportion in the comprehensive roughness have an obvious positive correlation with the areas of the vegetation. Consequently, on account of the different operation periods of the TGR, the floodplain roughness has a continuously increasing trend as the vegetated areas grow. In addition, the large number of waterway regulation projects in the middle reaches also increased the floodplain resistance.

Keywords: Resistance change; River channel morphology; Floodplain; Middle Yangtze River

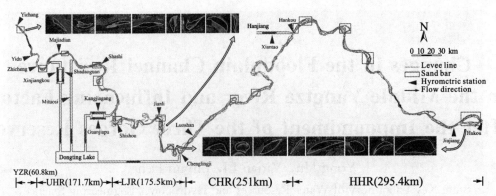

Figure 1. Sketch of the Middle Yangtze Reach with the locations of diversion inlets, 18 hydrometric stations, and 16 bars

Coupled 1D/2D Hydrodynamic Modelling and Hazard Risk Assessment

Boliang Dong, Junqiang Xia, Shanshan Deng, Meirong Zhou

State Key Laboratory of Water Resources and Hydropower Engineering Science,
Wuhan University, Wuhan 430072, China

Abstract

Modelling flood inundation processes and give precise assessments of corresponding flood hazard degrees are increasingly important for disaster prevention and mitigation. However, due to the influence of complex street layouts and various infrastructures, urban flooding processes are highly uncertain. In order to accurately simulate the whole urban flooding process, a coupled 1D/2D urban flooding hydrodynamic model was established in this study. The model was based on the finite volume method framework and involving subprocesses such as surface runoff inundation, flow exchange between surface runoff and sewer pipe flow, as well as sewer network drainage. Based on the high spatial-temporal resolution hydrodynamics results, mechanical-based stability criteria for human and vehicles were utilized to access the hazard risks. The established model was firstly validated by a laboratory experiment involving both surface runoff inundations over a typical urban street as well as the interaction between surface runoff and sewer pipe flow, then the model was applied to simulate an actual urban flooding process that occurred in Glasgow, UK. According to the analysis of simulation results, following conclusions have been drawn: (i) Comparisons between simulated results and observed data indicate that the developed model can accurately simulate the urban flooding process with NSE values greater than 0.7. (ii) For urban flash flooding hazards, the road network is the main path for flood routing and therefore becomes vulnerable areas for people and vehicles. Hazard degree distributions of people and vehicles present different characteristics, vehicles are more vulnerable to high water depth while the stability of the human body is more sensitive to flow velocity. (iii) Urban sewer system has a strong influence on the flood inundation process in terms of reducing water depth and flow velocity, delaying the arrival time of flood wave, shortening the inundation time, and therefore greatly mitigate the hazard degrees for people and vehicles.

Keywords: Urban flooding; 1D/2D dual drainage modeling; Flood risk assessment; Human stability

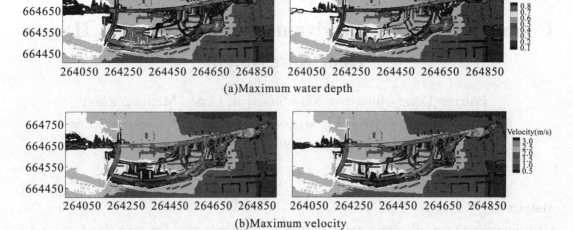

Figure 1. Comparison of simulated maximum water depth and velocity distributions for with and without sewer system drainage scenarios: without sewer drainage (left); with sewer drainage (right)

Data-driven Analysis and Forecasting for Sewer Flow of an Old Community in Ningbo, China

Biao Huang[1], Jiachun Liu[2], David Z Zhu[3]

[1,3] College of Civil and Environmental Engineering, Ningbo University, Ningbo 315211, China

[2,3] Department of Civil and Environmental Engineering, University of Alberta, Edmonton, Canada

Abstract

Smart monitoring, real-time control, and advanced management of sewer systems require reliable flow analysis and forecasting, where traditional hydrodynamics models are subject to significant uncertainties and applicability limitations. Data-driven models are increasingly being used in urban water systems around the world, with the development of internet-of-things, online monitoring, and big data analysis.

This study presents data-driven modeling methods for time series analysis and forecasting for a combined sewer system of an old community in Ningbo, China. Basic characteristics of sewer flow can be identified through statistical methods, including the difference between dry-weather and wet weather flows. Exploratory analysis was conducted first for the time series data of water level, flow rate and conductivity, and temporal structures were recognized (Fig. 1). Machine learning methods, such as Random Forest, were used to examine the importance of various parameters and predict future by transforming the time series dataset into a supervised learning problem first. The results indicate that for the combined sewer network under study, rainfall and day of month are two most critical factors influencing the flowrate. Deep learning models, are also developed in the present study by using Long Short-Term Memory and Gated Recurrent Unit neural networks, based on existing data from the monitoring sites. Both univariate and multivariate predictions were modeled using LSTM and GRU networks, which increase forecast horizons to one week and provide accurate estimations.

Keywords: Data driven analysis; Machine learning; Deep learning; Time series forecasting; Sewer flow; Smart monitoring

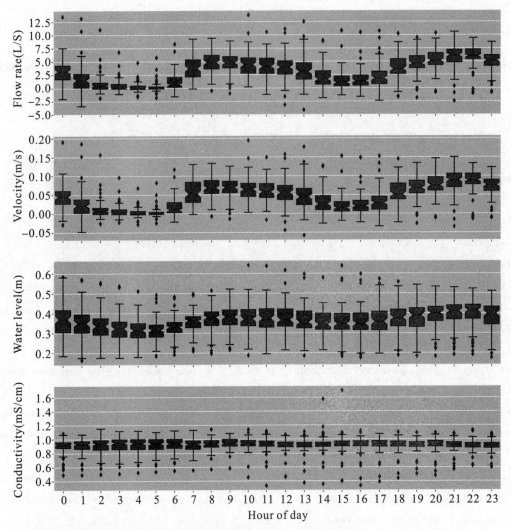

Figure 1. Statistical analysis of characteristics of dry-weather sewer flow at a monitoring site of an old community in Ningbo, China

Dynamic Simulation and Numerical Analysis of Coastal Flooding with Changes in Sea Level and Landscape

Dongmei Xie[1], Qingping Zou[2], Yongping Chen[3]

[1,3] Hohai University, Nanjing 210098, China
[2] Heriot-Watt University, Edinburgh, capital of Scotland, United Kingdom

Abstract

Coastal flooding due to extreme sea levels and waves during tropical and extratropical storms have caused extensive property damage and loss of lives throughout the world. Coastal inundation greatly depends on coastal topography, shoreline configuration, coastal landscape and sea states (Woodruff et al., 2013; Bilskie et al., 2014). Changes in sea level, storm intensity and tracks, and landscape may result in increased coastal flooding and altered inundation patterns (Woodruff et al., 2013; Bilskie et al., 2014). Previous studies mainly focus on investigate the contribution of different hydrodynamic processes on coastal inundation (e. g. Chen et al., 2013), or changes of flooding pattern with modified sea states for a given sea level rise scenario under present topographic conditions (e. g. Karim and Mimura, 2008). Recently, Bilskie et al. (2014) simulated the response of storm surge to changes in sea level and landscape, however, the nonlinear interactions between the sea states, sea level rise and landscape change and their impact on coastal flooding needs further investigation.

This study utilizes a physically based hydrodynamic storm surge and wind wave model to study coastal flooding with a rising sea level and changing landscape in Saco Bay, a mesotidal embayment in the northeastern United States. The nonlinear interactions between the sea states, sea level rise and landscape change and their impact on coastal flooding are also analyzed.

We employed the two-way coupled spectral wave and circulation model SWAN + ADCIRC to compute storm surge and wave conditions over the entire basin of the U. S. Atlantic Coast.

An unstructured grid was utilized to provide flexibility in resolving bathymetric features from ocean to coast. In Saco Bay, model resolution is down to 10 m to resolve the detailed bathymetric and topographic features (Fig. 1). The model was forced with wind and atmospheric pressure representative of the April 2007 nor' easter. Xie et al. (2019) is referred to for the detailed information on model setup and validation.

Numerical simulations were performed to explore the influences and sensitivities of inundation to nonlinear wave-current-bathymetry interaction, and changes in sea levels and landscape. During the 2007 April nor' easter, the coastal areas of Saco Bay were extensively flooded with increased water level due to storm surge. Waves contribute to water level through wave setup due to the momentum transfer from breaking waves to the mean flow. The distribution of wave setup is closely related to the local bathymetry and coastline geometry in the bay. By including the wave-current-bathymetry interaction, both the flooding extent and depth were enhanced. The increased sea level did not change the flooding patterns. We will examine the effect of nonlinear interaction between storm tides, increased sea level and changes in landscape on inundation in the full paper.

Changes in sea level and nearshore geomorphology will continue to affect the behavior of coastal inundations throughout the world. The study utilized a process-based approach to evaluate the impact of wave-current-bathymetry interaction, and changes in coastline morphology and sea level in the context of coastal inundation. More results and discussion on the behavior of coastal inundation with sea level rise and coastline change will be presented in the full paper. The nonlinear interaction between the change in sea states and coastal inundation will also be discussed.

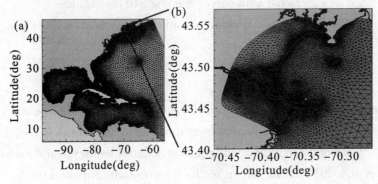

Figure 1. (a) Unstructured mesh for the east coast of United States; (b) zoom-in view of the unstructured grid in the Saco Bay

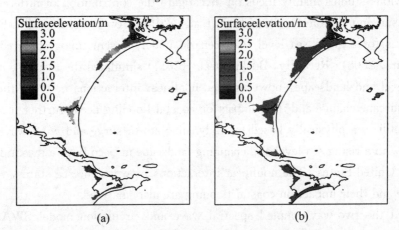

Figure 2. Coastal inundation map in the Saco Bay

Note: (a) Without sea level rise; (b) With wave effect.

Effect of a Single Air Bubble on the Migration Direction of a Cavitation Bubble

Jianbo Li[1], Yanwei Zhai[2]

[1] State Key Laboratory of Hydraulics and Mountain River Engineering, Sichuan University, Chengdu 610065, China

[2] China Three Gorges Corporation, Beijing 100038, China

Abstract

Studying the influence of a single air bubble on the migration direction has great significance for understanding the mechanism of air entrainment alleviating the cavitation erosion in hydraulic engineering. A rising air bubble interacts with a cavitation bubble which is generated by an underwater low-voltage electric discharge device. Only the cases where the cavitation bubble centroid, air bubble centroid, and hydrophone probe are located approximately on the same horizontal line are analyzed. As shown in the first frame of the figure, the black area in the center of the image represents the air bubble, and that on the right side of the image represents the two cross-electrodes. The results show that: there exists five different migration forms of cavitation bubble in the experiment, which are: (a) the cavitation bubble merges with the air bubble forming a gas-containing cavitation bubble before the first migration cycle of the cavitation bubble; (b) the non gas-containing cavitation bubble migrations in two directions simultaneously and appears to be dumbbell shaped after it migrations; (c) the non gas-containing cavitation bubble migrations away from the air bubble; (d) the non gas-containing cavitation bubble migrations towards the air bubble; and (e) the non gas-containing cavitation bubble does not exhibit any clear directions of migration. The relative distance, relative size and relative roundness between the air bubble and the cavitation bubble are important parameters which determine the migration direction of the cavitation bubble. Obviously, the air bubble affects the migration direction of the cavitation bubble. The migration direction of the cavitation bubble is always related to the direction of the collapse micro-jet. Therefore, it can be inferred that the existence of air bubble will affect the collapse energy release of cavitation bubble, thus reducing the collapse intensity of cavitation bubble. This research will provide better theoretical support for the air entrainment to alleviate cavitation and can also provide guidance for aeration measures in hydraulic engineering.

Keywords: Cavitation bubble; Air bubble; Migration direction

Figure 1. Different situations of the migration direction of the cavitation bubble with the influence of the air bubble

Effect of Viscosity in Faraday Waves of Two-layer Liquids

Dongming Liu, Yang Wu

State Key Laboratory of Hydraulic Engineering Simulation and Safety,
Tianjin University, Tianjin 300072, China

Abstract

Faraday waves in a two-layered viscous liquids system are discussed in this paper. Three kinds of silicone oil with different viscosities are chosen as the viscous liquids. In order to put the viscous liquid in the upper layer, fresh water is chosen to be the heavier and inviscid liquid. In the same way, we choose ethanol to be the lighter and inviscid liquid when we would like the viscous liquid in the lower layer. After conducting a series of experiments and numerical simulations for the upper layer resonance, the results indicate that the model called NEWTANK can simulate the whole process and meet the accuracy requirement. The numerical model NEWTANK solves spatially averaged NSEs on a non-inertial coordinate for external excitations. By means of the two-step projection method and the Bi-CGSTAB technique we can obtain the numerical solution. Under the assistance of the numerical model, several conditions can be simulated which are difficult to implement in the experiment, especially when we need large amplitude of eternal excitation to excite resonance of the lower layer instability. Two factors are found which may lead to the different performance of viscous liquid in different layers. One is the ratio of liquid density difference and the other is viscous liquid in the upper layer may easier to change the system's natural frequency. Further analysis shows that the viscosity of the liquid caused the variation in its natural frequency. Through the numerical model, a series of free sloshing cases were simulated. According to the experience, by changing the length of the tank we can get the frequency of different modes. For example, the first natural frequency of half the size of the tank is equal to the second natural frequency of the original tank. The result shows that with the increase of the viscosity, the natural frequency of the system will reduce which also illustrates the essential difference between the viscous liquid and the inviscid liquid in sloshing. When we intend to get upper layer resonance, the viscosity of the upper layer is significant and the lower layer is more important when we want to excite the lower layer resonance.

Keywords: Faraday wave; Two-layered liquid system; Viscosity

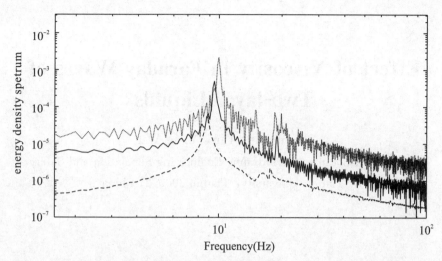

Figure 1. The dominant frequencies of different viscous silicon oil in the same free sloshing between viscosity kinematic coefficient $5.711\times10^{-5}\,m^2/s$ (red line), $1.098\times10^{-4}\,m^2/s$ (blue line) and $4.134\times10^{-4}\,m^2/s$ (black line)

Effects of Ambient Temperature on Air-water Flow Properties of Free Surface Flow

Xiaohui Zheng[1], Wei Liu[2], Wei Sang[3], Shanjun Liu[4], Hang Wang[5], Ruidi Bai[6]

[1,4,5,6] State Key Laboratory of Hydraulics and Mountain River Engineering, Sichuan University, Chengdu 610065, China

[2,3] Yongcheng Water Conservancy Bureau, Shangqiu 476610, China

Abstract

Self-aeration of open channel flow is an important factor for the enhancement of mass exchange of nitrogen, oxygen and carbon dioxide between the atmosphere and the river water. Sufficient dissolved oxygen (DO) in rivers and streams is vital to aquatic biota. Dams across the rivers affect the oxygen-water transfer processes in many ways. The storage and slow motions of the bulk water upstream the dam reduce the transfer rate and DO concentration. However, the mass exchange can be greatly enhanced during the water discharge through hydraulic structures (e. g. spillway) due to the intensive air-water interaction. Flow aeration is one common result of the air-water interplay, which often occurs in self-aerated free-surface flows, breaking waves, jet flows or hydraulic jumps. During the aeration, tremendous bubbles are entrained, and the complexity of the instantaneous two-phase flow behaviors has aroused the interests of researchers. To the author's knowledge, no information has been reported regarding the air-water flow properties within consideration of the change in ambient temperature. The ambient temperature affects the two-phase flow properties, in particular the water dynamic viscosity and surface tension. By definition, the Reynolds number does not only increase with increasing flow discharge, but also increases with reduced fluid kinematic viscosity. As a common knowledge, the flow characteristics of dynamic viscosity and surface tension decrease with increasing ambient temperature. For instance of different environmental conditions such as in the upper and bottom layers in the reservoir, cold and warm regions, or summer and winter seasons, the effects of ambient temperature can no longer be ignored, while the Reynolds number changes with different values of kinematic viscosity. In the present study, experiments were conducted in a concrete flume with a 1.0 m width, 0.40 m depth, 18 m length and a 21.8° sloping angle. The air entrainment characteristics of the self-aerated flows were systematically investigated and compared between two different flow temperatures of 0.3℃ and 24.8℃. Although the flow Reynolds number increased with decreasing water dynamic viscosity, experimental results indicated that no obvious differences were observed in term of interfacial velocity, air concentration, bubble frequency and bubble size spectrum. Since the Reynolds number has been widely involved in many empirical equations for the prediction of air-water flow development, a large deviation result may be obtained without considering the ambient temperature effects. Thus the ambient temperature should be carefully measured and recorded in model and prototype tests.

Keywords: Spillway chute; Self-aerated flow; Ambient temperature; Reynolds number; Air entrainment

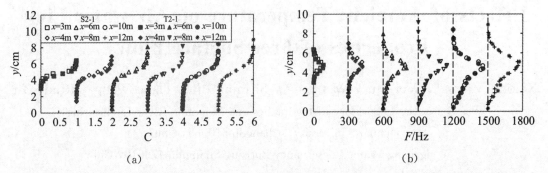

Figure 1. Development of air concentration (a) and bubble frequency (b) in the longitudinal direction and comparisons for different ambient temperatures

Note: Red color: $T_w = 0.3$ Celsius; Black color: $T_w = 24.8$ Celsius.

Effects of Hydrological Processes on Surface and Subsurface Nitrogen Losses from Purple Soil Slopes

Meixiang Xie[1], Pingcang Zhang[2]

[1]Department of rural water management, Nanjing Hydraulic Research Institute, Nanjing 210029, China

[2]Department of soil and water conservation, Changjiang River Scientific Research Institute, Wuhan 430010, China

Abstract

Relationships of hydrological processes via surface flow (SF) and subsurface flow (SSF) to nitrogen (N) losses from sloping farmlands have been rarely researched. In this study, laboratory experiments were conducted to investigate ammonia nitrogen (NH_4—N), nitrate nitrogen (NO_3—N) and total nitrogen (TN) losses from purple sloped soils due to SF, SSF and sediment (S). Effects of rainfalls and slope gradients on N losses were also studied. Three rainfall intensities (0.4±0.02, 1.0±0.04 and 1.8±0.11 mm·min^{-1}) and four slope gradients (5°, 10°, 15° and 20°) were designed in experiments. Larger SF discharges occurred with increasing rainfall intensities while SSF was prone to happen under low rainfall intensities. Although r^2 of regression results were low, both N loss concentrations and loads coincided positively with discharges except for a negative relation between N concentrations and SF discharges. In comparison, smaller SSF discharges produced substantial N loads with higher N concentrations especially for NO_3—N. NH_4—N, NO_3—N, and TN losses were dominated by S, SSF and SF, respectively. Furthermore, linear increases in loss loads with increasing discharges revealed that distributions of N loss loads were compatible with flow distributions in stormwater. 10° may be a critical slope gradient for SSF discharge and nutrient export.

Keywords: Nitrogen loss; Flow discharge; Purple soil; Slope gradient; Rainfall intensity

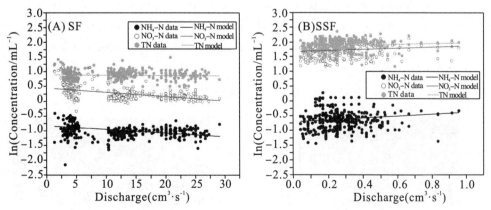

Figure 1. Relationships between the natural logarithm(ln) of concentration(mg·L^{-1}) of NH_4—N, NO_3—N and TN and flow discharge(cm^3·s^{-1}) shown as the observed data and the corresponding linear regression model for(A) SF and(B) SSF

Error Analysis of PIV Based Pressure Measurement Method

Zhongxiang Wang[1], Qigang Chen[2], Yanchong Duan[3], Qiang Zhong[4]

[1,2] School of Civil Engineering, Beijing Jiaotong University, Beijing 100083, China

[3] State Key Laboratory of Hydroscience and Engineering, Tsinghua University, Beijing 100084, China

[4] College of Water Resources and Civil Engineering, China Agricultural University, Beijing 100083, China

Abstract

The method of extracting pressure fields from velocity fields measured with particle image velocimetry(PIV) is an appropriate choice to non-intrusively measure pressure fields with high spatial and temporal resolutions. In this paper, the pressure-gradient field was firstly calculated from the velocity fields based on the Navier-Stokes equations by using an Eulerian approach. The pressure-gradient field was then integrated under given boundary conditions with an omni-directional virtual boundary scheme to obtain the pressure field. The accuracy, error sources, and major influencing factors on the accuracy of the proposed pressure-measurement method were analyzed with time-resolved direct numerical simulation(DNS) data and PIV measured velocity fields of open channel flows. A comparison between the reconstructed and simulated pressure fields from DNS showed that the mean bias error of this method is negligible, and the root mean square error is about 25%. Meanwhile, the process of calculating the pressure-gradient field was the major error source of the method. The temporal and spatial resolutions of the velocity fields significantly affect the calculation accuracy of the pressure-gradient field. The calculation error increases with the decrease of temporal resolution, and tends to be stable when the normalized time interval between consecutive velocity fields $\Delta t^+ = \Delta t u_*^2 / v$ (u_* is the friction velocity, v is the kinematic viscosity) is larger than 5. Meanwhile, the calculation error decreases first and then increases with the decrease of spatial resolution, and the minimum value is reached when the dimensionless velocity vector spacing $\Delta x^+ = \Delta x u_* / v$ is about 7. The results can provide theoretical basis and practical reference for measuring pressure field with PIV technology.

Keywords: Pressure field measurement; PIV; Velocity field; Error analysis

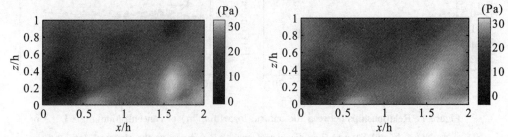

Figure 1. The pressure contours of DNS(left) and the pressure contours of measurement(right)

Evolvement Mechanism of Bed Morphology Changes Around a Finite Patch of Vegetation

Fujian Li, Chao Liu

State Key Laboratory of Hydraulics and Mountain River Engineering,
Sichuan University, Chengdu 610065, China

Abstract

Experiments were performed in channels with emergent model patches of vegetation distributed along the channel sidewall. The bed morphology and flow dynamics within and around the patches were analyzed. The bed morphology showed similarity under different continuous patch lengths and channel velocities. For the same patch pattern, the bed erosion near the leading edge and the side of the patch increased with channel velocity and a riverbed incised appeared on the opposite channel sidewall of the patch under a large channel velocity. The bed erosion was related to the interior flow adjustment length of the patch. The patch length and channel velocity had negligible influence on bed erosion. When vegetation was distributed as discontinuous patches, the distance between patches had high dependence on bed morphology. The erosion pattern of small intervals was similar with continuous patches. When the interval was large, each patch was eroded independently. Flow dynamics revealed the reason for the change of bed morphology. The diversion caused by patches made a decrease in velocity inside the patch and an increase in velocity outside the patch. For discontinuous patches, vortex formed between patches which caused reverse flows.

Keywords: Vegetation patch; Bed morphology; Flow dynamics

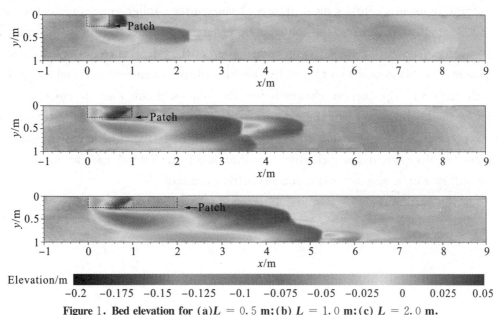

Figure 1. Bed elevation for (a) $L = 0.5$ m; (b) $L = 1.0$ m; (c) $L = 2.0$ m.

Flow was from left to right and dotted lines denote the patch

Experimental Studies of Boundary-controlledrip Current Systems

Sheng Yan[1,2], Zhili Zou[1]

[1] Dalian University of Technology, Dalian 116000, China
[2] Dalian Maritime University, Dalian 116000, China

Abstract

There are usually sandbars, groins or headlands on sandy beaches. Wave reflection can occur on the groins or headlands and lead to alongshore variability of wave heights. The longshore currents can be obstructed by these rigid lateral boundaries. So rip current systems often occur around these lateral boundaries and present a hazard to water users worldwide. The boundary-controlled rip current systems on a barred beach with the effects of intersecting waves and rip channels are studied by the laboratory experiment. In the reflection zone of the groin, the breaking of intersecting waves on the sand bar drives a limited number of rip current units, and the number of rip current units depends on the wave period. The wave heights, mean surface elevations and horizontal velocities of the wave field and flow field are measured. The measured data are used to analyze the mass conservation of each rip current unit and the lateral currents between rip current units are found to make a contribution to the rip current transport in the boundary-controlled rip current systems. The lateral currents between rip current units are generated due to a decreasing wave energy level from the groin. The influence of wave height on the flow characteristics is given by considering different incident wave heights. For different wave heights on barred beach without a rip current channel, the ratio of the sum of onshore Eulerian current transport, Stokes mass transport and surface roller transport to the rip current transport does not change much, so the contribution of lateral currents to rip current transports does not change much. It shows that the different contribution of the lateral currents of each unit to the rip current transports mainly depends on the relative position of the rip current unit and the groin. On the barred beach with the rip current channel, for the unit without the rip current channel the contribution of the lateral current to the rip current transport shows the opposite trend: the lateral current does not reduce the rip current transport, but it increases the rip current transport. For the units without the rip current channel, the contribution of the lateral current to the rip current transport is still similar to the case without a rip current channel.

Keywords: Rip current systems; Groins; Intersecting waves; Barred beach

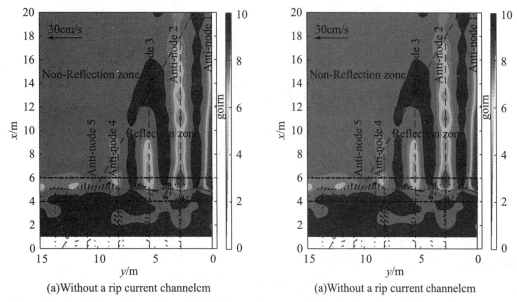

Figure 1. The measured mean flow fields and wave height contours inside and outside the reflection zone for regular waves with wave period $T=1.5$ s and incident wave height $H_i=4.16$ cm

Experimental Study on Cavitation Erosion Characteristics of Materials with Different Elasticities

Yanwei Zhai[1], Jianbo Li[2]

[1] China Three Gorges Corporation, Beijing 100000, China
[2] State Key Laboratory of Hydraulics and Mountain River Engineering,
Sichuan University, Chengdu 610065, China

Abstract

Using anti cavitation materials to reduce cavitation damage has always been a common method in hydraulic engineering, because anti cavitation materials can greatly improve the flow performance and service life of discharge structures and hydraulic machinery. The cavitation resistance of elastic materials has been lack of relevant research. In this paper, the cavitation cloud is generated by ultrasonic cavitation experiment system, and the cavitation effect on different elastic materials is observed to study the cavitation resistance of different elastic materials. The cavitation mass loss, surface morphology of different elastic materials and the cavitation resistance of different elastic materials has been studied. The main conclusions are as follows: (1) The cavitation effect of ultrasonic-induced cavitation bubble is similar to that of spark-induced cavitation bubble. Based on the research results of interaction mechanism between cavitation bubbles and different boundaries at meso level, the mechanism of air entrainment alleviating the cavitation erosion of different elastic materials at macro level can be revealed. (2) Through the ultrasonic cavitation experiment, it is found that the elastic material of elastic modulus of $1\sim11$ MPa shows no cavitation mass loss under the action of ultrasonic cavitation, and the macro surface morphology shows no obvious change. The micro surface only has some fine texture changes with the increase of experimental time, and there is no fracture or hole. The results show that the elastic material of this elastic modulus range exhibits good cavitation resistance. (3) With the increase of elastic modulus, the cavitation damage characteristics of different elastic materials are obviously different, which indicates that the correlation between cavitation resistance and elastic modulus decreases, which is related to the characteristics of material density and structure; The material with low density and sparse structure shows obvious cavitation mass loss and surface roughness increase. Besides, cavitation damage is more likely to occur at the surface defects of materials. The research results can provide relevant suggestions for the development and application of elastic cavitation resistant materials.

Keywords: Elastic material; Ultrasonic cavitation; Cavitation bubble; Cavitation erosion

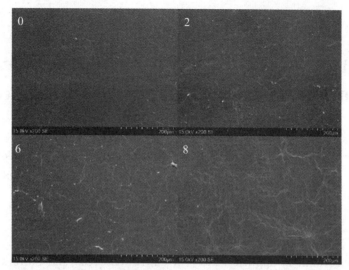

Figure 1. Microscopic surface morphology of elastic material with $E=9.7$ MPa, the number represents the cumulative test time

Flux Control Method for Uniform and Eco-friendly Fertilization Based on Differential Pressure Tank

Xinyu Hu, Xin Chen

Beijing Engineering Research Center of Safety and Energy Saving Technology for Water Supply Network System, China Agricultural University, Beijing 100083, China

Abstract

Differential pressure tank is applied widely in cultivated areas and suffers from nonuniformity fertilization, which negatively affects water and soil environment and the yield and quality of crops. This study proposes a flux control method to achieve uniform and eco-friendly fertilization on the basis of differential pressure tank. The flux control method divides the constant fertigation flux into two time-dependent fluxes; one flux flows into the tank to remove fertilizers uniformly, and another flux directly flows through the main pipe to mix with outlet fertilizers for an optimum fertilizer concentration in the fertigation system. The negative exponential decay of fertilizer concentration at the tank outlet is changed to a nearly linear decay by the flux control method. The method is applied directly for fertilizer in the case of large dispersion coefficient when the mixture of water and fertilizer is immediate and uniform. An extra parameter is required, and a simple predictor-corrector method is suggested for fertilizer in case of small dispersion coefficient. A source term of fertilizer dissolution is considered additionally in the flux control method when there is a certain degree of fertilizer dissolution. Two free parameters are introduced to describe the fertilizer dissolution, one parameter for the initial fertilizer dissolution rate and the other parameter for the decay rate of fertilizer dissolution. A grade control methodology is suggested for the convenient and easy application of flux control method. The methodology with four to six grades results in acceptable fertilizer concentration deviation around the optimal fertilization concentration. The applicability of flux control method is verified by the water-fertilizer numerical model. Overall, the traditional fertilization of differential pressure tank leads to more nitrogen and phosphorus losses due to excessive fertilization. The flux control method could offer positive to fertigation management and environmental protection by avoiding the excessive fertilizer losses found in traditional fertilization.

Keywords: Differential pressure tank; Flux control method; Fertilizer dissolution; Grade control methodology; Uniform and eco-friendly fertilization

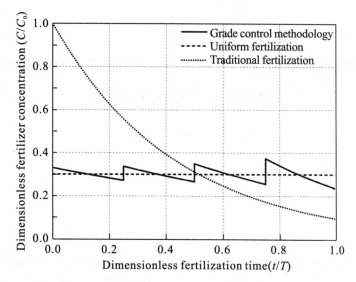

Figure 1. Comparison between uniform and eco-friendly fertilization and traditional fertilization

Hydrodynamics at the Channel Confluence with a Floodplain

Guanghui Yan, Saiyu Yuan

State Key Laboratory of Hydrology-Water Resources and Hydraulic Engineering,
Nanjing 210098, China

Abstract

The confluence of rivers is a key control node for the transportation of water, sand, pollutants and other substances in the river system. Near the confluence of the Yangtze River and the Poyang Lake, the Lake has a large area of floodplain which largely affects the confluence hydrodynamics. This kind of flow confluence with the effects of the floodplain has never considered before. The authors well designed a channel confluence flume which had a floodplain occupying a half of the channel width of the tributary. The effects of the floodplain on the confluent flows were investigated by comparing with the case without floodplain. Three-component velocities at the confluence were measured by acoustic Doppler velocimetry and turbulent flow characteristics such as mean velocity field, turbulent kinetic energy and Reynolds shear stress were analyzed. Hydrodynamic features such as a stronger helical flow, fluid upwelling, the distortion of the mixing layer with the presence of the floodplain were observed. An additional high-speed zone occurred, which should be attributed to the penetration of the main channel stream into the zone just downstream of the floodplain of the tributary. Two shear layers appeared at the confluence, which formed by the flow above the floodplain and left side of the floodplain sheared the mainstream. A stronger secondary flow movement occurred downstream of the confluence and continued for a long distance under the influence of the floodplain. The separation zone became wider which should be related to the higher flow velocity in the tributary with the floodplain; while its length was similar with that case with floodplain since the length of the separation zone shall be mainly related to the discharge ratio. These results are beneficial to understanding the relationship between the Yangtze River and the Poyang Lake as well as improvement of the ecology of the Yangtze River Basin.

Keywords: Confluence; floodplain; helical flow; separation zone

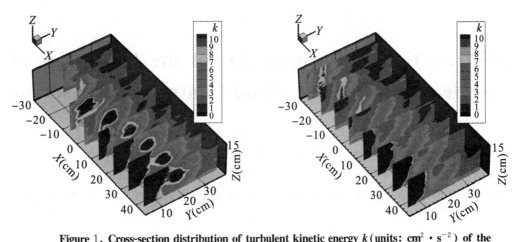

Figure 1. Cross-section distribution of turbulent kinetic energy k (units: $cm^2 \cdot s^{-2}$) of the main channel under two cases. The first section is located near the upstream junction corner, (a) represents the floodplain case; and (b) represents the comparison case, respectively

Identification and Application of the Thresholds of Pre-release Indexes for Flood Control of a Reservoir

Sizhong He [1], Cao Huang [2], Xu Deng [3], Chuqi Zhou [4], Wan Jiang [5]

[1,2,3,4,5] Changsha University of Science & Technology, Hunan 410022, China

[2] Changsha University of Science & Technology, Hunan Key Laboratory of Water Environment Treatment and Water Ecological Restoration in Dongting Lake, Hunan 410022, China

Abstract

The rolling updating pre-release operation of a reservoir coupled with flood forecasting and real-time demands is an important technology for scientifically coping with excessive floods. In a view of the problems of the reservoir pre-release operation, i.e. the various judgment criteria and the unclear guidance mechanism, the mathematical expressions of the pre-release indexes of a reservoir (the pre-release start threshold, the pre-release termination threshold and the pre-release depth) are interpreted according to the exploring of the start-stop mechanism of reservoir pre-release operation, and the thresholds identification method of the pre-release indexes is proposed by using the "simulation-optimization-simulation" technology. By taking the Shuifumiao Reservoir as an example, an hourly reservoir operation model for flood control is constructed based on the proposed pre-release indexes. The results show that: (1) the proposed model derived the smaller flood crests and the lower maximum water level of the reservoir than that derived by the flood routing calculation; (2) the overtime of the reservoir discharge exceeding the safety limited flow of the downstream, and the overtime of the water level exceeding the flood limited water level of the reservoir are shorter 17.4% and 16.2% respectively. By effectively coupling with flood forecasting, the result could provide a scientific basis for the reasonable "storage and discharge" of a reservoir to deal with the excessive floods.

Keywords: Flood control; Pre-release operation; Pre- release Index; Start-stop mechanism; Shuifumiao reservoir

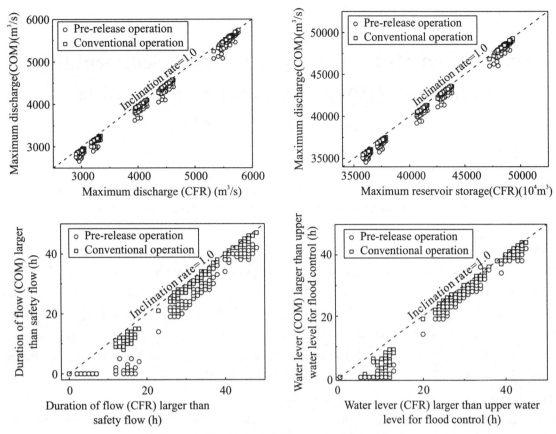

Figure 1. Results in indifferent operational methods

Note: COM(Fig. 1): Caculated by Other Methods; CFR(Fig. 1): Caculated by Flood Routing.

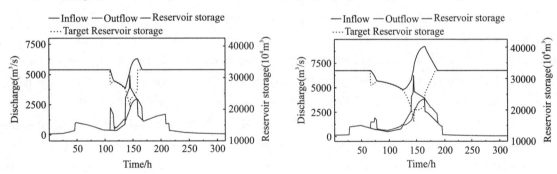

Figure 2. The process of pre-release operation

Imbalanced Stoichiometric Reservoir Sedimentation Regulates Methane Accumulation in China's Three Gorges Reservoir

Zhe Li, Lunhui Lu

CAS Key Lab of Reservoir Environment, Chongqing Institute of Green and Intelligent Technology, Chinese Academy of Sciences, Chongqing 400714, China

Abstract

The river-reservoir continuum drives a separation of sedimentation along the longitudinal hydrodynamic gradients, potentially causing imbalanced stoichiometric sedimentation. However, there are still uncertainties regarding the contribution of imbalanced stoichiometric sedimentation to methane emissions along this continuum. A two-year field survey and in situ experiments were conducted in China's Three Gorges Reservoir (TGR), a river-valley dammed reservoir. Sediments were trapped and collected to analyze for particulate carbon (C), nitrogen (N) and phosphorus (P) concentrations of different sizes in the summer flooded and winter dry seasons along the main stem. Large amount of sediments were deposited in the mid-part of the TGR, particularly in the summer flood season. Hydrodynamic gradients structured imbalanced stoichiometric sedimentation patterns. Averaged particulate C : N : P proportions in the sediment layer along the upper, mid and lower part of the TGR in the first 30 years of reservoir operation were estimated to be 68 : 7 : 1, 97 : 8 : 1, 151 : 13 : 1 respectively. The mid-part of the TGR, being the "control point" of methane accumulation based on seasons, exhibited the shift in stoichiometry and the stable isotope signatures indicating that there was likely a shift in sources of POM. P, specifically in smaller size particles (<10 μm), seemed to be the potential key driver that regulates particulate C : N : P in the sediment of the TGR. Its sedimentation, primarily in the mid-part of the TGR, contributed to the significant decreases in particulate C/P ratios, and could possibly regulate long-term CH_4 accumulation in the reservoir.

Keywords: Sedimentation rate; River-reservoir continuum; Organic carbon; Size of sediments; Greenhouse gas

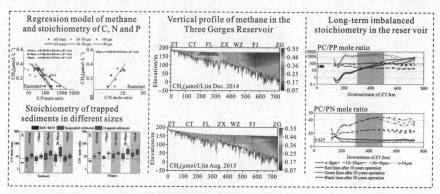

Figure 1. Vertical profiles of methane in the Three Gorges Reservoir is regulated by the imbalanced stoichiometric sedimentation of particulate C, N and P

In-situ Ecological Remediation of Urban Black-odor Water Bodies Based on Biological Carriers

Wenyao Jia, Xiaoyuan Qi, Kai He, Bingjun Liu

Sun Yat-sen University, School of Civil Engineering, Zhuhai 519082, China

Abstract

On the basis of investigating the black and odorous water body in-situ ecological restoration engineering technology, this article attempts to build the reactor system with basalt fiber(BF, Basalt Fiber) as the biological carrier and discusses the feasibility of biological contact oxidation technology with BF in the in-situ restoration of urban black and odorous water bodies. In this experiment, two closed reactors were operated. The test process is divided into the pre-hanging bio-membrane stage and stably running stage. In the rehanging bio-membrane stage, aeration devices are installed on the partition and under the partition to investigate the effect of the aeration position on the removal of pollutants in the water body and the BF carrier membrane. The changes in the chemical oxygen demand(COD), ammonia nitrogen(NH_4^+—N), and total phosphorus(TP) concentrations in the overlying water of the two reactors are monitored continuously. The growth of biofilms is also observed. The result shows that different aeration positions have no significant effect on the removal of pollutants in the overlying water. The removal rate of COD and TP in the two reactors can reach more than 85%. The removal rate of NH_4^+—N does not exceed 40%, and the nitrification is not good. Aeration under the baffle helps to speed up the filming of BF carrier. After replacing the bottom sludge and the overlying water body, the two sets of reactors enter the stably running stage, and the aeration devices in the two sets of reactors are placed on the baffle. The removal effect of COD and TP has no significant difference compared with the rehanging bio-membrane stage, and the removal effect of NH_4^+—N is not good. Add smoldered nitrifying bacteria. After 11 days of operation, the NH_4^+—N in the two reactors can reach more than 50%, and the nitrification is improved. In summary, the biological contact oxidation technology based on basalt fiber as the biological carriers is suitable for the in-situ treatment of the black and odorous water body and has good prospects for future application.

Keywords: Basalt fiber; Contact oxidation process; In-situ Aeration; Nitrifying bacteria

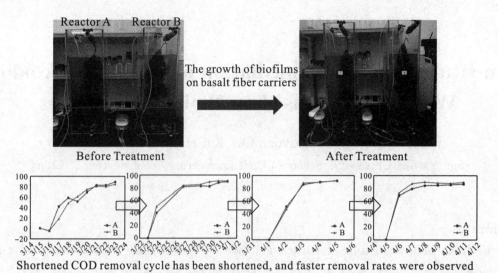

Figure 1. Schematic diagram of the biological contact oxidation technology based on basalt fiber

ISPH Simulation of Wave Breaking with k-ε Turbulence Model

Dong Wang[1], Philip L.-F. Liu[2]

[1] College of Environmental Sciences and Engineering, Dalian Maritime University, Dalian 116026, China

[1,2] Department of Civil & Environmental Engineering, National University of Singapore, Singapore

Abstract

Generation and transport of turbulence associated by flow separation and wave breaking are dynamic, complex and multi-scale processes. An Incompressible Smoothed Particle Hydrodynamics (ISPH) method solving the 2D RANS (Reynolds Averaged Navier-Stokes) equations with the k-ε turbulence closure is constructed. In the present model, the concept of "massless ISPH" utilizing the definition of "particle density" (number of computational particles within unit volume) is stressed. The detailed discretization of the equations based on the variable "particle density" is provided for completeness. The skills of this numerical model are tested by applying to two laboratory experiments: (1) A non-breaking solitary wave propagating over a bottom-mounted barrier and (2) a solitary wave breaking on a 1 on 50 slope. In the former case flow separation occurs behind the barrier as the wave crest passes by and a vortex is generated, which later interacts with free surface causing breaking. In the latter case wave breaking and bottom friction both generate significant turbulence. For both cases, the effects of initial seeding of turbulent kinetic energy, required in the k-ε model, are studied and it is concluded that initial values of $O(10^{-10})$ to $O(10^{-8})$ m^2/s^2 should be used. An adaptive wall boundary condition for k-ε turbulence model is employed to avoid the unrealistic production of turbulence near the wall boundary. The numerical dissipation of the present ISPH method has been discussed in this paper. The sensitivity study shows that the numerical dissipation will gradually reduce with the growing of the particle resolution until it is fine enough to capture the physical viscous effect while finally overcomes the numerical dissipation problem, causing a continuous balance of the two numerical issues. The numerical results, in terms of free surface profile, mean velocity field, vorticity field, turbulent kinetic energy and turbulent shear stress, are compared with experimental data. Very reasonable agreement is observed. This paper presents the first comprehensively validated 2D ISPH model with the k-ε turbulence closure, which can be applied to transient free surface wave problems.

Keywords: 2D ISPH; k-ε model; Wave breaking; Turbulent kinetic energy

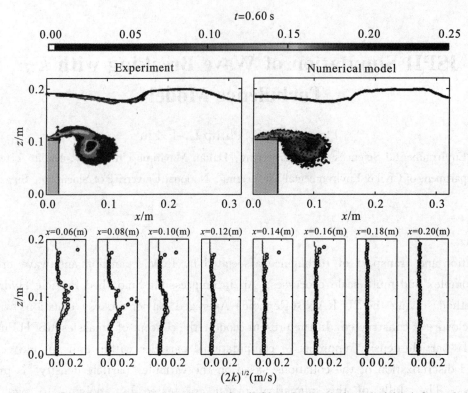

Figure 1. Comparisons between experimental data(left) and numerical results(right) for turbulence intensity(m/s) (top panels) and the corresponding vertical cross-sections (lower panels). In the lower panels, solid lines and circles represent the numerical and experimental results, respectively

Mathematical Description of Formation and Emergency Response Process of Barrier Lake Based on Calculus Theory

Jingwen Wang[1], Guangming Tan[1], Caiwen Shu[1], Shasha Han[2,3]

[1] State Key Laboratory of Water Resources and Hydropower Engineering Science,
Wuhan University, Hubei 430072, China

[2] Key Laboratory of Lower Yellow River Channel and Estuary Regulation, Ministry of Water Resources, Yellow River Institute of Hydraulic Research, Zhengzhou 450003, China

[3] State Key Laboratory of Hydroscience and Engineering,
Tsinghua University, Beijing 100084, China

Abstract

Barrier lake is a secondary disaster caused by earthquake and rainstorms as the landslides and debris flow accumulation associated with these two common natural disasters would dam and cut off the downstream of a river. The wholesome and mathematical analysis about the process of barrier lake from initial formation to artificial breach are laking in current research. In addition, there are few studies on the inplementing mechanisms and evaluating benefits of artificial measures under different dam break modes. This paper aimed to make a deeper understanding of the formation and emergency response of barrier lake from a mathematical side using the on-the-spot investigation and calculus theory, and then discussed the inplementing mechanisms and evaluating benefits of two engineering measures. Results showed that the formation of barrier lake leads to a sudden variation in the flow change rate(normal to infinite). However, after implementing a series of emergency measures, it returned to normal. This whole rescue process was regarded as the accumulation of disposal effects from the perspective of calculus. The volume change of mainstreams could be expressed by a differential equation of the lake surface area and water level variation. In addition, the corresponding mathematical description of flow discharges was also given when the engineering measures had been adopted. For example, after the excavation of artificial diversion channels, a dam breach process was divided into the breach development period and breach stability period, then their mathematical expressions of flow discharge were given respectively according to some practical phenomena. Furthermore, as another effective emergency measure, when engineering blasting was chosen to widen and deepen the cross-section, the velocity and stress distribution on diversion channels were uneven. On this condition, the flow discharge of certain sections and the stress on diversion channels were also been described mathematically according to the calculus theory. Overall, this paper tried to mathematicize and theorize the existing emergency measures, which helps to better understand their implementation principles and application requirements, therefore takes effective strategies for future emergency response. At the same time, it also provides a theoretical basis for establishing an emergency response calculation model which responds the dam breach immediately and accurately in the future.

Keywords: Barrier lake; Emergency response; Calculus theory; Diversion channel; Engineering blasting

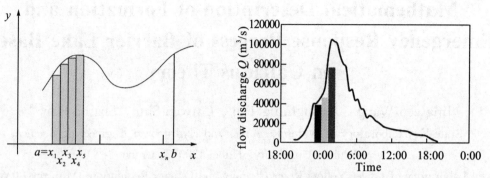

Figure 1. The differential theory in the barrier lake research

Note: (a) The area segmentation under a f curve; (b) The area segmentation under a mainstream flow discharge curve of Yigong barrier lake.

Microplastics in Freshwater River Sediments in Zhoushan, China

Yichen Sun, Cao Lu, Di Wu, Bin Zhou, Qiang Li

Department of Port and Transportation Engineering, Zhejiang Ocean University, Zhoushan 316022, China

Abstract

Microplastic(MPs) contamination is an emerging environmental problem of global concern. The role of rivers as a major transport pathway for all sizes of plastic debris into the ocean is widely recognized. However, only a few studies have directly assessed occurrence and fate of MPs in rivers. In order to better understand the pollution of MPs on urban rivers, the investigation of the distribution and abundance of MPs in the sediments from the river in Dinghai District of Zhoushan City were made. MPs were separated from the sediments by density flotation and classified according to their shapes and sizes under a stereomicroscope. The total of MPs were identified by Fourier-transform infrared(FTIR) spectrometry. The mean concentration of MPs at the 7 sites was (7.0 ± 0.28) N/kg. It was found that the concentration of MPs is much lower compared to other river systems, revealing the population and industrial have significant impacts for the abundance of MPs. Four types of MPs were mainly identified in the studied sediments, including fragments, particles, films and fibers. In addition, the dominant size of MPs was smaller than 5 mm which accounted for 92.95%. Polypropylene(PP), polyethylene(PE) and polyvinyl chloride(PVC) were the dominant polymer-types of MPs analyzed. This study can provide a valuable reference to further understanding the MPs pollution in inland freshwater ecosystem.

Key words: Microplastic pollution; River sediments; Abundance; Particle sizes

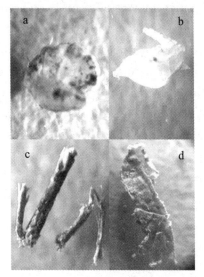

Figure 1. Different shapes of microplastic samples collected from the sediments

Note: a. Granules; b. Films; c. Fiber; d. Fragments.

Migration of Erosion Centers Along the Yichang-Chenglingji Reach Downstream of the Three Gorges Dam

Hualin Wang, Shan Zheng, Guangming Tan

State Key Laboratory of Water Resources and Hydropower Engineering Science,
Wuhan University, Wuhan 430072, China

Abstract

The operation of the Three Gorges Dam has caused profound morphology adjustment downstream of channel reaches since 2002. The location of the erosion centers (zones with the greatest erosion rates) and its migration with time and in space has not been thoroughly understood. In this study, based on the data of water discharge, sediment load, bed material size and cross-sectional surveys during 2002—2018, we studied the morphological adjustments of a channel reach between Yichang and Chenglingji with a river length of ~408 km downstream of the TGD. The channel reaches were divided into 32 sub-reaches according to the planform. The vertical, lateral and bankfull channel changes were calculated. Based on the calculated channel erosion/deposition volume, the erosion/deposition rates at the sub-reaches were classified into different levels indicating the relative degree of erosion/deposition processes. A clustering analysis method in spatial analysis was introduced to study the clustering pattern of the erosion centers. The results indicate that the channel experienced strong erosion during the initial operation of the TGD from 2003 to 2007, erosion slowed after the reservoir started to operate with the pool level of 175 m from 2008 to 2012. Spatial clustering analysis showed that, the erosion centers migrated downstream with decaying rates from 2003 to ~2012. After ~2012, cascade reservoirs started to operate upstream of the TGD, and erosion enhanced again with the increasing rate of downstream migration of erosion centers. The migration rate slowed again a few years later. The average migration rate of the erosion centers was ~7.5 km/yr. during 2003—2018. The erosion center was about ~140km downstream of the dam in 2018. After the erosion centers migrated to the downstream, the erosion rate at the channel may decrease or even aggradation may occur. Inter-annual transition between degradation and aggradation was common at the downstream channel reaches, where the erosion center has not yet reached. However, the degradation/aggradation rates generally decayed as time elapsed, implying the adjustments trend towards quasi-equilibrium. (Use "Arial" font, size 10pt for the abstract texts)

Keywords: Three Gorges Dam; Yichang-Chenglingji Reach; Morphological evolution; Erosion centers; Spatial clustering

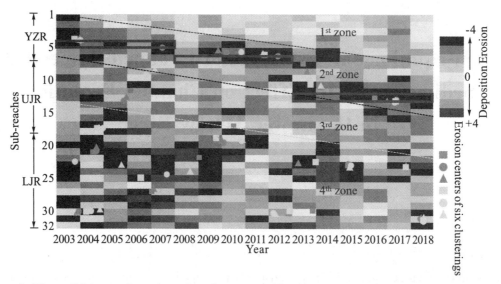

Figure 1. The spatial-temporal metrics and erosion centers of bankfull area. The blue and red represent erosion and deposition, respectively. The darker the color, the more intense the corresponding variation

Non-equilibrium Transport of Grouped Suspended Sediment and Its Effect on Channel Evolution in the Lower Yellow River

Yifei Cheng, Junqiang Xia, Meirong Zhou, Shanshan Deng

Wuhan University, Wuhan 430072, China

Abstract

The understanding of non-equilibrium transport of grouped sediment is the basis of modelling and predicting channel evolution, and the latter is important in river training decision-making. Since the operation of Xiaolangdi reservoir in 1999, the transport of grouped suspended sediment and the channel geometry in the lower reaches have greatly changed, and moreover discrepancies have been observed in channel evolution of different reaches. Based on the measured hydrological data from 1999 to 2018, the transport characteristics of different grain size fractions during the flood seasons has been analyzed, finding that the sediment discharge of each grain size fraction has dramatically decreased after the operation of the reservoir, among which the sediment discharge of medium grain size decreased most by 85%. Each grain size fraction shows different recovery capacities along the lower reach. The sediment discharge of fine grain size seldom increased along the channel, while that of medium and coarse grain size could increase to stations of Lijin and Aishan respectively. The conditions of riverbed sediment supply of different reaches have been investigated. The riverbed in the lower reaches has become more coarsely grained with the continuous erosion, which results in the inadequacy of fine and medium grain size supply in the braided and transitional reaches. While the mildly coarsened meandering reach could still supply the medium grain size fraction to the suspended load. Variations in channel dimensions of each reach has been calculated in the same period, indicating that undercutting generally occurs in the lower reaches accompanied with the widening in the braided reach. Resultantly, the geomorphic coefficient of each reach has considerably decreased, with the maximum decrease of 53% in the braided reach. To quantify the combined effects of non-equilibrium transport characteristics of grouped sediment and the channel boundary condition on the channel evolution, an empirical function has been established between the cumulative evolution volume(V) of each reach and the corresponding incoming sediment coefficient of different grain size fractions and the post-flood geomorphic coefficient in the last year. Results show that when the incoming sediment coefficient is small, the channel erosion induced by unit change of sediment coefficient could be more efficient, which is the same case for the response to the geomorphic coefficient. The determination coefficients of each function are greater than 0.85, suggesting channel evolution in the lower reaches could be relatively accurately modelled.

Keywords: Grouped suspended sediment; Non-equilibrium transport; Riverbed sediment supply; Channel evolution; Lower Yellow River

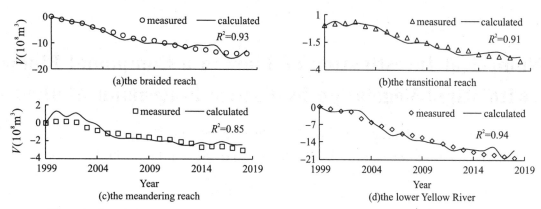

Figure 1. Comparison between calculated and measured channel evolution in the lower reaches

Numerical Investigation of Flow in a Compound Flume with Rigid Vegetation by Lattice Boltzmann Method

Hefang Jing, Weiwei Zhao, Weihong Wang

School of Civil Engineering, North Minzu University, Yinchuan 750021, China

Abstract

Vegetation usually grows on the floodplains of natural rivers. The vegetation on the floodplain will not only affect the water flow structure on the floodplain, but also have a great influence on the flow structure in the main channel. As a result, the vegetation will affect the sediment transport, bed deformation, water environment change and navigation. Therefore, it is of great practical value to study the problem of vegetated water flow in compound channels.

In this study, rigid vegetation, which is replaced by glass rods, is arranged in different conditions on the floodplain in the compound flume with a two-stage floodplain at the right bank. Laser Doppler velocimeter(LDV) is employed to measure the flow velocity and turbulence intensity under a series cases. In order to investigate the flow characteristics in the compound flume with vegetation, a three-dimensional lattice Boltzmann model (D3Q27) is developed. In the model, multi-time ralaxation (MTR) technology is adopted and the drag force of vegetation is considered. Comparisons between the simulated and measured data indicate that the D3Q27 model is able to simulate the flow in a compound open channel, and the arrangements of vegetation in the multi-stage floodplain led to different flow structures both in the main channel and the floodplain.

Table 1 presents the twelve cases in which water depth and vegetation arrangements are given. Fig. 1 shows the sketch of the vegetation arrangements for parallel and staggered. In Fig. 2, when the vegetation is staggered arranged, the transverse distribution of the simulated flow velocity on a typical horizontal section, in which water depths are 18 and 30, respectively. Fig. 3 shows the velocity distribution along transverse direction under the condition of Case D-4-3 on horizontal sections $Z=$ 17, 30 and 37, respectively.

Keywords: Lattice Boltzmann method; D3Q27 model; Compound flume; Rigid vegetation

Table 1. The water depth and vegetation set up of varied simulated cases

Cases	Water depth	Vegetation arrangement	Cases	Water depth	Vegetation arrangement
D-1-1	18	Without vegetation	D-3-1	30	Without vegetation
D-1-2	18	parallel	D-3-2	30	parallel
D-1-3	18	Staggered	D-3-3	30	Staggered
D-2-1	25	Without vegetation	D-4-1	40	Without vegetation
D-2-2	25	parallel	D-4-2	40	parallel
D-2-3	25	Staggered	D-4-3	40	Staggered

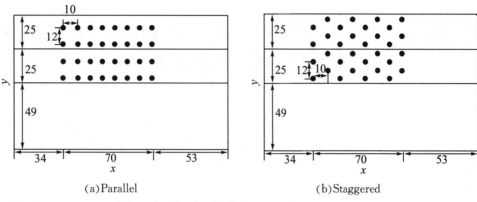

(a) Parallel   (b) Staggered

Figure 1. The sketch of the vegetation arrangement

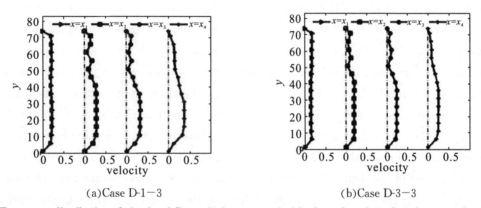

(a) Case D-1-3   (b) Case D-3-3

Figure 2. Transverse distribution of simulated flow velocity on a typical horizontal section when the vegetation is staggered

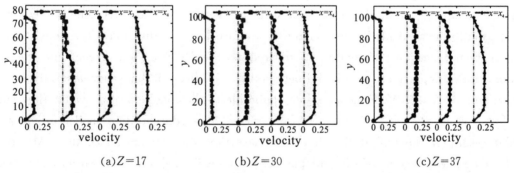

(a) $Z=17$ (b) $Z=30$ (c) $Z=37$

Figure 3. Velocity distribution along transverse direction under the condition of Case D-4-3

Numerical Investigation of Two-phase Flow at a Chute Offset Aerator

Jijian Lian, Panhong Ren, Dongming Liu

State Key Laboratory of Hydraulic Engineering Simulation and Safety, School of Civil Engineering, Tianjin University, Tianjin 300072, China

Abstract

Chute aerators are usually adopted to prevent spillways from cavitation damage by entraining air into the water flow. Therefore, it is of great significance to study the aeration characteristics of aerators in spillways, especially for high dams. However, due to the complexities of water-air interactions caused by the high flow velocity, the theoretical model has not yet established. Hydraulic model tests play an indispensable role in the aerated flow, but the model tests have the scale effect that is difficult to ignore. It is a great challenge to study the air-water two-phase flow, especially for the high flow velocity and it is necessary to develop the numerical simulation to study the complex aeration characteristics. In this study, a set of numerical models consisting of the mixed model, the concentration model and the turbulence model is conducted. Then, this numerical model is verified by a chute offset aerator model test. Through the comparison of air concentration between numerical simulation and experimental data, the applicability of the model has been preliminarily proved. Subsequently, the length between the jet take-off and the pressure impingement point and the characteristic length of the black-core water are further verified. Good agreement between numerical simulation and formula calculation further proves the feasibility of this numerical model to study the aeration characteristics of the aerator. Compared to the VOF model, except for the effect of convection, the entrained air is also influenced by the effect of buoyancy and diffusion in the concentration model. The VOF model is just applicable to the condition of air-water flow with clear free surface. When the air and water is fully mixed, the concentration model provides a more reasonable result for the typical air-water two-phase flow downstream an aerator. Moreover, the study of the bubble advection velocity in high-speed flows is limited to the laboratory instrumentation. Thus, the bubble advection velocity is analysed qualitatively in this study. The numerical result of this model has relatively high accuracy, which can provide reference for the study of aeration problem and be helpful for the design of the aerator.

Keywords: Numerical simulation; The mixed model; Air concentration; The pressure impingement point; The jet black-water core

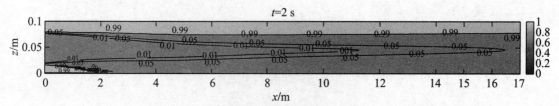

Figure 1. Overview of the air concentration distribution downstream the chute offset aerator

Numerical Investigation on Flow Around Two Side-by-side Patches of Emergent Vegetation

Jian Wang, Jingxin Zhang

School of Naval Architecture, Ocean and Civil Engineering, Shanghai Jiao Tong University, Shanghai 200240, China

Abstract

Aquatic vegetation plays a significant role in water bodies due to its hydrodynamic, ecological and biomechanical processes and the interplay between them. Knowledge of flow-vegetation interaction will promote a better understanding of these processes and assist the protection of fluvial ecosystems and river management. Flow through and around an isolated patch of vegetation has been well documented both experimentally and numerically. However, the coexistence of two or more vegetation patches is ubiquitous in natural rivers, few studies have covered the hydrodynamics of flow with more than one patch. Thus, the flow around two side-by-side circular patches of emergent vegetation was investigated numerically in this study, with the vegetation being modeled as rigid cylinders. The flow characteristics of wake influenced by the solid volume fraction of the patches and the gap distance between the patches were extensively analyzed. Compared with the wake behind an isolated patch, the wake interference between two side-by-side patches was more complex. For the patches with low solid volume fraction, no patch-scale vortex street occurred in the wake behind two patches, the development of the separated shear layers was suppressed at a small gap distance. For the patches with high solid volume fraction, the 'biased flow' pattern wake and 'parallel vortex street' pattern wake occurred in the flow with two side-by-side cylinders were also observed in the wake behind two patches. Overall, the flow regimes of the wake behind two side-by-side patches were divided into two regimes, i.e. the near wake interference regime and the far wake interference regime. In the near wake interference regime, the time-averaged streamwise velocity profile in the near wake zone of two patches was different from that of an isolated patch, while in the far wake interference regime, the time-averaged streamwise velocity profile in the near wake zone of two patches was similar to that of an isolated patch. The wake interference patterns regarding different solid volume fractions and gap distances were summarized, which can be applied to the analyses of flow with more than two patches of vegetation.

Keywords: Emergent vegetation; Side-By-Side patches; Wake interference; Near wake; Far wake

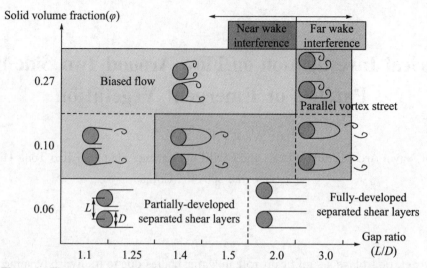

Figure 1. Flow regimes of the wake behind two side-by-side circular patches of cylinders

Numerical Simulation of Fluid-structure Interactions of Bridge Piers with Local Scours

Jing Chen[1], Yang Xiao[2], Wei Xu[3], Chentao Li[4], Zixuan Wang[5]

[1,2,4,5] College of Water Conservancy and Hydropower Engineering, Hohai University, Nanjing 210098, China

[2] Institute of Water Science and Technology, Hohai University, Nanjing 210098, China

[3] Department of Engineering Mechanics, Hohai University, Nanjing 210098, China

Abstract

The motion of the vibrating bridge piers and interactions between vortex that sheds from the lower shear layer of the piers and sediment bed are the primary mechanisms that cause the formation of the local scours around bridge piers, which can jeopardize stability of the piers and may induce entire failures of the bridges. Therefore, there is a great need to investigate the effects of local scours on flow fields around piers and dynamic responses of piers, on which basis safety of the bridges can be comprehensively assessed. With this concern, this study numerically simulates fluid-structure interactions between currents and piers with local scours, whereby the coupled fluid and structural fields of the current-pier-scour system can be achieved. In particular, distributions of flow velocity, shear stress, vorticity and turbulence kinetic energy of the flow fields around piers can be characterized with different parameters of local scours. Besides, distributions of dynamic responses such as stresses, strains, displacements, and accelerations can be obtained to estimate real ultimate bearing capacities of bridge piers with local scours. By means of inverse analysis, the limit parameters of local scours can be determined for safety assessments and predictions of the bridges.

Keywords: Bridge pier; Local scour; Fluid-structure interaction; Flow field; Dynamic response

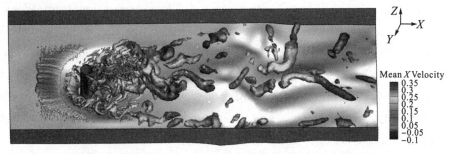

Figure 1. Visualization of the main horseshoe vortices and of the vortical content of the near wake in the mean flow using a Q iso-surface($Q=0.5$) colored with the average flow direction velocity

Numerical Simulation of Longdaohe River Water Quality Improvement

Suiliang Huang

State Key Laboratory of Subtropical Building Science, South China University of Technology, Guangzhou 510640, China

Abstract

According to the construction design, Longdaohe River water quality improvement demonstration project, a tributary of Wenyuhe River and located in the suburban area of Beijing where an international exposition centre is under construction, mainly includes construction of secondary river channel, wetland river channel in the upper part of the River and in the lower part construction of meandering secondary river channel with deep pools and shoals and ladder pools, vegetation buffer zone etc. (Wang et al., 2010; Liu et al., 2012). In the middle of the River, there connects Luomaxi lakes, where dredging of riverbed is considered, as shown in Fig. 1. The upstream and downstream of Longdaohe River are prismatic open channels with varied cross-sectional shapes.

After collecting its riverbed topography, information of operation modes, hydrology, water qualities and decontamination capacity data of Longdaohe demonstrative project of water quality improvement, a two-dimensional numerical model (Huang, 2007; Huang, 2009; Xiao et al., 2018) was set up on the basis of generating fine grids to simulate spacial-temporal variations of water qualities with different disposal levels (different degradation coefficients) and different diversion flow rates in 330 days. The water quality indexes are represented by chemical oxygen demand (COD), ammonia nitrogen (NH_3—N) and total phosphorus (TP). Taking TP as an example, the main conclusions drawn from the simulation are as follow:

When the incoming discharge is $0.5 \ m^3 \cdot s^{-1}$ and TP concentration from Wenyuhe River is $2.06 \ mg \cdot L^{-1}$ and the comprehensive degradation coefficient is $5 \times 10^{-6} \ s^{-1}$, the calculated average outlet concentration is reduced to $0.30 \ mg \cdot L^{-1}$ and the reduction rate is up to 85.3%, as shown in Figs. 2 and 3. However, when the comprehensive degradation coefficient is $5 \times 10^{-7} \ s^{-1}$ and $5 \times 10^{-8} \ s^{-1}$, as shown in Fig. 1, TP concentration at the outlet is reduced much less, with an average reduction rate of 12.1% and 2.2%, respectively. The corresponding outlet concentration is $1.81 \ mg \cdot L^{-1}$ and $1.99 \ mg \cdot L^{-1}$, respectively.

The simulation also shows that most of the water surface of Longdaohe River demonstration project, especially most water is located in the lower part of Longdaohe River, and degradation of COD, NH_3—N and TP in the river is mainly completed in this part.

When the incoming TP concentration is $2.06 \ mg \cdot L^{-1}$ and the water diversion flow rate is $0.25 \ m^3 \cdot s^{-1}$, TP concentration at outlet is $0.14 \ mg \cdot L^{-1}$ and the removal rate is 93.2%. While the flow rate is increased to $0.5 \ m^3 \cdot s^{-1}$, the outlet concentration is $0.30 \ mg \cdot L^{-1}$, and removal rate is 82.2%.

Figure 1. Locations of Wenyuhe River and Longdaohe River (Length of Longdaohe River is 7871 m)

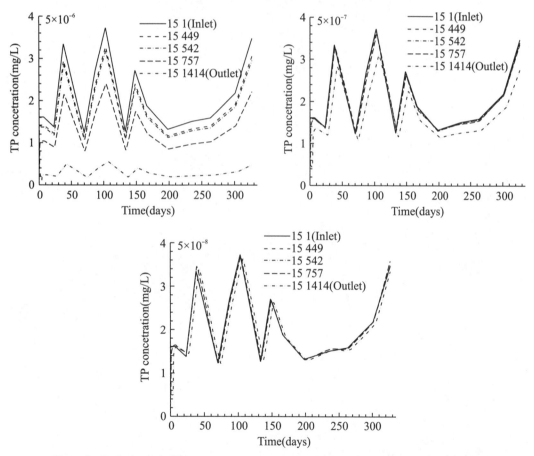

Figure 2. Variations of TP concentrations with time at different cross-sections for different comprehensive degradation coefficients of $5\times10^{-6}\text{s}^{-1}$, $5\times10^{-7}\text{s}^{-1}$, $5\times10^{-8}\text{s}^{-1}$

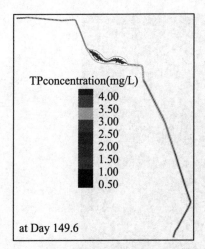 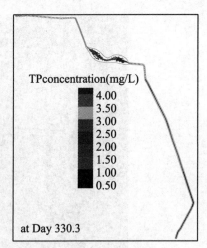

Figure 3. Distributions of TP concentrations in Longdaohe River

Possible Impact of Climate Change on Tropical Cyclone Activities Based on a New Trajectory Model

Kaiyue Shan, Xiping Yu

State Key Laboratory of Hydroscience and Engineering, Department of Hydraulic Engineering, Tsinghua University, Beijing 100084, China

Abstract

The establishment of a tropical cyclone(TC) trajectory model that can represent the basic physics and is practically applicable considering both accuracy and computational cost is essential to the climatological studies of global TC activities. In this study, we propose a simple deterministic model by developing semi-empirical formula for the beta drift under known conditions of the environmental steering flow. The horizontal shear in the environmental flow was assumed to be the control variable in a climatologically effective expression for the beta drift. In order to verify the proposed model, all historical TC tracks in the western North Pacific and the North Atlantic ocean basins during the period 1979—2018 are simulated and statistically compared with the relevant results derived from observed data. The proposed model is shown to well capture the spatial distribution patterns of the TC occurrence frequency in the two ocean basins. Prevailing TC tracks as well as the latitudinal distribution of the landfall TC number in the western North Pacific ocean basin are also shown to agree better with the results derived from observed data, as compared to the existing models that took different strategies to include the effect of the beta drift. It is confirmed that the present model is advantageous in terms of not only the accuracy but also the capacity to accommodate the varying climate. Then, the TC trajectory model is applied to assess the potential changes in TC activities over the western North Pacific and the potential threat to China at the end of the 21st century. It is suggested that the TC occurrence in southern China will decrease, while the more landfalling TCs further north will lead potential threat to Zhejiang Province, Jiangsu Province, and northern coastal provinces. Reducing greenhouse gas emissions can effectively reduce the threat of TCs to high-latitude regions.

Keywords: Tropical cyclone; Trajectory model; Climate change

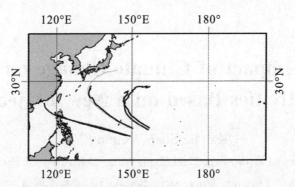

Figure 1. The prevailing tracks of the three types averaged from the observed data (black lines) and the computed results with the present model in the western North Pacific Ocean basin during the period 1979—2018. For the computed results, the green line, the red line and the blue line represent the straight-moving type, the curved-to-landfall type and the curved-to-ocean type, respectively

Research on the Influence of Rough Strip Energy Dissipator on Flow Characteristics of Channel Bend

Honghong Zhang[1,2], Zhenwei Mu[1,2], Fan Fan[1,2]

[1] College of Hydraulic and Civil Engineering, Xinjiang Agricultural University,
Urumqi 830052, China

[2] Xinjiang Key Laboratory of Hydraulic Engineering Security and Water Disasters Prevention,
Urumqi 830052, China

Abstract

The curved channel flow, characterized by complexity and diversity, is a popular subject in hydraulic research. Taking the "635" spillway regulation project in Xinjiang as an example, the rough strip energy dissipator(R-SED) is set to the bottom of bend to increases the flow resistance through roughness, R-SED is one kind of energy dissipation and flow stabilization measure applied to the channel bend. In this paper, the R-SED body structure is optimized based on the characteristics of backwater on the concave bank and falling on the convex bank, so that the longitudinal section is a trapezoidal shape(the height of the R-SED on the concave side is greater than that of the convex side). Based on 21 sets of experimental results of different layout types, combining with ANSYS CFD numerical simulation technology, the influence of R-SED on the secondary flow structure and intensity, cross-section circulation, vorticity characteristics and other microstructures are investigated. At the same time, the macroscopic changes such as the energy dissipation rate and lateral velocity distribution of different flow layers are analyzed, and then the influence principle and energy dissipation mechanism of R-SED on the bend are revealed. The results show that, after adding the R-SED, the secondary flow structure basically disappears, the main flow area expands, the circulation intensity of typical sections significantly become weak, and energy dissipation is aggravated by the increase of eddy density, with the maximum vorticity up to 159.1 per second. In 21 sets of experiments, when the flow rate Q reaches 15 L/s, the energy dissipation rate η is generally higher; when the R-SED height (h_1-h_2) is 1.5~0.75 cm, the strip distance ΔL is 25 cm, and the rotation angle θ is 20°, η can reach the maximum value of 49.5%. Under different R-SED layouts, the upper layer, middle layer and lower layer of typical sections have different velocity distribution rules. The R-SED can redistribute the flow velocity of different current layers and increase the number of hydraulic jumps, thereby aggravating the turbulence of the bend flow. Combined with the characteristics of the energy dissipation structure, the R-SED energy dissipating part of the bend can be divided into three zones: the hydraulic jump energy dissipation zone, the mainstream zone and the junction zone of the surrounding water body, and the vortex energy dissipation zone on both sides of mainstream area. The R-SED has been well used in the prototype of the Xinjiang "635" spillway, which can act as an effective engineering measure to the problem of bend flow.

Keywords: Curved channel flow; Rough strip energy dissipator; Secondary currents; U underflow energy dissipation; Vortex density; Energy dissipation mechanism

Figure 1. Comparison of flow regime under R-SED condition and non-R-SED condition and R-SED structural layout

Note: (a) non-R-SED condition of curved section; (b) R-SED condition of curved section; (c) R-SED layout and section shape.

Research on the Producing Mechanism and Exchange Processes of Total Dissolved Gas

Bin Zhang, Xiaoli Fu

Department of Civil Engineering, Tongji University, Shanghai 200092, China

Abstract

Dams are often operated to discharge over the spillways, once the water flows through the dam and entrains a large amount of air, it then falls into the depths of the stilling pool. Under the action of static pressure, the entrained air bubbles are compressed to make their density far greater than normal, and they quickly exchange mass with the water body and destroy the balance between dissolved air and water, resulting in gas supersaturation, which can be harmful to fish and even caused the death of fish and other aquatic animals due to bubble disease. The total dissolved gas (TDG) concentration on the stilling basin where the most complicated flow pattern happens is relevant with spillway flow, spillway jet, powerhouse operation and tailwater channel water depth. Previous studies have paid more attention to the mechanism and certain characteristics of supersaturated gas. Different from previous researches, in order to explore the generation of mechanism and change of supersaturated gas, this paper describes physical processes governing the producing mechanism and TDG exchange at forebays, spillways, powerhouses, stilling basins and tailwater channels based on studies at field observations. The results illustrate that the production and exchange process of supersaturated gas mainly occurs in the spillway and the stilling basin. In the spillway and the stilling basin, the generation and change mechanism of supersaturated gas is mainly affected by the flow rate of the spillway, the flipper and the water depth in the stilling basin. In the concentrated habitats and spawning breeding grounds of important fish or aquatic organisms (especially in the fish breeding period and migration period), or for the water conservancy hubs where conditions are available, certain methods can be adopted, such as installing a jet in the spillway to reduce the adverse effects of high supersaturated gas in the downstream area. Similarly, on the premise that the distribution of supersaturated gas can be predicted, reasonable scheduling methods can also provide a reference for the proposed water conservancy project, so that it can cooperate with the design of reasonable discharge facilities to improve the water environment and restore the TDG concentration to a normal state.

Keywords: Gas supersaturation; Spillway release; Field observation; Optimal operation

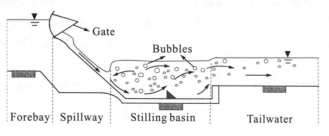

Figure 1. The TDG sketch

Source Characteristics and Exacerbated Tsunami Hazard of the 2020 Mw 7.0 Samos Earthquake in Eastern Aegean Sea

Gui Hu[1], Linlin Li[1,2], Wanpeng Feng[1,2], Yuchen Wang[3], Çagil Karakaş[4], Yunfeng Tian[5], Xiaohui He[1,2]

[1] Guangdong Provincial Key Laboratory of Geodynamics and Geohazards, School of Earth Sciences and Engineering, Sun Yat-sen University, Zhuhai 519082, China

[2] Southern Marine Science and Engineering Guangdong Laboratory(Zhuhai), Zhuhai 519082, China

[3] Earthquake Research Institute, The University of Tokyo, Tokyo, Japan

[4] Schlumberger-SNTC Research, Stavanger, Norway

[5] National Institute of Natural Hazards, Ministry of Emergency Management, Beijing 100085, China

Abstract

On October 30 2020 11:51 UTC, a Mw 6.9 normal faulting earthquake occurred off the northern coasts of Samos Island, Eastern Aegean, Greece. Over 120 people were killed and more than 1000 people were injured during the seismic sequence. The quake produced a moderate tsunami that swamped the coastal areas of Izmir(Turkey) and Samos(Greece). Finding the source of such a tsunami has been puzzling as a normal faulting earthquake with Mw 6.9 would not be adequate to generate metric-scale waves. Identifying the seismogenic fault responsible for the mainshock has been difficult due to the lack of near-field observations. In this study, we infer the tsunami source characteristics and tsunami resonance effect based on simulation of six slip models, and analysis of aftershock distribution and tsunami spectra of the selected tide gauges. We provide 3 slip models from joint inversion of two Sentinel-1 coseismic interferograms with a maximum Line of Sight(LOS) change of 8 cm on the coastal areas at the Samos island. We obtain a north-dipping fault model, which can better explain the geodetic observations, tsunami wave, and post-event survey. The spatial distribution of the aftershocks helps to constrain the fault depth and dimension. The spectral analysis of tsunami waveforms suggests that the tsunami period band is within 4.6~21.3 min and the primary wave period is ~14.2 min. Using the period as an indirect constraint, we show that the source dimension of our slip model can produce tsunami waveforms with a similar wave period. We also find high-energy waves of the Samos earthquake that lasted 20 h, and fundamental oscillation periods of Siğacik Bay are remarkably close to some dominating tsunami periods. We emphasize that tsunami resonance in Siğacik bay is one of the key factors responsible for the amplified tsunami wave heights observed in this bay.

Keywords: 2020 Samos earthquake; Tsunami; Finite fault model; Tsunami resonance

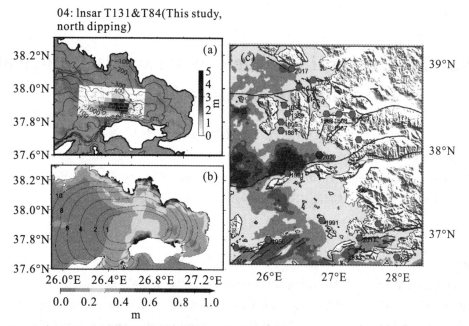

Figure 1. (a) The fourth fault slip model, M4 for the tsunami of 2020 Samos earthquake on the discretized fault; (b) Maximum tsunami wave heights with arrival time (thin black line) in minutes; (c) historical tsunamis (green polygon) between A.D. 142 and A.D. 2020 in Samos island and adjacent areas, and active faults colored by faulting type with red for normal fault, green for strike-slip, lightgreen for strike-slip dextral, and orange for strike-slip sinistral

Spatial-temporal Characteristics of Hydrological Extremes in the Han River Basin

Yingying Feng[1], Maochuan Hu[1,2]

[1] School of Civil Engineering, Sun Yat-sen University, Guangdong 510275, China

[2] Guangdong Engineering Technology Research Center of Water Security Regulation and Control for Southern China, Sun Yat-sen University, Guangdong 510275, China

Abstract

Understanding changes of hydrological extremes is of high significance for water resource management. In this study, a suite of extreme indices derived from daily precipitation and streamflow at 15 stations were analyzed to assess changes in the hydrological extremes from 1960 to 2018 in the Han River Basin, which is the second largest river basin in Guangdong Province, China. Due to the impact of monsoon and typhoon, this region always suffers flood in wet season(April-September) and drought in dry season(October-next March). Thus, the wet extreme indices in wet season and dry extreme indices in dry season were focused. The Mann-Kendall(MK1) and a modified Mann-Kendall test(MK2) as well as Sen's slope estimator were used to analyze significant trends($p<0.05$) in a time series with and without serial correlation and their magnitudes. The results indicate that(1) some extreme indices were highly autocorrelated at some station. MK2 improved the accuracy of trend analysis.(2) The trends of annual wet season's maximum daily precipitation(RX1), total precipitation (PRCPTOT) and consecutive wet days(CWD) were various at different stations. And the magnitudes of changes in precipitation extremes increased from southwest to northeast in this basin(Fig. 1). (3) There was significant decrease in annual dry season's consecutive dry days(CDD) at Zijin station. The intensity and magnitudes of changes in CDD decreased from the southwest and northeast to the central region;(4) There was a downward trend in annual wet season's highest daily streamflow(SX1) at all discharge stations. Meanwhile, a downward trend in annual dry season's consecutive low flow days(CDS) was found at Hengshan and Chaoan Stations. Overall, annual wet season's and dry season's extreme indices decreased in the downstream of Han River Basin. The extreme indices of Mei River Basin(upstream of Han River) and Ting River Basin(upstream of Han River) show different trends during the annual wet season and dry season. There was high random variability of extreme indices compared with their linear changes, thus the trends of most extreme indices at many stations had no statistical significance. This study provides information for the Han River Basin Authority to take proactive measures for sustainable water management.

Keywords: Precipitation; Streamflow; Extreme indices; Trend; Modified Mann-Kendall test

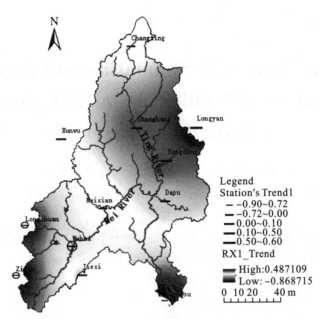

Figure 1. Change rate of annual maximum 1-day precipitation (R×1) during wet season

Study on Numerical Simulation and Emergency Control Strategy of Sudden Water Pollution Scenarios in Dongzhang Reservoir

Linlin Yan[1,2], Jijian Lian[1,2,3], Ye Yao[1,2]

[1] State Key Laboratory of Hydraulic Engineering Simulation and Safety, Tianjin University, Tianjin 300072, China

[2] School of Civil Engineering, Tianjin University, Tianjin 300072, China

[3] School of Water Conservancy & Hydropower, Hebei University of Engineering, Handan 056038, China

Abstract

It has been a long term concern problem about the water quality safety subject of the water source area. The water pollution risk level of water source area rises with the economy developing and living standard improving along the water diversion project. The water resource allocation project in Pingtan and Minjiang Estuary under construction is one of 172 major water conservancy projects in China. As a regulating reservoir of the project, it is more important to ensure the water quality safety of Dongzhang reservoir. So, in this study, we identified and evaluated its potential pollution risks and types by LS Matrix Method. It is classified according to the location of different pollution sources, and then the potential pollution types of each pollution source are distinguished. Second, we establish and calibrate a two-dimensional water quality model coupling with the hydrodynamic model based on the available data. Third, we set different water pollution scenarios. The main pollution types simulated by each working condition are industrial pollution represented by aniline pollution, and domestic pollution, including ammonia nitrogen, total phosphorus, and permanganate index. The effects of different wind conditions and upstream flow on the diffusion characteristics of pollutants were evaluated by analyzing the characteristic values of peak concentration, peak time and intrusion time. Also, we identified adverse environmental factors. Last, a sudden water pollution risk assessment framework and emergency control plan has been established based on the above work. In the event of a low-risk pollution accident, there is no need to close the gate and no local pollution treatment is required. In the event of a medium-risk pollution accident, there is no need to close the gate, but we need to carry out partial treatment for the pollution. In the event of a high-risk pollution accident, the gate must be closed in time and local pollution treatment must be carried out. This study can provide theoretical support and decision-making reference for the early response mechanism of sudden water pollution incidents control in the Dongzhang Reservoir along the water transfer project.

Keywords: Dongzhang Reservoir; Simulation of accidental water pollution; Emergency regulation

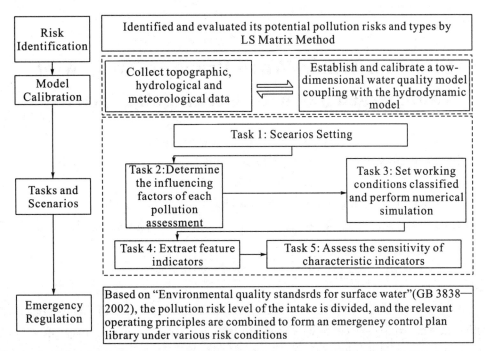

Figure 1. Research ideas and technical routes of study on numerical simulation and emergency control strategy of sudden water pollution scenarios in Dongzhang reservoir

Study on Water Wave Propagation Law of Submerged Jet with Rectangular Orifice

Shuguang Zhang[1,2], Jijian Lian[1,3]

[1] State Key Laboratory of Hydraulic Engineering Simulation and Safety, Tianjin University, Tianjin 300072, China

[2] School of Civil Engineering, Tianjin University, Tianjin 300072, China

[3] School of Water Conservancy & Hydropower, Hebei University of Engineering, Handan 056038, China

Abstract

Submerged jets are ubiquitous in drainage systems, fishways and drainage structures. Surge usually occurs at the free surface downstream of the jet stream, which may affect downstream shipping in serious cases. However, the characteristics of such waves are quite different from those of wind and waves on the sea, and the research results on wave characteristics in the ocean cannot be directly applied to the downstream channel of the junction. In order to investigate the water wave propagation law of submerged jet with rectangular orifice, a series of physical model tests are designed in this paper. The time-history information of the downstream water level variation at different water levels is obtained by measurement. Previous scholar often used threshold method to remove the value near zero, which will lead to errors. In this paper, the Hample coupled with wavelet threshold method is used to process data. Fig.1 shows water level time-history diagram and wave height distribution. We can find that after filtering with Hample coupled with wavelet threshold method, the spectrum characteristics and propagation law of wave height can be analyzed more accurately. And then, the propagation law of submerged jet water wave in time domain and frequency domain is obtained. The results showed that: 1) The correlation between water surface processes at different measuring points is weak, and the correlation between adjacent wave heights at the same measuring point is weak. 2) With the rise of the upstream water level, the wave height at each measuring point gradually increases, and the wave height decreases exponentially along the path, and the wave height at each measuring point approximately corresponds to the Weibull distribution. 3) In the process of flood discharge surge wave propagation, the low frequency large wave has a tendency to be transformed into a slightly high frequency small wave due to wave breakage and tanning. The related research results obtained in this paper reveal the propagation law of submerged jet water wave, which can provide reference for the navigation safety design of the downstream hub.

Keywords: Submerged jets; Wave; Wavelet threshold; Propagation law

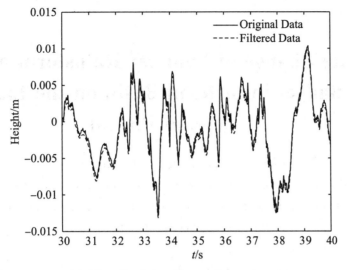

(a) Water level time-history diagram(part)

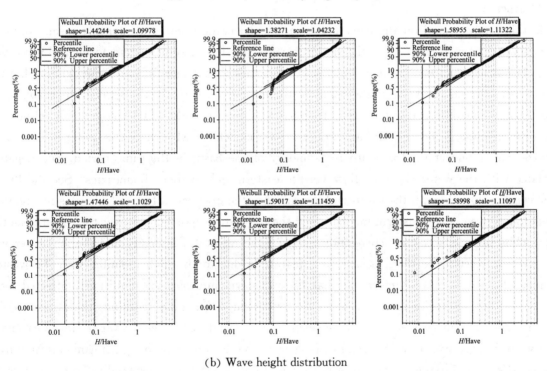

(b) Wave height distribution

Figure 1. Water level time-history diagram and Wave height distribution

The Climate Change of Summer Rainstorm and Water Vapor Budget in Sichuan Basin on the East Side of Tibetan Plateau

Dongmei Qi, Yueqing Li, Changyan Zhou

Institute of Plateau Meteorology, China Meteorology Administration, Chengdu, China;
Heavy Rain and Drought-Flood Disasters in Plateau and Basin Key Laboratory of Sichuan
Province, Chengdu 610000, China

Abstract

In this study, based on the daily precipitation observation data and monthly mean era-interim reanalysis data from 1979 to 2016, the variation characteristics of summer rainstorms and water vapor budget in the sichuan basin are analyzed. The results show that the spatial and temporal distribution of rainstorms in the sichuan basin is the result of the interaction between the special topography of the sichuan basin and different water vapor transports at low latitudes. The amount and frequency of summer rainstorms in the sichuan basin are positively correlated with the water vapor inflow and the net water vapor budget via the southern boundary of the basin in the same period, and negatively correlated with the water vapor outflow via the eastern and northern boundaries. Specifically, the water vapor inflow through the southern boundary has the greatest influence. There is no significant correlation between the rainstorm intensity and the net water vapor budget over the eastern/southern basin, while there is a positive correlation between the rainstorm intensity and the net water vapor budget in the western/northern basin. This feature is due to the high mountains on the western/ northern edge of the basin, indicating that the rainstorm intensity is significantly affected by the dynamics of different regional terrains and local topography. Generally speaking, the variations of the rainstorm amount and frequency in the sichuan basin depend on the water vapor transport and the water vapor budget over different regions. However, the rainstorm intensity is mainly affected by the dynamic effect of different regional terrains and local topography, especially over the western/ northern basin. The summer rainstorm is not only directly related to the regional circulations and water vapor transports in the eastern china and the sichuan basin, but also closely related to the atmospheric circulation and its associated weather system anomalies over the key areas of the air-sea interaction in tropical regions. It reveals the new mechanism of multi-scale interactions between the special topography of the sichuan basin and different water vapor transports at low latitudes. When there is the consistent easterly(westerly) airflow over the key area(130°~180°E, 0°~10°N) from the east of philippines and eastern indonesia to the north of papua new guinea, and the anticyclone (cyclone) system on its north establishes and maintains, the anomalous southeasterly(easterly) airflow and water vapor divergence(convergence) are maintained in the key areas of eastern and southern china. Due to the effect of the special topography of plateau-basin, the distributions of water vapor budget, convergence and divergence in the sichuan basin are special. The anomalous southeasterly(northeasterly) airflow and water vapor convergence(divergence) maintain in the sichuan basin, resulting in much more(fewer) rainstorms in summer.

Keywords: Sichuan Basin; Summer rainstorm; Water vapor budget; Circulation anomalies; Physical mechanism

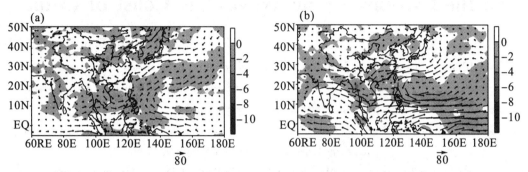

Figure 1. The anomaly of integrated water vapor flux(vectors, units: kg·m^{-1}·s^{-1}) and its divergence(colored areas, units: 10^{-5} kg·m^{-2}·s^{-1}) in the more rainstorm years(a), and fewer rainstorm years(b) in Sichuan Basin

The Influence of the BSISO Oscillation in the Summer Season on the Extreme Ocean Waves Off Coast of China

Xincong Chen[1], Xi Feng[2], Xiangbo Feng[3], Jingfu Peng[4], Yunshu Wu[5], Xinqi Zhang[6]

[1,2] Affiliation Institution 1, Nanjing, China, Key Laboratory of Coastal Disaster and Defence(Hohai University), Ministry of Education & College of Harbor, Coastal and Offshore Engineering, Hohai University, Nanjing 210098, China

[3] Affiliation Institution 2, Reading, United Kingdom, NCAS-climate & Department of Meteorology, Reading University, United Kingdom

[4,5,6] Affiliation Institution 3, Nanjing, China, College of Dayu, Hohai University, Nanjing 210098, China

Abstract

The present study assesses the impact of the Boreal Summer Intra-Seasonal Oscillation(BSISO) on the extreme waves in the China adjacent seas using the latest version of ECMWF analysis(ERA5) during the summer monsoon months June through August(JJA). Composite analysis of anomalies of extreme significant wave heights (SWH) and the relative excessing proportionality of the SWH anomalies for 8 phases of BSISO has been conducted to investigate their linking to BSISO signal. Suppressed extreme wave height anomalies are noticeable during phase 2~4 and enhanced extreme wave height anomalies are distinct during phase 6~8. Whilst phase 1 and phase 5 are found to be neutral/transitional period for the intra-seasonal variability of extreme wave anomalies in the study domain. In addition, this study separated the storm waves caused by cyclones from the extreme waves samples and composite analysis revealed that the cyclones are the main contributors to the intra-seasonal oscillations in extreme wave anomalies. The conclusion drawn by this study aims to provide theoretical basis for the accurate prediction of extreme waves, particularly for nearshore and offshore regime along China coasts.

Keywords: ERA5; BSISO; Tropical cyclone; Extreme wave; Composite analysis

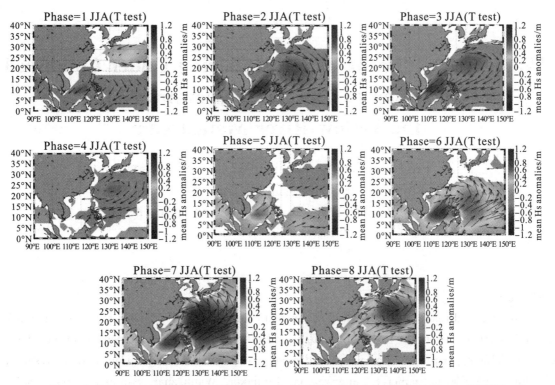

Figure 1. **Mean wave field for combined waves during JJA associated with eight phases of BSISO with significance tests (blank area indicates p-value > 0.05). The contour indicates the mean wave height averaged over all the BSISO days during JJA from 1979-present and the quiver indicate the major direction of waves at each grid point, the length of which represents the associated wave height**

Water and Sediment Benefits Sharing Under Combined Operation of Cascade Reservoirs Along the Yellow River Mainstream

Fengzhen Tang[1], Yuanjian Wang[2], Xin Wang[3]

[1, 2, 3] Yellow River Institute of Hydraulic Research, Zhengzhou 450003, China

[1, 2, 3] Key Laboratory of Lower Yellow River Channel and Estuary Regulation, MWR, Zhengzhou 450003, China

Abstract

The unbalanced relation of water and sediment in the Yellow River Basin makes it imperative for water and sediment resources to be used fairly and efficiently under combined operation of cascade reservoirs. This paper proposed an integrated model to optimize the regulation mode of reservoirs in the Yellow River Basin, namely, Longyangxia, Liujiaxia, Wanjiazhai, Sanmenxia and Xiaolangdi. Multiple objectives were considered, including flood control, water supply, hydropower and scouring-silting. According to this model, cooperation of reservoirs during the first rainy season in 2019 can bring more incremental benefit, about $33.54 million, compared with independent operation of each reservoir. Based on the historical streamflow and sediment process in 2006—2019, the relationship between the incremental benefit and the flow duration curve(sediment duration curve) was built. It was found that the systematic incremental benefit and flow duration curve are closely related, and the systematic incremental benefit is larger when the Yellow River Basin is wetter. According to the direct gains of each stakeholder, it is obvious seen that Longyangxia contributes a lot to the grand coalition for its upstream location, while it doesn't get any incremental benefit. Therefore, the Gately point method was employed to search the possible benefit sharing solutions, ensuring the benefit of each stakeholder higher than independent operation, so that each stakeholder is likely to join the grand coalition. The results show that five reservoirs all need to be operated for downstream reservoirs' or downstream river channel's beneficial uses under cooperation. Longyangxia, Liujiaxia and Xiaolangdi reservoirs obviously take the most important roles in the whole cooperation, thus, the incremental benefit is necessary to be compensated more to these three reservoirs in the benefit reallocation. The downstream river channel, without regulating capacity, always shares benefits with the upper reservoirs. This model is able to be used to optimize the operation mode of reservoirs in the Yellow River Basin, to improve the whole benefits of cascade reservoirs, and provides reference for stakeholders of different reservoirs in the Yellow River Basin to reallocate incremental benefits under cooperation.

Keywords: Yellow River Basin; Cascade reservoirs; Combined operation; Cooperative game theory; Water and sediment benefits sharing

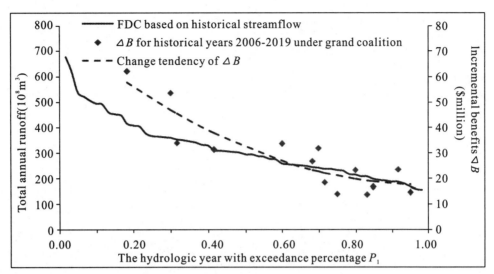

Figure 1. Relationship between ΔB and FDC at the Tongguan station based on the historical streamflow in 2006—2019

Note: FDC is obtained from historical streamflow in 1961—2019. The horizontal axis represents the specific hydrologic years arrayed by the exceedance percentage.